STUDIES IN INTERFACE SCIENCE

Introduction to Molecular-Microsimulation of Colloidal Dispersions

STUDIES IN INTERFACE SCIENCE

SERIES EDITORS
D. Möbius and R. Miller

Vol. 1 Dynamics of Adsorption at Liquid Interfaces. *Theory, Experiment, Application.* By S.S. Dukhin, G. Kretzschmar and R. Miller
Vol. 2 An Introduction to Dynamics of Colloids. By J.K.G. Dhont
Vol. 3 Interfacial Tensiometry. By A.I. Rusanov and V.A. Prokhorov
Vol. 4 New Developments in Construction and Functions of Organic Thin Films. Edited by T. Kajiyama and M. Aizawa
Vol. 5 Foam and Foam Films. By D. Exerowa and P.M. Kruglyakov
Vol. 6 Drops and Bubbles in Interfacial Research. Edited by D. Möbius and R. Miller
Vol. 7 Proteins at Liquid Interfaces. Edited by D. Möbius and R. Miller
Vol. 8 Dynamic Surface Tensiometry in Medicine. By V.M. Kazakov, O.V. Sinyachenko, V.B. Fainerman, U. Pison and R. Miller
Vol. 9 Hydrophile-Lipophile Balance of Surfactants and Solid Particles. *Physicochemical Aspects and Applications.* By P.M. Kruglyakov
Vol. 10 Particles at Fluid Interfaces and Membranes. *Attachment of Colloid Particles and Proteins to Interfaces and Formation of Two-Dimensional Arrays.* By P.A. Kralchevsky and K. Nagayama
Vol. 11 Novel Methods to Study Interfacial Layers. By D. Möbius and R. Miller
Vol. 12 Colloid and Surface Chemistry. By E.D. Shchukin, A.V. Pertsov, E.A. Amelina and A.S. Zelenev
Vol. 13 Surfactants: Chemistry, Interfacial Properties, Applications. Edited by V.B. Fainerman, D. Möbius and R. Miller
Vol. 14 Complex Wave Dynamics on Thin Films. By H.-C. Chang and E.A. Demekhin
Vol. 15 Ultrasound for Characterizing Colloids. *Particle Sizing, Zeta Potential, Rheology.* By A.S. Dukhin and P.J. Goetz
Vol. 16 Organized Monolayers and Assemblies: Structure, Processes and Function. Edited by D. Möbius and R. Miller
Vol. 17 Introduction to Molecular-Microsimulation of Colloidal Dispersions. By A. Satoh

Introduction to Molecular-Microsimulation of Colloidal Dispersions

A. Satoh

Faculty of System Science and Technology,
Akita Prefectural University, Japan

2003
ELSEVIER
Amsterdam – Boston – London – New York – Oxford – Paris
San Diego – San Francisco – Singapore – Sydney – Tokyo

ELSEVIER SCIENCE B.V.
Sara Burgerhartstraat 25
P.O. Box 211, 1000 AE Amsterdam, The Netherlands

First edition 2003

Library of Congress Cataloging in Publication Data
A catalog record from the Library of Congress has been applied for.

British Library Cataloguing in Publication Data
A catalogue record from the British Library has been applied for.

ISBN: 0 444 51424 4

♾ The paper used in this publication meets the requirements of ANSI/NISO Z39.48-1992 (Permanence of Paper).
Printed in The Netherlands.

To Toshiko, Shin, and Sana

Preface

This book is an introductory textbook on molecular-microsimulation methods for colloidal dispersions. It is expected to meet the needs of young researchers, such as postgraduate and advanced undergraduate students in physics, chemistry, mechanical and chemical engineering, who require a systematic understanding of the theoretical background to simulation methods, together with a wide range of practical skills for developing their own computational programs.

A functional fluid (colloid) is a colloidal dispersion in which particles with functional properties are stably dispersed in a base liquid. These kinds of fluids are very attractive from an application point of view because they behave as if the fluid itself had functional properties under the conditions of an applied magnetic or electric field. Typical functional fluids that can be made by modern dispersion techniques are magnetic fluids (ferrofluids), electro-rheological (ER) fluids, and magneto-rheological (MR) suspensions. Ferrofluids and MR suspensions respond to an applied magnetic field whereas ER fluids respond to an applied electric field. In typical applications, these fluids show their important properties in a flow field while being subjected to an appropriate external field. Under these conditions it is possible to bring about a controlled change in the internal structure which can result in a controlled change of a useful physical property, such as viscosity.

Molecular simulation methods and microsimulation methods have been receiving widespread attention as a means to investigate physical phenomena in colloids for both the thermodynamic equilibrium and non-equilibrium state, particularly when subjected to a flow field. This is because colloidal particles show very complicated behaviors in various conditions of magnetic fields, electric fields and flow fields. Molecular-microsimulation methods are therefore indispensable for the investigation of the static and dynamic behavior of colloidal particles at the microscopic level. In many cases of a fluid problem involving a colloidal dispersion, the theoretical or the experimental approaches are very difficult, and then molecular-microsimulation methods play a most important role in the investigation of physical phenomena at the microscopic level.

In the 1980s and early '90s, it was not possible to conduct a large-scale computer simulation of a physical problem unless a supercomputer was used. Nowadays, the situation regarding computing facilities has been changed so dramatically that even a high-performance personal computer may be sufficient to conduct a computer simulation for some physical problems. Thus it is possible for a researcher to have the choice to conduct a large-scale simulation on a machine at a large computing facility, or conduct a medium-scale simulation on a personal computer in the researchers own laboratory. Today, therefore, computer simulations are a widely accepted approach in many fields of research and are no longer confined to a few specialist researchers.

Computer simulation methods are generally classified into two groups, (1) numerical analysis methods, which are based on finite difference methods, finite element methods, etc., and (2) molecular simulation or microsimulation methods, such as molecular dynamics, Monte Carlo, Stokesian dynamics and Brownian dynamics. In this textbook, the latter microsimulation (molecular simulation) methods are explained in detail from the perspective of practical applications in colloid physics and colloid engineering.

Simulation methods suitable for thermodynamic equilibrium problems, such as Monte Carlo, are based on equilibrium statistical mechanics, and so in Chapter 2 we outline the relevant background of the theory. The Monte Carlo methods are then explained in detail in Chapter 3, with some examples of their application to ferromagnetic colloidal dispersions. Monte Carlo methods are not applicable to non-equilibrium phenomena such as a flow field because they are not able to include a necessary approximation for the friction effect of colloidal particles. For this case, molecular dynamics methods are applicable, and they are explained in Chapter 7, in which theories for spherical and rodlike particles are presented. In the case of a non-dilute colloidal dispersion, the hydrodynamic interaction between particles is the main factor for determining the motion of particles subjected to a flow field. Hence, theories for single particle motion and two-particle motion in a linear flow field are addressed in Chapter 5, in which a spherical and an axisymmetric particle system are considered. In this chapter, the velocity and angular velocity of particles are related to the forces and torques acting on the ambient fluid using two different formulations, i.e. the mobility formulation and the resistance formulation. In Chapter 6, the theory for a two-particle system is extended to the multi-body hydrodynamic interaction among particles by means of two different approximations, i.e. the additivity of forces approximation and the additivity of velocities approximation. In Chapter 8 we explain the Stokesian dynamics method, which is a microsimulation technique applicable to non-dilute colloidal dispersions and is based on the additivity approximation. If colloidal particle size is around or below micron-orders, then the Brownian motion of particles cannot be treated as negligible and the Brownian dynamics methods explained in Chapter 9 must be used. The methodology in Chapter 11 and the examples of Stokesian and Brownian dynamics methods in Chapter 12 should be a very helpful guide to understand in more detail the actual tactics employed in the application of the microsimulation methods. Higher-order approximations of multi-body hydrodynamic interactions are addressed in Chapter 13. Other useful microsimulation methods, that may be promising in treating hydrodynamic interactions among particles, such as dissipative particle dynamics and lattice Boltzmann methods, are outlined in Chapter 14.

We have made an effort to produce a book that is suitable for both self-study and reference. Therefore, where necessary, detailed mathematical description has been given in a self-contained manner, together with sufficient reference formulae, in the Appendices or in the instructive Exercises. Hence, we believe that the reader should be able to follow the mathematical manipulation presented in this book by using the material provided. Many of the exercises at the end of each chapter are designed to encourage the understanding of the subjects addressed therein. In some instances, mathematical manipulation has been placed in an Appendix in order to retain the logical clarity in the body of a Chapter.

There are also some useful subroutine programs and several full programs to be found in the Appendices which will assist the reader to develop a program to meet their own requirements. These programs, written in the FORTRAN language, have been designed to be user-friendly rather than over-elaborate so that the reader is able to discern the concepts and then has the option to modify, and improve them where necessary.

Acknowledgments

Finally, I deeply appreciate the assistance given to me by Dr. G. N. Coverdale in proof-reading the manuscript. Also, I would like to express my gratitude to Dr. R. Miller for providing me with the opportunity to write this book.

Kisarazu City, Chiba Prefecture, Japan, December 2002 A. Satoh

Contents

CHAPTER 1

WHAT KINDS OF MOLECULAR-MICROSIMULATION METHODS ARE USEFUL FOR COLLOIDAL DISPERSIONS ?

A colloidal dispersion is a suspension of fine particles (dispersoid) which are stably dispersed in a base liquid (dispersion medium). The dimensions of the dispersed particles are generally within the range of 1 nm to 10 μm. If the particles are smaller than about 1 nm, the dispersion behaves like a true solution, and at the other end of the scale the particles tend to sink due to the force of gravity. Hence a dispersion of particles outside this range is not regarded as being in the colloidal state. The main objective of the present book is to discuss molecular-microsimulation methods for colloidal dispersions, whereas general matters concerning colloid science should be referred to in other textbooks [1-3]. In this book, we restrict our attention to solid particles as colloidal ones which are not distorted in shape in a flow field.

One possible technique for generating functional or intelligent materials is to systematically combine known materials to make a new material with the desired functional properties. Such functional or intelligent materials are expected to exhibit highly attractive properties under certain circumstances. This method of generating new materials is similar to the concept of developing micromachines in the mechanical engineering field. Such intelligent materials may be called "artificially-controlled materials." According to this concept, new functional colloids such as ferrofluids, ER fluids, and MR suspensions have been generated in the colloid physics-engineering field. These functional fluids are generated by making functional particles stably dispersed in a base liquid. They behave as if the liquid itself had functional properties in an external field such as an applied magnetic or electric field. Another example of expanding the concept of the synthesis of functional colloids in the materials field of the surface quality change leads to the possibility of developing higher quality magnetic recording materials (tapes).

In order to investigate a physical phenomenon theoretically, the governing or basic equations are first constructed, and then solved analytically or numerically. Where possible, the theoretical or simulation results are compared with experimental observations in order to deepen our understanding of the physical phenomena. When we construct governing equations for a physical phenomenon, we don't usually need the information of the microstructure of molecules or atoms which constitute the system, rather we generally stand on a larger scale. For example, let us consider the flow problem of water around a circular cylinder under the condition of an ordinary pressure. The physical phenomenon in this case is governed by the Navier-Stokes equation which is well known as a basic equation for flow problems in the fluid mechanics field. In the derivation of the Navier-Stokes equation, we never need the detailed information concerning a water molecule which is composed of one oxygen and two hydrogen atoms. Rather, water is regarded as a continuum, and the Navier-Stokes equation is derived on such a continuum scale.

How does the molecular simulation or microsimulation deal with physical phenomena? In molecular-microsimulations, we stand on the microscopic level of the constituents of a system, such as atoms, molecules or sometimes ultrafine particles, and follow the motion of the constituents to evaluate microscopic or macroscopic physical quantities. Hence, in investigating the above-mentioned flow problem around a cylinder, the equations for the translational and rotational motion of water molecules have to be solved, and the desired macroscopic quantities are evaluated from an averaging procedure over the corresponding microscopic ones, which may be a function of molecular positions and velocities. In this example, the flow field is evaluated by averaging the velocity of each molecule and illustrates the fact that we cannot simulate a physical phenomenon unless the molecular structure and interaction energies between molecules are known. It is, therefore, clear that molecular-microsimulation methods are completely different from numerical analysis methods such as finite-difference or finite-element methods. The former methods stand on the microscopic level of molecules or fine particles, and the latter methods on the macroscopic level of a continuum, in which the governing equations are discretized and these algebraic equations are solved numerically.

Molecular-microsimulations for colloidal dispersions may be classified mainly into four groups based on the simulation method. Monte Carlo methods are highly useful for thermodynamic equilibrium in a quiescent flow field. If the particle Brownian motion is negligible and a dispersion is dilute, molecular dynamics methods are applicable. If the particle Brownian motion can be neglected and a dispersion is not dilute, we have to take into account hydrodynamic interactions among particles. In this case, Stokesian dynamics methods have to be used instead of molecular dynamics methods. Finally, if colloidal particles are not sufficiently smaller than micron-order and the particle Brownian motion has to be taken into account, then Brownian dynamics methods are indispensable. In the following sections we just survey the essence of each simulation method.

1.1 Monte Carlo Methods

The Monte Carlo method is named after the famous gambling town, Monte Carlo, which is the Principality of Monaco. In this method, a series of microscopic states is generated successively through the judgment of whether or not a new state is acceptable using random numbers. This procedure is similar to a gambling game where success or failure is determined by dice. Monte Carlo methods are used for simulating physical phenomena for a system in thermodynamic equilibrium, and are therefore based on the theory of equilibrium statistical mechanics.

We now consider a canonical ensemble of the particle number N, system volume V, and temperature T. For this statistical ensemble, the probability density function, that is, the canonical distribution $\rho(\mathbf{r})$ is written as

$$\rho(\mathbf{r}) = \frac{1}{Q}\exp\left\{-\frac{1}{kT}U(\mathbf{r})\right\}, \tag{1.1}$$

in which the abbreviation $\mathbf{r}$ is used for $\mathbf{r}_1, \mathbf{r}_2, \ldots, \mathbf{r}_N$ for simplicity. With this distribution, the statistical average $\langle A \rangle$ of a certain quality A is expressed as

$$\langle A \rangle = \int A(\mathbf{r})\rho(\mathbf{r})d\mathbf{r}, \tag{1.2}$$

in which k is Boltzmann's constant, $d\mathbf{r} = dx_1 dy_1 dz_1 \cdots dx_N dy_N dz_N$, Q is the configuration integral, and U is the total interaction energy of a system. The expressions for Q and U are written as

$$Q = \int \exp\left\{-\frac{1}{kT}U(\mathbf{r})\right\}d\mathbf{r}, \tag{1.3}$$

$$U = \sum_{i=1}^{N}\sum_{\substack{j=1 \\ (j>i)}}^{N} u_{ij}. \tag{1.4}$$

In Eq. (1.4), u_{ij} is the interaction energy between particles i and j, and the contribution of three-body interactions to the system energy has been neglected.

Since the configuration integral Q is $3N$-fold for a three-dimensional system, analytical evaluation is generally impracticable and numerical integration methods such as the Simpson method are also quite unsuitable for multi-fold integrals. However, if it is taken into account that the integrand is an exponential function, we can expect that certain particle configurations make a further more important contribution to the integral than others. Thus, if such important microscopic states can be sampled with priority in a numerical procedure, then it may be possible to evaluate the ensemble average, Eq. (1.2), in a reasonable time. This is exactly the concept of importance sampling. It is the Monte Carlo method that generates microscopic states successively with the concept of important sampling. As will be shown in Sec. 3.3, Metropolis' transition probability is almost universally used for the transition of one microscopic state to the next. With this transition probability, the probability of realizing a microscopic state is determined by the canonical distribution for the case of the canonical ensemble. Hence Eq. (1.2) can be simplified as

$$\langle A \rangle \approx \sum_{n=1}^{M} A(\mathbf{r}^n)/M , \tag{1.5}$$

in which M is the total sampling number, and $\mathbf{r}^n$ is the n-th sampled microscopic state.

On the other hand, the quantity $\overline{A}$, which is measured in the laboratory, is the time average of A and expressed as

$$\overline{A} = \lim_{t\to\infty} \frac{1}{(t-t_0)} \int_{t_0}^{t} A(\tau)d\tau . \tag{1.6}$$

The relationship between the statistical ensemble $\langle A \rangle$ and the time average $\overline{A}$ will be discussed later in Sec. 2.1.

1.2 Molecular Dynamics Methods

In this book, we consider molecular dynamics methods from the viewpoint of microsimulation for colloidal dispersions. Thus general matters of molecular dynamics methods should be referred to in appropriate textbooks [4-6].

Hydrodynamic interactions among colloidal particles are negligible for a dilute dispersion. Then, if the particle Brownian motion can be neglected, the motion of an arbitrary particle i in a system is governed by the following equation:

$$m_i \frac{d^2\mathbf{r}_i}{dt^2} = \mathbf{F}_i - \xi \mathbf{v}_i , \tag{1.7}$$

in which m_i is the mass of particle i, $\mathbf{r}_i$ is the position vector, $\mathbf{F}_i$ is the sum of forces acting on particle i by the ambient particles, and ξ is the friction coefficient. Equation (1.7) is valid for spherical particles. In many cases of flow problems in colloidal dispersions, the inertia term on the left-hand side in Eq. (1.7) is negligible, and the equation of motion reduces to

$$\mathbf{v}_i = \mathbf{F}_i / \xi . \tag{1.8}$$

Hence the particle velocity can be evaluated from Eq.(1.8) under an appropriate initial condition. Additionally, if the following finite difference approximation to $\mathbf{v}_i = d\mathbf{r}_i / dt$ is used:

$$\mathbf{r}_i(t+\Delta t) = \mathbf{r}_i(t) + \Delta t \mathbf{v}_i(t) , \tag{1.9}$$

then the particle position at the next step is obtained from this equation. This is essentially nothing else but the molecular dynamics approach.

1.3 Stokesian Dynamics Methods

Hydrodynamic interactions among particles cannot be neglected unless a colloidal dispersion is dilute. The equation of motion of particles for a non-dilute dispersion is expressed as a generalization of Eq. (1.8):

$$\begin{bmatrix} \mathbf{F}_1 \\ \mathbf{F}_2 \\ \vdots \\ \mathbf{F}_N \end{bmatrix} = \eta \mathbf{R} \begin{bmatrix} \mathbf{v}_1 \\ \mathbf{v}_2 \\ \vdots \\ \mathbf{v}_N \end{bmatrix}, \tag{1.10}$$

in which N is the particle number of a system, η is the viscosity of the base liquid, and **R** is the resistance matrix, which is composed of resistance tensors and characterizes hydrodynamic interactions among particles. Since the resistance matrix **R** is usually dependent only on the particle position, the inverse matrix $\mathbf{R}^{-1}$ can be evaluated independently of the particle velocities. Thus the particle velocity can be obtained from the mathematical manipulation of the product of $\mathbf{R}^{-1}$ and the forces. Then the particle position at the next time step is calculated from Eq. (1.9). It is clear from this brief statement that Stokesian dynamics methods do not take into account the particle Brownian motion, but they do take into account hydrodynamic interactions among particles.

1.4 Brownian Dynamics Methods

The Brownian motion of particles is clearly induced by the motion of solvent molecules. Since solvent molecules are generally much smaller than colloidal particles, the characteristic time for the motion of solvent molecules is much shorter than that for the motion of colloidal particles. Hence, simulations based on the time scale of the motion of solvent molecules may be unrealistic from a computation time point of view for evaluating physical quantities. To circumvent this difficulty, solvent molecules are regarded as a continuum and the influence of the motion of solvent molecules is combined into the equations of motion as a stochastic term. The following Langevin equation is a typical expression in this approach:

$$m_i \frac{d^2\mathbf{r}_i}{dt^2} = \mathbf{F}_i - \xi \mathbf{v}_i + \mathbf{F}_i^B, \tag{1.11}$$

in which $\mathbf{F}_i^B$ is the random force inducing the particle Brownian motion and has stochastic properties. It is the Brownian dynamics method that describes the particle motion according to an equation with a stochastic term such as Eq. (1.11). Equation (1.11) is applicable only to a dilute dispersion since hydrodynamic interactions among particles are not taken into account. If we take into account such interactions, more complicated Brownian dynamics algorithms have to be considered

References

1. W. B. Russel, D. A. Saville, and W. R. Schowater, "Colloidal Dispersions," Cambridge University Press, 1989
2. T. G. M. van de Ven, "Colloidal Hydrodynamics," Academic Press, London, 1989
3. R. J. Hunter, "Foundations of Colloid Science," Vol. 1, Clarendon Press, Oxford, 1986
4. M. P. Allen and D. J. Tildesley, "Computer Simulation of Liquids," Clarendon Press, Oxford, 1987
5. D. W. Heermann, "Computer Simulation Methods in Theoretical Physics," 2nd Ed., Springer-Verlag, Berlin, 1990
6. J. M. Haile, "Molecular Dynamics Simulation: Elementary Methods," John Wiley & Sons, New York, 1992

CHAPTER 2

STATISTICAL ENSEMBLES

When we use simulation methods to investigate the behavior of a colloidal dispersion in equilibrium, it is practically possible to neglect the solvent molecules and treat only the colloidal particles. Even from this simplified approach, one can sufficiently understand the essence of a physical phenomenon in equilibrium. Monte Carlo methods are very useful for this kind of simulation. They are based on statistical mechanics, and therefore it is important to understand the essence of statistical mechanics before we start to explain the theory of Monte Carlo methods in detail. Thus, in this chapter, we outline the basis of equilibrium statistical mechanics which is for a system in thermodynamic equilibrium [1-9]. It is noted that a state in thermodynamic equilibrium does not change with time macroscopically, but particles in a system change in their positions and velocities with time microscopically.

2.1 The Concept of Statistical Ensembles

We consider a system of the particle number N, system volume V, and temperature T, which are specified and are constant. As shown in Fig. 2.1 (a), microscopic states of the system change successively with time. Hence, a certain microscopic quantity $A(t)$ which depends on the microscopic state of the system is measured every certain time interval and the time average of such data is calculated using Eq. (1.6) to obtain the corresponding macroscopic quantity $\overline{A}$.

We now discuss the concept of a statistical average. As shown in Fig. 2.1(b), we consider a large number of microscopic states representing a collection of all the possible states, which may appear and are determined by all the particle configurations in a physical system. A collection of such microscopic states, which satisfy the condition of a given value of (N,V,T) in the present case, is called the statistical ensemble. If the probability that an arbitrary microscopic state of the statistical ensemble appears is known, the statistical average $\langle A \rangle$ of a physical quantity A can be evaluated. This averaging procedure is called the ensemble average.

We explain the concept of the ensemble average more concretely on a mathematical basis. It is assumed that the state of the system is prescribed by the particle positions $(\mathbf{q}_1,\mathbf{q}_2,\ldots,\mathbf{q}_N)$, and the particle momenta $(\mathbf{p}_1,\mathbf{p}_2,\ldots,\mathbf{p}_N)$. From now on, the abbreviations $\mathbf{q}$ and $\mathbf{p}$ are used for

so that the Gibbs free energy G can be defined, using the free energy F, as

$$G = F + PV\,. \tag{2.37}$$

Next we show the expressions for typical thermodynamic quantities using the grand canonical partition function Ξ. Since the grand potential Ω is expressed as

$$\Omega = F - G = E - TS - N\mu\,, \tag{2.38}$$

the differential of the grand potential, $d\Omega$, can be obtained as

$$d\Omega = dE - d(TS) - d(N\mu) = -SdT - PdV - Nd\mu\,. \tag{2.39}$$

In deriving this equation, Eq. (2.34) has been used. Hence, the following expressions can be derived straightforwardly.

$$S = -\left(\frac{\partial \Omega}{\partial T}\right)_{V\mu} = k\ln\Xi + kT\left(\frac{\partial}{\partial T}\ln\Xi\right)_{V\mu} = k\ln\Xi + \frac{E}{T} - \frac{\mu}{T}\langle N\rangle\,, \tag{2.40}$$

$$E = kT^2\left(\frac{\partial}{\partial T}\ln\Xi\right)_{V\mu} + \langle N\rangle\mu\,, \tag{2.41}$$

$$P = -\left(\frac{\partial \Omega}{\partial V}\right)_{T\mu} = kT\left(\frac{\partial}{\partial V}\ln\Xi\right)_{T\mu}\,, \tag{2.42}$$

$$\langle N\rangle = -\left(\frac{\partial \Omega}{\partial \mu}\right)_{VT} = kT\left(\frac{\partial}{\partial \mu}\ln\Xi\right)_{VT}\,. \tag{2.43}$$

The last equation on the right-hand side in Eq. (2.40) has been obtained by differentiating Eq. (2.28). Also, note that the average particle number $\langle N\rangle$ in these equations is used since the number of particles, N, fluctuates for the case of the grand canonical ensemble.

Finally we show the expression for the chemical potential for an ideal gas. If the system energy is taken as $U(\mathbf{r})=0$, and the exponential term is expressed in a Maclaurin expansion, then Eq. (2.29) reduces to a simplified form. Using this simplified expression, the differential on the right-hand side in Eq. (2.43) can be carried out straightforwardly, and the following relation is obtained:

$$\frac{\langle N\rangle}{V} = \frac{e^{\mu/kT}}{\Lambda^3}\,. \tag{2.44}$$

With this expression and the equation of state, $PV=\langle N\rangle kT$, the chemical potential for an ideal gas can be derived as

$$\mu = kT\ln\left(\frac{P}{kT}\Lambda^3\right). \tag{2.45}$$

2.2.4 The isothermal-isobaric ensemble

The pressure ensemble or isothermal-isobaric ensemble is a statistical ensemble which is specified by the particle number of the system, N, temperature T, and pressure P. This statistical ensemble is highly useful for treating a physical system where the system volume fluctuates, although the pressure is kept constant. We consider a microscopic state, r, which has the system volume V and energy $E_V(r)$. In this statistical ensemble, the probability density $\rho(E_V(r))$ can be expressed as

$$\rho(E_V(r)) = \exp\left\{-\frac{E_V(r)+PV}{kT}\right\}\bigg/ Y, \tag{2.46}$$

in which Y corresponds to the partition function Z for the canonical ensemble and is written as

$$Y = \sum_V \sum_r \exp\left\{-\frac{E_V(r)+PV}{kT}\right\}. \tag{2.47}$$

For a classical-mechanical system, the probability density $\rho(\mathbf{r},\mathbf{p},V)$ and Y are expressed as

$$\rho(\mathbf{r},\mathbf{p},V) = \frac{\exp\left\{-\dfrac{H(\mathbf{r},\mathbf{p})+PV}{kT}\right\}}{N!h^{3N}Y}, \tag{2.48}$$

$$Y = \frac{1}{N!h^{3N}}\iiint \exp\left\{-\frac{H(\mathbf{r},\mathbf{p})+PV}{kT}\right\} d\mathbf{r}d\mathbf{p}dV = \int \exp\left(-\frac{PV}{kT}\right) Z dV. \tag{2.49}$$

If the Hamiltonian $H(\mathbf{r},\mathbf{p})$ is expressed in Eq. (2.8), Y can be written as

$$Y = \frac{1}{N!\Lambda^{3N}}\iint \exp\left\{-\frac{U(\mathbf{r})}{kT}\right\}\exp\left(-\frac{PV}{kT}\right) d\mathbf{r}dV. \tag{2.50}$$

For this statistical ensemble, thermodynamic quantities can be related to the statistical-mechanical ones through the relationship between the Gibbs free energy G and Y:

$$G = -kT\ln Y. \tag{2.51}$$

If we use the definition of the Gibbs free energy, Eq. (2.37), the Helmholtz free energy, Eq. (2.20), and the first law of thermodynamics, Eq. (2.10), then the differential in the Gibbs free energy, dG, can be expressed as

$$dG = -SdT + VdP. \tag{2.52}$$

Hence the following expressions can be derived straightforwardly from Eqs. (2.51) and (2.52):

$$S = -\left(\frac{\partial G}{\partial T}\right)_P = k\ln Y + kT\left(\frac{\partial}{\partial T}\ln Y\right)_P, \tag{2.53}$$

$$\langle V\rangle = \left(\frac{\partial G}{\partial P}\right)_T = -kT\left(\frac{\partial}{\partial P}\ln Y\right)_T, \tag{2.54}$$

$$\tilde{H} = kT^2 \left(\frac{\partial}{\partial T} \ln Y \right)_P . \tag{2.55}$$

In Eq. (2.55), $\tilde{H}$ is called the enthalpy and defined by the expression, $\tilde{H}=E+P\langle V\rangle$, which is related to the Gibbs free energy through the equation, $G=\tilde{H}-TS$.

2.3 Thermodynamic Quantities

The most basic and representative thermodynamic quantities are the temperature and pressure. The temperature has a relationship with the thermal motion of particles, that is, the particle velocities. We here show an expression for the pressure P in terms of the ensemble average of the particle positions and the forces acting on the particles. This is well known as the virial equation of state and expressed as

$$P = \frac{N}{V} kT + \frac{1}{3V} \left\langle \sum_i \sum_{\substack{j \\ (i<j)}} \mathbf{r}_{ij} \cdot \mathbf{f}_{ij} \right\rangle , \tag{2.56}$$

in which $\mathbf{r}_{ij}=\mathbf{r}_i-\mathbf{r}_j$, $\mathbf{f}_{ij}$ is the force exerted on particle i by particle j. The first term on the right-hand side in Eq. (2.56) is due to the momenta of particles, and the second term is due to the interactions between particles. The term of the ensemble average in Eq. (2.56) may be written in different form using the radial distribution function and this will be shown in Sec. 10.1. The derivation process of Eq. (2.56) may give useful help in understanding the contribution of colloidal particles to physical quantities such as viscosity. Hence we show the detailed derivation of Eq. (2.56) in Appendix A8.

References

1. K. Huang, "Statistical Mechanics," 2nd Ed., John Wiley & Sons, New York, 1987
2. J.W. Gibbs, "Elementary Principles in Statistical Mechanics," Ox Bow Press, Woodbridge, Coneticatt, 1981
3. F.C. Andrews, "Equilibrium Statistical Mechanics," 2nd Ed., John Wiley & Sons, New York, 1975
4. E.A. Jackson, "Equilibrium Statistical Mechanics," Prentice-Hall, Englewood Cliffs, N.J., 1968
5. R. Kubo, "Statistical Mechanics," North-Holland Publishing Company, Amsterdam, 1965
6. F. Mandl, "Statistical Physics," 2nd Ed., John Wiley & Sons, New York, 1988
7. D.A. McQuarrie, "Statistical Mechanics," Happer & Row, New York, 1976
8. B.J. McClelland, "Statistical Thermodynamics," Chapman & Hall, London, 1973

9. J.P. Hansen and I.R. McDonald, “Theory of Simple Liquids,” 2nd Ed., Academic Press, London, 1986

10. G.A. Bird, “Molecular Gas Dynamics,” Chap.3, Clarendon Press, Oxford, 1976

Exercises

2.1 By taking into account that the kinetic energy K is $\frac{1}{2m}\sum_{i=1}^{N} p_i^2$ and the square of the momentum $\mathbf{p}_i$ of particle i is expressed as $p_i^2 = p_{ix}^2 + p_{iy}^2 + p_{iz}^2$, carry out the integration of Z_K in Eq. (2.16) to obtain Eq. (2.17).

2.2 By substituting Z shown in Eq. (2.16) into Eq. (2.24), prove that $E=\langle K+U\rangle$, that is,

$$E = \frac{1}{N!h^{3N}Z}\int (K(\mathbf{p})+U(\mathbf{r}))\exp\left\{-\frac{(K(\mathbf{p})+U(\mathbf{r}))}{kT}\right\} d\mathbf{r}d\mathbf{p}. \tag{2.57}$$

2.3 For an ideal gas, the grand partition function Ξ is written, from Eq. (2.29), as

$$\Xi = \sum_N \frac{V^N}{N!\Lambda^{3N}}\exp\left(\frac{\mu N}{kT}\right). \tag{2.58}$$

Derive Eq. (2.44) by substituting Eq. (2.58) into Eq. (2.43). Also, using the equation of state, $PV=\langle N\rangle kT$, and Eq. (2.44), prove that the chemical potential for an ideal gas is expressed by Eq. (2.45).

2.4 Particles of a system in thermodynamic equilibrium have a certain velocity distribution. We now consider a system in equilibrium which is specified by a constant N, V, and T. When we focus our attention on a certain particle i, the probability that the momentum of the particle can be found within the range of $\mathbf{p}_i$ to $(\mathbf{p}_i+d\mathbf{p}_i)$ is expressed as $\rho_p(\mathbf{p}_i)d\mathbf{p}_i$. By assuming that the Hamiltonian H is expressed by Eq. (2.8), prove that the probability density $\rho_p(\mathbf{p}_i)$ can be written as

$$\rho_p(\mathbf{p}_i) = \frac{1}{(2\pi mkT)^{3/2}}\exp\left(-\frac{p_{ix}^2+p_{iy}^2+p_{iz}^2}{2mkT}\right). \tag{2.59}$$

If the probability that the velocity can be found within the range of $\mathbf{v}_i$ to $(\mathbf{v}_i+d\mathbf{v}_i)$ is expressed as $f_p(\mathbf{v}_i)d\mathbf{v}_i$, prove that the probability density $f_p(\mathbf{v}_i)$ can be written, from Eq. (2.59), as

$$f(\mathbf{v}_i) = \left(\frac{m}{2\pi kT}\right)^{3/2}\exp\left\{-\frac{m}{2kT}\left(v_{ix}^2+v_{iy}^2+v_{iz}^2\right)\right\}. \tag{2.60}$$

This velocity distribution f is called the Maxwellian distribution or Maxwell's velocity distribution. It is noted that the Maxwellian distribution can also be derived from the molecular gas dynamics theory [10].

2.5 The Maxwellian distribution, Eq. (2.60), is the probability density function for velocity components. However, the probability density for the speed of particles (or the magnitude of particle velocities) is very helpful to understand the actual velocity distribution more straightforwardly than the component probability density function. We assume that the probability of the particle speed being found within the range of v_i to (v_i+dv_i) is expressed as $\chi(v_i)dv_i$. By changing the variables from the velocity components (v_{ix},v_{iy},v_{iz}) of particle i to (v_i,θ,φ) through the relation of $v_{ix}=v_i\sin\theta\cos\varphi$, $v_{iy}=v_i\sin\theta\sin\varphi$, and $v_{iz}=v_i\cos\theta$, prove that the probability density function $\chi(v_i)$ is expressed as

$$\chi(v_i) = 4\pi\left(\frac{m}{2\pi kT}\right)^{3/2} v_i^2 \exp\left(-\frac{m}{2kT}v_i^2\right). \tag{2.61}$$

The speed $v_{mp}= (2kT/m)^{1/2}$, which gives the maximum of χ, is called the most probable thermal speed.

CHAPTER 3
MONTE CARLO METHODS

The Monte Carlo method is a highly powerful technique to investigate physical phenomena from a microscopic point of view for a molecular system. This method is also very useful even for colloidal dispersions in thermodynamic equilibrium. Although colloidal dispersions are composed of dispersion medium (or base liquid) and dispersed phase (or colloidal particles), the interactions among particles play the most important role in aggregation phenomena of particles; the interactions between a base liquid and colloidal particles are less important than particle-particle interactions in aggregation phenomena of some colloids such as ferromagnetic colloidal dispersions. Hence, it is physically reasonable to use Monte Carlo methods in which solvent molecules are neglected and colloidal particles alone are treated for understanding the essence of physical phenomena. You can see, therefore, that the Monte Carlo method is directly applicable to a colloidal dispersion by regarding colloidal particles like molecules in a molecular system. It is noted, however, that special techniques (for example, the cluster-moving method) are indispensable in simulations for a strongly-interacting system, which will be explained later.

In the Monte Carlo method, microscopic states of a system are generated using random numbers. [1-9]. A sequence of microscopic states made by random numbers corresponds to (but is not equal to) the trajectory of state points made dynamically by molecular dynamics methods for a molecular system in equilibrium. The probability of a certain microscopic state appearing has to obey the probability density for the statistical ensemble of interest. The Metropolis method [10] on which Monte Carlo methods are overwhelmingly based doesn't need an explicit expression for the probability density (exactly, the partition function) in carrying out simulations; actually, such an explicit expression for the probability density is not generally known in almost all cases. According to the type of physical system of interest, Monte Carlo methods are classified into microcanonical, canonical, grand canonical Monte Carlo methods, etc. But the basic concept of each Monte Carlo method is the same and based on importance sampling and Markov chains. In the following sections, we explain first such basic concepts and then the Metropolis method, which may be another important concept for Monte Carlo simulations. After that, Monte Carlo algorithms will be shown for each statistical ensemble. As pointed out before, it is noted that Monte Carlo methods are basically for simulating physical phenomena in thermodynamic equilibrium, and are not suitable for non-equilibrium phenomena which are important subjects in engineering fields.

3.1 Importance Sampling

We consider a system composed of N particles; the position vector of particle i is denoted by $\mathbf{r}_i$, and all the position vectors $\mathbf{r}_1$, $\mathbf{r}_2$, ..., $\mathbf{r}_N$ are denoted simply by the vector $\underline{\mathbf{r}}$ for simplicity. As already explained, the ensemble average $\langle A \rangle$ of a quantity A which depends only on the particle positions $\underline{\mathbf{r}}$ can be evaluated by means of the probability density for the statistical ensemble of interest. For example, it is expressed, by taking into account the relation in Eq. (2.16) for the canonical ensemble, as

$$\langle A \rangle = \frac{1}{Z_U} \int A(\underline{\mathbf{r}}) \exp\left\{ -\frac{1}{kT} U(\underline{\mathbf{r}}) \right\} d\underline{\mathbf{r}} . \tag{3.1}$$

It is generally impossible to evaluate the configuration integral Z_U (Eq. (2.16)), so that a numerical approach such as Simpson's formula is frequently adopted. However, since the application of Simpson's formula to a multi-fold integration is inappropriate (or extremely inefficient), a stochastic approach such as Monte Carlo methods is employed.

In simple Monte Carlo methods, the multi-fold integration on the right-hand side in Eq. (3.1) is evaluated by scanning randomly the integration area with uniform random numbers. If the configuration which is scanned in the n-th order is denoted by $\underline{\mathbf{r}}^n$, then Eq. (3.1) can be written as

$$\langle A \rangle \approx \frac{\sum_{n=1}^{M} A(\underline{\mathbf{r}}^n) \exp\left\{ -U(\underline{\mathbf{r}}^n)/kT \right\}}{\sum_{n=1}^{M} \exp\left\{ -U(\underline{\mathbf{r}}^n)/kT \right\}} , \tag{3.2}$$

in which M is the total sampling number. However, this kind of multi-fold integration may not be efficient, since the convergence of the summation with respect to n in Eq. (3.2) is very slow with increasing the value of M, for actual physical problems. The following importance sampling method [1,2,5], therefore, may become the key concept for the successful application of Monte Carlo methods to physical problems.

The concept of importance sampling is that the particle configurations which make a more important contribution to the integration are sampled more frequently. This can be expressed as an equation using a weight function $w(\underline{\mathbf{r}})$:

$$\langle A \rangle = \frac{1}{Z_U} \int \frac{A(\underline{\mathbf{r}})}{w(\underline{\mathbf{r}})} \exp\left\{ -\frac{1}{kT} U(\underline{\mathbf{r}}) \right\} \cdot w(\underline{\mathbf{r}}) d\underline{\mathbf{r}} . \tag{3.3}$$

In this equation, $w(\underline{\mathbf{r}})$ has to be the probability density for the statistical ensemble of interest. For the canonical ensemble, the canonical distribution may be used as $w(\underline{\mathbf{r}})$:

$$w(\underline{\mathbf{r}}) = \exp\{-U(\underline{\mathbf{r}})/kT\}/Z_U \quad . \tag{3.4}$$

Hence, if particle configurations are sampled with a weight $w(\underline{\mathbf{r}})$ instead of a uniform sampling, Eq. (3.3) reduces to the following simplified expression:

$$\langle A \rangle \approx \sum_{n=1}^{M} A(\underline{\mathbf{r}}^n)/M \, . \tag{3.5}$$

Equation (3.5) may seemingly enable us to evaluate the multi-fold integration, that is, the ensemble average $\langle A \rangle$. However, the difficulty in the evaluation of the multi-fold integration is not solved in a meaningful way, since the weight function $w(\underline{\mathbf{r}})$ in Eq. (3.4) cannot be used unless the exact expression for Z_U is known. We remind the reader that we are discussing Monte Carlo approaches because the exact expression for Z_U cannot be derived analytically. Our difficulty is circumvented by means of introducing the concept of ergodic Markov chains by Metropolis et al. [10]. In the next section we outline Markov chains.

3.2 Markov Chains

Monte Carlo methods are based on the important concept of Markov chains [2, 11-14]; a Markov process in which both time and state space are not continuous but discrete, is called a Markov chain.

We now consider a stochastic process where the particle configuration changes with a sequence of $\underline{\mathbf{r}}^0$, $\underline{\mathbf{r}}^1$, ..., $\underline{\mathbf{r}}^n$, in which $\underline{\mathbf{r}}^n$ stands for the configuration of all particles at time n. If the states $\underline{\mathbf{r}}^0$, $\underline{\mathbf{r}}^1$, ..., $\underline{\mathbf{r}}^{n-1}$ are known, then the probability of state $\underline{\mathbf{r}}^n$ appearing obeys the conditional probability $p(\underline{\mathbf{r}}^n \mid \underline{\mathbf{r}}^{n-1},...,\underline{\mathbf{r}}^1,\underline{\mathbf{r}}^0)$. With this definition of the conditional probability, a Markov chain can be expressed in mathematical form. That is, if the following condition is satisfied for an arbitrary value n,

$$p(\underline{\mathbf{r}}^n \mid \underline{\mathbf{r}}^{n-1},\cdots,\underline{\mathbf{r}}^1,\underline{\mathbf{r}}^0) = p(\underline{\mathbf{r}}^n \mid \underline{\mathbf{r}}^{n-1}), \tag{3.6}$$

the stochastic process $\underline{\mathbf{r}}^0$, $\underline{\mathbf{r}}^1$, ..., $\underline{\mathbf{r}}^n$ is called a Markov chain. Equation (3.6) clearly states that the probability of a transition from the configuration at time (n-1) to that at time n is independent of the configurations before the time (n-1). This is called the Markov property. Thus, a Markov chain may be defined using the Markov property such that the stochastic process which has the Markov property is a Markov process (Markov chain). It is noted that the probability $p(\underline{\mathbf{r}}^n \mid \underline{\mathbf{r}}^{n-1})$ is the transition probability that a microscopic state $\underline{\mathbf{r}}^{n-1}$ transfers to $\underline{\mathbf{r}}^n$. Now we consider the transition from a microscopic state $\underline{\mathbf{r}}_i$ to $\underline{\mathbf{r}}_j$; the subscripts i

and j do not represent a difference in time points, but in microscopic states. In this case, the transition probability p_{ij} is written as

$$p_{ij} = p(\underline{\mathbf{r}}_j \mid \underline{\mathbf{r}}_i). \tag{3.7}$$

In ordinary simulations, p_{ij} is independent of the time point, or stationary. In the following, we treat such a Markov chain.

The important point in applying Markov chains to Monte Carlo simulations is to generate Markov chains such that the probability of a microscopic state appearing obeys the probability density for the statistical ensemble of interest. For example, for a system of a given constant temperature, volume, and number of particles, the probability of the state $\underline{\mathbf{r}}_i$ appearing, $p_i(=p(\underline{\mathbf{r}}_i))$, is required to be

$$p_i \propto \exp\{-U(\underline{\mathbf{r}}_i)/kT\}. \tag{3.8}$$

Some constraint conditions have to be placed on the transition probability in order that microscopic states appear according to the probability density of interest, irrespective of a given initial state. Before we show which type of Markov chain satisfies the above-mentioned properties, some general statements concerning Markov chains are given.

(1) General statement A

A probability distribution (the probability of state i appearing is denoted by ρ_i) is called invariant or stationary for a given Markov chain if the following conditions are satisfied:

$$\left.\begin{array}{ll} 1.\ \rho_i > 0 & \text{(for all } i\text{)}, \\ 2.\ \sum_i \rho_i = 1, & \\ 3.\ \rho_j = \sum_i \rho_i p_{ij}. & \end{array}\right\} \tag{3.9}$$

(2) General statement B

A Markov chain in which every state can be reached from every other state is called an irreducible Markov chain. If a Markov chain is not irreducible, but absorbable, the sequences of microscopic states may be trapped into some independent closed states and never escape from such undesirable states. This means that Eq. (3.8) cannot be satisfied under such circumstances.

(3) General statement C

That a state $\underline{\mathbf{r}}_i$ is recurrent means that $\underline{\mathbf{r}}_i$ reappears in the future; hence the state $\underline{\mathbf{r}}_i$ can appear many times in a Markov chain. In contrast, that a state $\underline{\mathbf{r}}_i$ is transient means that the probability of the state $\underline{\mathbf{r}}_i$ appearing decreases gradually and finally the state $\underline{\mathbf{r}}_i$ never appears in a stochastic process. We show this statement more exactly using mathematical equations. The probability that a state $\underline{\mathbf{r}}_j$ is first attained from a state $\underline{\mathbf{r}}_i$ at n-th step is denoted by $f_{ij}^{(n)}$, and the following expression are also taken into account:

$$\left.\begin{aligned} f_{ij}^{(0)} &= 0, \\ f_{ij} &= \sum_{n=1}^{\infty} f_{ij}^{(n)}, \\ \mu_i &= \sum_{n=1}^{\infty} n f_{ii}^{(n)}. \end{aligned}\right\} \tag{3.10}$$

In these equations, f_{ij} means the probability that the state $\underline{\mathbf{r}}_j$ is reachable from the state $\underline{\mathbf{r}}_i$ in finite steps. If f_{ii}=1, the state $\underline{\mathbf{r}}_i$ is called recurrent or persistent. In the persistent case, the recurrent time is important and equal to μ_i. If f_{ii}<1, the state $\underline{\mathbf{r}}_i$ is called transient.

(4) General statement D

The probability of the transition from the states $\underline{\mathbf{r}}_i$ to $\underline{\mathbf{r}}_j$ in m steps is denoted by $p_{ij}^{(m)}$ (thus, $p_{ij}=p_{ij}^{(1)}$). A state $\underline{\mathbf{r}}_i$ is periodic and has a period d if the following conditions are satisfied:

$$p_{ii}^{(nd)} > 0 \quad (n = 1,2,\cdots), \qquad p_{ii}^{(m)} = 0 \quad (m \neq nd). \tag{3.11}$$

A state can return to the original state exactly at time points d, $2d$, …, in which the period d is an integer. A state $\underline{\mathbf{r}}_i$ is called aperiodic if such a value of d (>1) does not exist.

(5) General statement E

If a state $\underline{\mathbf{r}}_i$ is aperiodic and persistent with a finite recurrence time, the state $\underline{\mathbf{r}}_i$ is called ergodic. A Markov chain which has ergodic states alone is called an ergodic Markov chain.

(6) General statement F

An ergodic Markov chain has a stationary appearance probability. In this case, ρ_i>0 for all values of i, and the absolute appearance probability is irrespective of the initial condition. We now show the proof for this statement in the following. We consider a Markov chain with the total M microscopic states, and use the notation $p_{ij}^{(m)}$ for the transition probability of the state i (or $\underline{\mathbf{r}}_i$) transferring to the state j (or $\underline{\mathbf{r}}_j$) after m steps. Note that the transition probability p_{ij} has to satisfy the following condition:

$$\left.\begin{aligned} & p_{ij} > 0, \\ & \sum_{j=1}^{M} p_{ij} = 1. \end{aligned}\right\} \tag{3.12}$$

If all states are ergodic, an arbitrary state j can be reached from an arbitrary state i at a finite step m. This is expressed in recurrent form as

$$p_{ij}^{(m)} = \sum_{k=1}^{M} p_{ik}^{(m-1)} p_{kj} \, . \tag{3.13}$$

Hence, if we take the limit of $m \to \infty$, the absolute appearance probability of the state j, ρ_j, should be obtained. That is,

$$\lim_{m \to \infty} p_{ij}^{(m)} = \rho_j \, . \tag{3.14}$$

This means that ρ_j is independent of state i. Also, it is clear that the following conditions are satisfied:

$$\left.\begin{aligned} & \rho_j > 0, \\ & \sum_{j=1}^{M} \rho_j = 1. \end{aligned}\right\} \tag{3.15}$$

If Eq. (3.13) is substituted into Eq. (3.14), the following relation can be obtained:

$$\rho_j = \sum_{i=1}^{M} \rho_i p_{ij} \, . \tag{3.16}$$

Since Eqs. (3.15) and (3.16) are exactly equal to Eq. (3.9), an ergodic Markov chain has a stationary appearance probability, as already pointed out. Consequently, it is seen that, if an appropriate transition probability by which an ergodic Markov chain is generated is used, the microscopic states arise according to a certain unique probability density, and such a probability density is independent of the initial state. Finally, we show an example of the classification of Markov chains:

Markov chains
- irreducible Markov chains
  - ergodic Markov chains
  - periodic Markov chains
- absorbable Markov chains

3.3 The Metropolis Method

As already pointed out, an explicit expression for the weight function $w(\underline{\mathbf{r}}_i)$ has to be known to evaluate the average $\langle A \rangle$ in Eq. (3.5) in simulations based on the concept of importance sampling. Now we change the way of thinking. If an appropriate transition probability is used in simulations to conduct importance sampling and the state i (or $\underline{\mathbf{r}}_j$) ultimately arises with the following probability $\rho_i(=\rho(\underline{\mathbf{r}}_i))$, for example, for the canonical ensemble:

$$\rho_i = \frac{\exp\{-U(\underline{\mathbf{r}}_i)/kT\}}{\sum_i \exp\{-U(\underline{\mathbf{r}}_i)/kT\}}, \tag{3.17}$$

then importance sampling with the weight function in Eq. (3.4) may be considered to have been conducted. Hence the average $\langle A \rangle$ can now be evaluated from Eq. (3.5) in simulations. Since an ergodic Markov chain has a stationary appearance probability, this probability can become equal to Eq. (3.17) using an appropriate transition probability.

The transition probability p_{ij} which gives a stationary appearance probability ρ_i has to satisfy the following minimum conditions:

$$\left.\begin{array}{ll} (1)\, p_{ij} > 0 & \text{(for all } i \text{ and } j\text{)}, \\ (2)\, \sum_j p_{ij} = 1 & \text{(for all } i\text{)}, \\ (3)\, \rho_i = \sum_j \rho_j p_{ji} & \text{(for all } i\text{)}. \end{array}\right\} \tag{3.18}$$

Instead of condition (3) in Eq. (3.18), the following condition of detailed balance or microscopic reversibility is generally used:

$$(3)'\, \rho_i p_{ij} = \rho_j p_{ji} \qquad \text{(for all } i \text{ and } j\text{)}. \tag{3.19}$$

The microscopic reversibility condition is only sufficient but not necessary for the condition (3) in Eq. (3.18). This is straightforwardly verified by summing Eq. (3.19) with respect to index j with the condition (2) in Eq. (3.18). There is considerable freedom in determining the transition probability, explicitly since it cannot uniquely be specified from Eq. (3.18).

Metropolis et al. [10] proposed the following transition probability which satisfies the conditions (1), (2), and (3)′:

$$\left.\begin{array}{l} p_{ij} = \begin{cases} \alpha_{ij} & (i \neq j \text{ and } \rho_j/\rho_i \geq 1), \\ \alpha_{ij}\rho_j/\rho_i & (i \neq j \text{ and } \rho_j/\rho_i < 1), \end{cases} \\ p_{ii} = 1 - \sum_{j(\neq i)} p_{ij}, \end{array}\right\} \tag{3.20}$$

in which α_{ij} has to satisfy the following conditions:

$$x_k' = x_k + (2R_1 - 1)\delta r_{max}$$
$$y_k' = y_k + (2R_2 - 1)\delta r_{max}$$
$$z_k' = z_k + (2R_3 - 1)\delta r_{max}$$

5. Compute the potential U' for this new configuration
6. If $\Delta U = U' - U \leq 0$, accept the new configuration, take $\mathbf{r}_k = \mathbf{r}_k'$ and $U = U'$, and return to step 3
7. If $\Delta U > 0$, take a new random number R_4 from the above uniform number sequence
   7.1 If $\exp(-\Delta U/kT) \geq R_4$, accept the new configuration, take $\mathbf{r}_k = \mathbf{r}_k'$ and $U = U'$, and return to step 3
   7.2 If $\exp(-\Delta U/kT) < R_4$, reject the new configuration, regard the old configuration as a new one, and return to step 3

An important point to be noted in the algorithm is, as stated in step 7.2, that, if the new configuration is rejected, particle k is retained at its old position and the old configuration is recounted as a new state in the Markov chain. Step 6 ensures that the system is driven towards a minimum potential energy state with a dispersed distribution according to Boltzmann's factor.

3.4.2 The grand canonical Monte Carlo algorithm

The Monte Carlo method for the grand canonical ensemble, which is specified by a given (V,T,μ), is called the grand canonical MC, grand canonical ensemble MC, or $VT\mu$ MC method. Since the number of particles in a system is not constant, but fluctuates for this ensemble, the MC algorithm becomes more complicated than the canonical MC one. If the configuration for a system of N particles is denoted by $\underline{\mathbf{r}}^N$, the probability density is simplified to $\rho(\underline{\mathbf{r}}^N)$ after the integration of Eq. (2.27) with respect to the momenta. With this probability density, the ensemble average $\langle A \rangle$ of a microscopic quantity $A(\underline{\mathbf{r}}^N)$ is expressed as

$$\begin{aligned} \langle A \rangle &= \sum_N \int \rho(\underline{\mathbf{r}}^N) A(\underline{\mathbf{r}}^N) d(\underline{\mathbf{r}}^N) \\ &= \frac{1}{\Xi} \sum_N \frac{1}{N!\Lambda^{3N}} \exp(\mu N/kT) \int A(\underline{\mathbf{r}}^N) \exp\{-U(\underline{\mathbf{r}}^N)/kT\} d\underline{\mathbf{r}}^N . \end{aligned} \tag{3.24}$$

The grand canonical MC method is composed of the following three main parts:

(1) Movement of particle: move a particle
(2) Destruction of particle: delete a particle from the system

(3) Creation of particle: add a particle to the system

Now we show an expression for ρ_j/ρ_i which is necessary for computing the transition probability in simulations.

The procedure (1) is exactly the same as in the canonical MC method, since the number of particles, N, is constant. Hence Eq. (3.23) can be used without any changes.

In the procedure (2), a particle is chosen randomly and deleted from the system, whereby state i composed of N particles is transferred to a new state j composed of (N-1) particles. In this case, the ratio ρ_j/ρ_i can be derived, using Eqs. (2.27) and (2.8), as

$$\begin{aligned}\rho_j/\rho_i &= N\Lambda^3 \exp(-\mu/kT)\exp\left[-\left\{U(\underline{\mathbf{r}}^{N-1})-U(\underline{\mathbf{r}}^{N})\right\}/kT\right] \\ &= N\Lambda^3 \exp(-\mu/kT)\exp\left\{-\left(U_j-U_i\right)/kT\right\}.\end{aligned} \tag{3.25}$$

In the procedure (3), state i composed of N particles is transferred to a new state j composed of (N+1) particles by adding a particle to the system at a random position. The ratio ρ_j/ρ_i for this case can be expressed, using Eqs. (2.27) and (2.8), as

$$\rho_j/\rho_i = \frac{1}{(N+1)\Lambda^3}\exp(\mu/kT)\exp\left\{-\left(U_j-U_i\right)/kT\right\}. \tag{3.26}$$

Next we consider an appropriate ratio of attempting these three main procedures. To satisfy the condition of the microscopic reversibility, the procedures of creation and destruction have to be attempted with the same probability, whereby Eq. (3.21) can be satisfied. In contrast, there is considerable freedom concerning the ratio of attempting the procedures of creation/destruction and movement. It is known that the fastest convergence to an equilibrium state can be attained if these three main procedures are attempted with an equal probability [15]. The problem with the simple grand canonical MC algorithm is that the attempt of adding a particle to the system becomes less successful as the number density of particles increases. That is, the attempts of the creation of a particle are rejected with a higher rate for a denser system. This drawback may be improved by introducing the concept of ghost (or virtual) particles [16,17]. We consider next an improved Monte Carlo method based on this concept of ghost particles.

For a given volume, the system has substantially the maximum number of particles which can be packed into this volume. Let this maximum particle number be N_{max}. Simulations treat the system of N_{max} particles in the volume V; there are N real particles and (N_{max}-N) ghost particles in the system. The interactions between real particles, of course, have to be taken into account in the simulations, but ghost particles have no influence on the real and other ghost particles. With this concept, a particle is created by making a ghost particle real and the destruction of a particle is conducted by making a real particle virtual. We now derive an expression for ρ_j/ρ_i which corresponds to Eqs. (3.25) or (3.26).

Since the configuration integral can be written as

$$\int \exp\left\{-U(\underline{\mathbf{r}}^N)/kT\right\} d\underline{\mathbf{r}}^N = \frac{1}{V^{N_{max}-N}} \int \exp\left\{-U(\underline{\mathbf{r}}^{N_{max}};N)/kT\right\} d\underline{\mathbf{r}}^{N_{max}}, \tag{3.27}$$

the probability density $\rho(\underline{\mathbf{r}}^{N_{max}};N)$ is expressed as

$$\rho(\underline{\mathbf{r}}^{N_{max}};N) = \frac{1}{N!\Lambda^{3N}} \exp(\mu N/kT) \exp\left\{-U(\underline{\mathbf{r}}^{N_{max}};N)/kT\right\} V^N/\Xi', \tag{3.28}$$

in which Ξ' is written as

$$\Xi' = \sum_{N=1}^{N_{max}} \frac{V^N}{N!\Lambda^{3N}} \exp(\mu N/kT) \int \exp\left\{-U(\underline{\mathbf{r}}^{N_{max}};N)/kT\right\} d\underline{\mathbf{r}}^{N_{max}}. \tag{3.29}$$

The ensemble average $\langle A \rangle$ based on this probability density is expressed as

$$\langle A \rangle = \sum_{N=1}^{N_{max}} \int \rho(\underline{\mathbf{r}}^{N_{max}};N) A(\underline{\mathbf{r}}^N) d\underline{\mathbf{r}}^{N_{max}} = \sum_{N=1}^{N_{max}} \int \rho(\underline{\mathbf{r}}^N) A(\underline{\mathbf{r}}^N) d\underline{\mathbf{r}}^N. \tag{3.30}$$

This is in agreement with Eq. (3.24).

The expression for ρ_j/ρ_i in deleting a particle from the system can be derived, using Eq. (3.28), as

$$\begin{aligned} \rho_j/\rho_i &= \frac{N\Lambda^3}{V} \exp(-\mu/kT) \exp\left[-\left\{U(\underline{\mathbf{r}}^{N_{max}};N-1) - U(\underline{\mathbf{r}}^{N_{max}};N)\right\}/kT\right] \\ &= \frac{N\Lambda^3}{V} \exp(-\mu/kT) \exp\left\{-(U_j - U_i)/kT\right\}. \end{aligned} \tag{3.31}$$

For the case of adding a particle to the system,

$$\begin{aligned} \rho_j/\rho_i &= \frac{V}{(N+1)\Lambda^3} \exp(\mu/kT) \exp\left[-\left\{U(\underline{\mathbf{r}}^{N_{max}};N+1) - U(\underline{\mathbf{r}}^{N_{max}};N)\right\}/kT\right] \\ &= \frac{V}{(N+1)\Lambda^3} \exp(\mu/kT) \exp\left\{-(U_j - U_i)/kT\right\}. \end{aligned} \tag{3.32}$$

The main part of the grand canonical MC algorithm with the concept of ghost particles is as follows:

1. Specify an initial configuration $\underline{\mathbf{r}}^{N_{max}}$ for N_{max} particles, including ghost particles
2. Choose N_{real} particles from N_{max} particles randomly
3. Compute the potential energy U_i $(=U(\underline{\mathbf{r}}^{N_{max}};N))$
4. Movement of a particle:

4.1 Choose one particle from N_{max} particles

4.2 Generate a new configuration j by moving the selected particle using a random number

4.3 Compute the potential energy U_j for the new configuration

4.4 If $\Delta U = U_j - U_i \leq 0$, accept the new configuration, regard U_j as U_i, and go to step 5

4.5 If $\Delta U > 0$, take a random number R from a uniform random number sequence ranging between 0 and 1

4.5.1 If $\exp(-\Delta U/kT) \geq R$, accept the new configuration, regard U_j as U_i, and go to step 5

4.5.2 If $\exp(-\Delta U/kT) < R$, reject the new configuration, retain the old one, and go to step 5

5. Destruction of a particle:

5.1 Choose a real particle randomly which should be attempted to be deleted

5.2 Compute the potential energy U_j $(=U(\underline{\mathbf{r}}^{N_{max}}; N-1))$

5.3 Compute the value of ρ_j/ρ_i in Eq. (3.31)

5.4 If $\rho_j/\rho_i \geq 1$, make the particle virtual, regard U_j as U_i, and go to step 6

5.5 If $\rho_j/\rho_i < 1$, take a random number R from the uniform random number sequence

5.5.1 If $\rho_j/\rho_i \geq R$, make the particle virtual, regard U_j as U_i, and go to step 6

5.5.2 If $\rho_j/\rho_i < R$, keep the selected particle real, and go to step 6

6. Creation of a particle:

6.1 Choose a ghost particle randomly which should be attempted to be added to the system

6.2 Compute the potential energy U_j $(=U(\underline{\mathbf{r}}^{N_{max}}; N+1))$

6.3 Compute the value of ρ_j/ρ_i in Eq.(3.32)

6.4 If $\rho_j/\rho_i \geq 1$, make the selected particle real, regard U_j as U_i, and go to step 4

6.5 If $\rho_j/\rho_i < 1$, take a random number R from the uniform random number sequence

6.5.1 If $\rho_j/\rho_i \geq R$, make the particle real, regard U_j as U_i, and go to step 4

6.5.2 If $\rho_j/\rho_i < R$, keep the selected particle virtual, and go to step 4

In this algorithm, each procedure of movement, destruction, and creation is conducted with the same probability in a series of processes. It is noted that, if the selected particle in step 4 is virtual, the movement of the particle is always attained since ghost particles have no interactions with the other particles.

3.4.3 The isothermal-isobaric Monte Carlo algorithm

The Monte Carlo method for the isothermal-isobaric ensemble which is specified by a given (N,P,T) is called the isothermal-isobaric MC, isothermal-isobaric ensemble MC, or *NPT* MC method. Since the pressure is specified and constant, the system volume V fluctuates. If the probability density for the configuration $\underline{\mathbf{r}}$ and volume V is denoted by $\rho(\underline{\mathbf{r}};V)$, this is expressed as

$$\rho(\underline{\mathbf{r}};V) = \exp(-PV/kT)\exp\{-U(\underline{\mathbf{r}};V)/kT\}/Y_U \,, \tag{3.33}$$

in which

$$Y_U = \iint \exp(-PV/kT)\exp\{-U(\underline{\mathbf{r}};V)/kT\}d\underline{\mathbf{r}}dV \,. \tag{3.34}$$

Hence the ensemble average $\langle A \rangle$ of an arbitrary quantity $A(\underline{\mathbf{r}};V)$, which depends on $\underline{\mathbf{r}}$ and V, can be written as

$$\begin{aligned}\langle A \rangle &= \iint A(\underline{\mathbf{r}};V)\rho(\underline{\mathbf{r}};V)d\underline{\mathbf{r}}dV \\ &= \iint A\exp(-PV/kT)\exp\{-U(\underline{\mathbf{r}};V)/kT\}d\underline{\mathbf{r}}dV \Big/ Y_U \,.\end{aligned} \tag{3.35}$$

The fluctuation in the system volume is not desirable in simulations, so some techniques may be necessary to circumvent this unpleasant situation. Such a method is shown below.

The following method is very similar to the extended system method which is used in molecular dynamics methods. We consider a new coordinate system which is independent of the volume fluctuation. This may be attained by the coordinate transformation such that the particle positions $\underline{\mathbf{r}}(=(\mathbf{r}_1,\mathbf{r}_2,\ldots,\mathbf{r}_N))$ are transformed to $\underline{\mathbf{s}}(=(\mathbf{s}_1,\mathbf{s}_2,\ldots,\mathbf{s}_N))$ by the following equation:

$$\underline{\mathbf{s}} = \underline{\mathbf{r}}/L \,, \tag{3.36}$$

in which L is the side length of the system volume V, which is equal to $V^{1/3}$ for a cube. The volume V is nondimensionalized by V_0 which may be a volume at the close-packed configuration for a given particle number N; the dimensionless volume V^* is expressed as $V^*=V/V_0$. With the variables $\underline{\mathbf{s}}$ and V^*, Eq. (3.35) can be written as

$$\langle A \rangle = \iint A(L\underline{\mathbf{s}};V)\exp(-PV_0V^*/kT)\exp\{-U(L\underline{\mathbf{s}};V)/kT\}V^{*N}d\underline{\mathbf{s}}dV^* \Big/ Y_U' \,, \tag{3.37}$$

in which Y_U' is expressed as

$$Y_U' = \iint V^{*N}\exp(-PV_0V^*/kT)\exp\{-U(L\underline{\mathbf{s}};V)/kT\}d\underline{\mathbf{s}}dV^* \,. \tag{3.38}$$

Hence, the probability density $\rho(\underline{\mathbf{s}};V^*)$ concerning the probability of the state $(\underline{\mathbf{s}},V^*)$ appearing under circumstances of given values of (N, P, T) and V_0, can be derived as

$$\rho(\underline{\mathbf{s}};V^*)=\exp\left[\ -\left\{PV_0V^*+U(L\underline{\mathbf{s}};V)-NkT\ln V^*\right\}/kT\ \right]/Y'_U\ . \tag{3.39}$$

The expression of ρ_j/ρ_i for the transition from the states $(\underline{\mathbf{s}}_i;V_i^*)$ to $(\underline{\mathbf{s}}_j;V_j^*)$ can be obtained from Eq. (3.39):

$$\rho_j/\rho_i=\exp\left[-\left\{PV_0(V_j^*-V_i^*)+U_j-U_i-NkT\ln(V_j^*/V_i^*)\right\}/kT\right]. \tag{3.40}$$

The main part of the Monte Carlo algorithm using this coordinate transformation is as follows:

1. Specify an initial volume and configuration , and regard this state as state i
2. Compute the potential energy U_i
3. Change the volume from V_i^* to V_j^* using a random number R_1 taken from a uniform random number sequence ranging from 0 to 1; that is, $V_j^*=V_i^*+(2R_1-1)\delta V_{max}^*$
4. Choose a particle randomly and move this selected particle to an arbitrary position using other random numbers, and regard this new state (with a new configuration and a new volume) as state j
5. Compute the potential energy U_j for state j
6. Compute $\Delta H'=\{PV_0(V_j^*-V_i^*)+U_j-U_i-NkT\ln(V_j^*/V_i^*)\}$
7. If $\Delta H'\leq 0$, accept the new state j, regard state j as state i, and return to step3
8. If $\Delta H'>0$, take another random number R_2
   - 8.1 If $\exp(-\Delta H'/kT)\geq R_2$, accept the new state j, regard state j as state i, and return to step 3
   - 8.2 If $\exp(-\Delta H'/kT)<R_2$, reject state j, retain the old volume V_i^* and old configuration $\underline{\mathbf{s}}_i$, and return to step 3

It is noted that, in step 8.2, the old state i has to be regarded as a state of a Markov chain. In this algorithm, the procedures of moving particles and changing the volume are combined together, but separate procedures are possible.

Finally, we state some points to be noted in computing potential energies. Potential energies have to be computed for the original coordinate $\underline{\mathbf{r}}$, not for the transformed one $\underline{\mathbf{s}}$. This means that the interaction energies of all pairs of particles have to be recomputed after the volume change. Hence, generally speaking, the isothermal-isobaric MC algorithm is computationally much more expensive than the canonical MC algorithm. However, in some cases, for example, the Lennard-Jones potential, this difficulty does not appear seriously since the potential energies after the volume change can be computed using the old configuration and potential energies. We exemplify this situation using equations below.

If the particle-particle interaction energy is given by the Lennard-Jones potential (Eq. (A3.1)), the potential energy U_i of the system in state i can be written, using the transformed coordinate $\underline{\mathbf{s}}$, as

$$U_i = 4\varepsilon \sum_k \sum_{\substack{l \\ (k<l)}} \left(\frac{\sigma}{L_i s_{kl}} \right)^{12} - 4\varepsilon \sum_k \sum_{\substack{l \\ (k<l)}} \left(\frac{\sigma}{L_i s_{kl}} \right)^{6}, \tag{3.41}$$

in which $s_{kl} = | \mathbf{s}_k - \mathbf{s}_l |$. If the first and second terms on the right-hand side are denoted by $U^{(12)}$ and $U^{(6)}$ (not including the negative sign), respectively, then the potential energy U_j of the system in state j after the volume change from V_i to V_j (or L_i to L_j) can be written as

$$U_j = U_i^{(12)} (L_i / L_j)^{12} - U_i^{(6)} (L_i / L_j)^{6}. \tag{3.42}$$

Hence, if $U^{(12)}$ and $U^{(6)}$ are separately saved as variables in simulations, the new potential energy U_j can be evaluated from the ratio L_i/L_j and the old values of $U_i^{(12)}$ and $U_i^{(6)}$ without directly calculating the interaction energies between particles in the new configuration.

3.5 The Cluster-Moving Monte Carlo Algorithm

Aggregation phenomena of dispersed particles in ferrofluids and in usual colloidal suspensions are of great interest from the standpoint of computer simulations. If the physical phenomenon of particle aggregation proceeds far beyond the time range of computer simulations, it is impossible for conventional MC algorithms to reproduce the correct aggregate structure of particles. It is the cluster-moving Monte Carlo method [18,19] that has been developed for improving drastically the convergence rate from an initial state to equilibrium. We here show the essence of the cluster-moving Monte Carlo method, and then in the next section the cluster analysis method [20-22] will be explained, which is a sophisticated technique for defining the cluster formation.

We consider a dispersion in which the fine particles are dispersed. If the attractive interaction between particles is much greater than the thermal energy, particles should aggregate to form clusters. For example, for a ferrofluid [23] in which ferromagnetic particles are dispersed, such particles should aggregate to form chainlike clusters along an applied magnetic field if the strength of magnetic interactions to the thermal energy, λ, is much greater than unity (see Eq. (3.58) for λ). However, the ordinary MC method predicts only short clusters, even for a strong interaction case such as λ=15, as shown in Fig. 3.4(a); such short clusters do not grow with advancing MC steps (a MC step means N attempts of moving N

particles of a system). Therefore, we may say that the ordinary MC method cannot capture a physically reasonable aggregate structure. We will mention this point again in Sec. 3.7.

Why does the aggregation of particles not progress from the structure of Fig.3.4(a) in the process of the conventional MC algorithm? In order for clusters to grow, the particles have to separate from clusters and combine with growing clusters. For very strong forces between particles such as λ=15, however, such separation of particles does not frequently occur. This is why the ordinary algorithms cannot capture a correct aggregate structure. Hence it is reasonable that the cluster-moving MC algorithm, with the movement of clusters, shows an outstanding convergence rate from an initial state to equilibrium.

In the cluster-moving MC method, particle clusters which are formed during the process of a simulation are allowed to move as unitary particles. For a strongly-interacting system, the aggregate structure of particles is hardly changed during the simulation process by the conventional algorithm, i.e. a lot of clusters may be formed but they will remain dispersed without joining each other. The cluster-moving MC algorithm, however, regards such scattered clusters as unitary particles, and moves the clusters with a transition probability which is equal to p_{ij} in Eq. (3.20) in the Metropolis method; the acceleration of the particle aggregation process can significantly be achieved. Using the Metropolis transition probability in both the movement of particles and the movement of clusters ensures that the algorithm returns to the conventional one unless particles significantly aggregate.

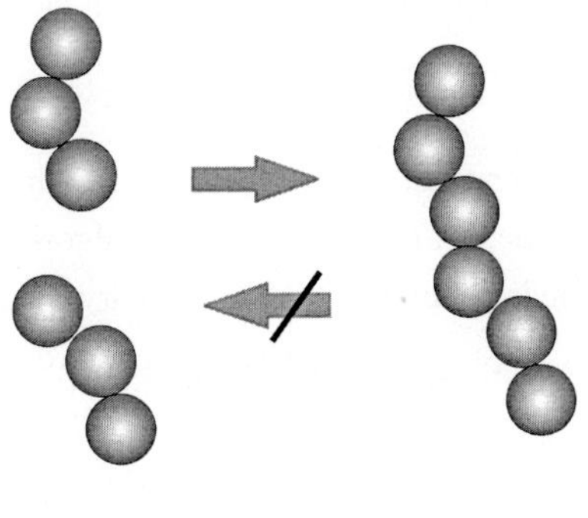

Figure 3.2 Characteristics of cluster-moving MC algorithm.

We now consider the theoretical background of the cluster-moving MC algorithm. As shown in Fig.3.2, the present algorithm includes the step of joining two clusters into one, but not that of dissociating one cluster into two. This means that Eq. (3.21) cannot be satisfied. Hence, it should be noted that the cluster-moving MC algorithm has to be switched to a conventional one when a system has attained equilibrium. After the further attainment to equilibrium by the switched algorithm, the mean values of quantities of interest may be evaluated. If the trial of the cluster movement is not attempted so frequently, the equilibrium with the cluster-moving MC algorithm is essentially equivalent to that achieved by the algorithm without the cluster movement; this is because the transition probability based on the Metropolis method is used in moving clusters. It may be possible to develop a modified cluster-moving MC method which satisfies the condition in Eq. (3.21) by introducing the procedure that clusters are dissociated into small clusters.

The main part of the cluster-moving algorithm, that is, the transition from states i to j, is as follows:

1. Check the formation of clusters
2. Select a cluster in order or randomly
3. Compute the interaction energy of the selected cluster with the other clusters, U
4. Move the cluster randomly and compute the interaction energy of the cluster at the new position, U'
5. If $\Delta U=U'-U\leq 0$, then accept the new position and return to step 2
6. If $\Delta U>0$, then generate a uniform random number R $(0<R<1)$
    6.1 If $\exp(-\Delta U/kT)\geq R$, then accept the new position and return to step 2
    6.2 If $\exp(-\Delta U/kT)<R$, then reject the movement, retain the old position, and return to step 2 (the old state is regarded as a new state)

The procedure from step 2 is repeated for all the clusters; single particles are regarded as clusters for convenience. Practically, the cluster-moving algorithm is used in such a way that the above-mentioned procedure is integrated into the ordinary Metropolis MC algorithm.

To verify the validity of the cluster-moving MC method, two-dimensional simulations were carried out for ferromagnetic dispersions. We used the following assumptions and parameters in simulations. The magnetic interaction energy between two particles is expressed by Eq. (3.57). The magnetic moment of the particles points in the field direction under circumstances of a strong magnetic field. The cutoff radius r_{coff} for the interaction energies, which will be explained in Sec. 11.3, is $r_{coff}=5d$ (d is the particle diameter). The number of particles is N=400 and the size of a square cell as a simulation region is adjusted such that the volumetric packing fraction of particles φ_V was taken as φ_V =0.046. All the results shown in the following were obtained by simulating up to 20,000 MC steps, in which the procedure with the cluster movements (the cluster-moving MC algorithm) is up to 10,000 MC steps and the remaining procedure is by the Metropolis algorithm.

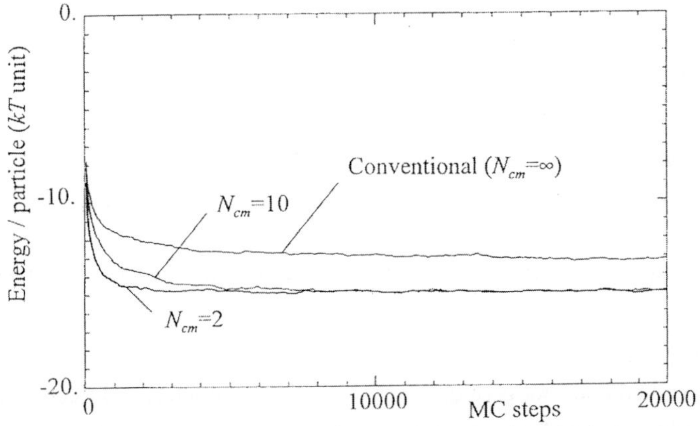

Figure 3.3 Evolution of interaction energies.

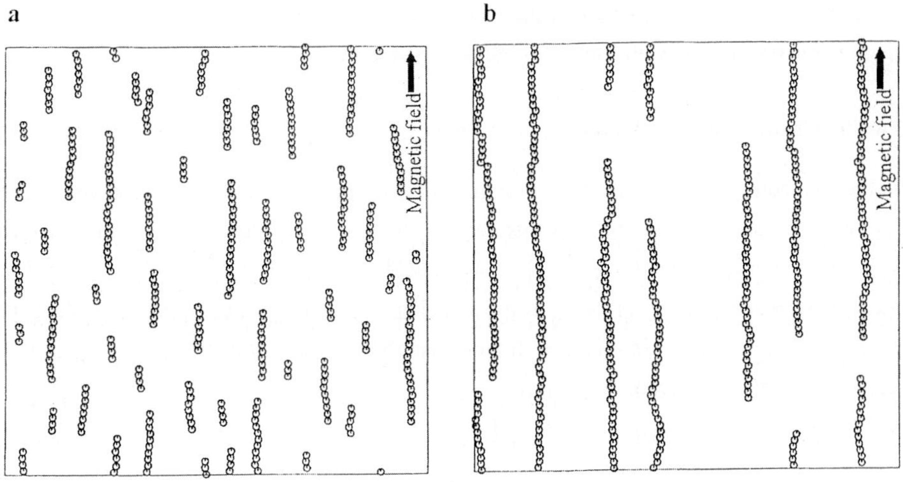

Figure 3.4 Aggregate structures of ferromagnetic particles: (a) for N_{cm}=∞,and (b) for N_{cm}=2.

There are some methods by which the cluster formation is quantified, and we here adopted the simplest method which is based on the particle-particle separation. That is, if the separation between two particles is smaller than a criterion value r_c, then the two particles are regarded as forming a cluster. A more sophisticated method for defining clusters will be explained in the next section. In the present simulations, such a criterion value was taken as r_c=1.3d.

The influence of N_{cm} on the convergence of the system is shown in the following for three cases of N_{cm}, where N_{cm} means that cluster movements are attempted at every N_{cm} MC steps. Figure 3.3 shows the evolution of the interaction energy per particle for λ=15. The curve by the ordinary algorithm ($N_{cm}=\infty$) indicates a very slow convergence rate, so that it is impossible for the conventional algorithm to capture the correct aggregate structure of particles. The aggregate structure after 20,000 MC steps by the ordinary algorithm is given in Fig. 3.4(a). This structure is not essentially different from that after 1,000 MC steps. It is seen from Eq. (3.57) that particles should aggregate to form long chainlike clusters along the field direction for λ=15, since the minimum of the potential curve for this case is about 30kT. Hence the ordinary MC method does not give the correct aggregate structure. In contrast the curve for the cluster-moving MC method gives outstanding convergence compared to the ordinary one, and the larger N_{cm} is, the faster the equilibrium is attained. Figure 3.4(b) shows the aggregate structure predicted by the cluster-moving MC algorithm. Chainlike clusters with almost infinite length are formed along the field direction, which is highly reasonable from a physical point of view. It is concluded that the cluster-moving method is highly useful for simulating aggregation phenomena for a strongly-interacting system.

3.6 The Cluster Analysis Method

In the preceding section, the definition in terms of particle-particle separations is used in defining the cluster formation. In this section, we outline a more sophisticated method for defining the cluster formation [20-22]. We consider a system of N particles, in which many clusters are assumed to be formed. If the i-th cluster is composed of N_i particles, the equation of $\sum_i N_i = N$ is valid. In this case the total number of ways of distributing all particles into each cluster, $W(\{N_i\})$, can be written as

$$W(\{N_i\}) = W(N_1, N_2, \cdots) = N!\Big/\prod_j N_j!. \tag{3.43}$$

The entropy $S(\{N_i\})$ can be obtained from Sec. 2.2.1 for a given distribution of $\{N_i\}$as

$$S(\{N_i\}) = k\ln W(\{N_i\}) = k(\ln N! - \sum_j \ln N_j!)\,, \tag{3.44}$$

in which k is Boltzmann's constant. If there are no interactions between particles, the system approaches a cluster partition such as to maximize the entropy S, that is, all particles move individually without cluster formation. However, if there are interactions between particles, the energies of forming clusters have influence on the cluster distribution.

Since cluster j is composed of N_j particles, the energy of forming the cluster, U_j, is evaluated by summing the interaction energies between the particles forming the cluster:

$$U_j = \frac{1}{2}\sum_{k=1}^{N_j}\sum_{\substack{l=1 \\ (l\neq k)}}^{N_j} u_{kl}, \tag{3.45}$$

in which u_{kl} is the interaction energy between particles k and l. Hence, by summing all the cluster energies, the energies of forming clusters in the system, $U(\{N_i\})$, can be obtained as

$$U(\{N_i\}) = \sum_j U_j = \frac{1}{2}\sum_j \sum_{k_j=1}^{N_j}\sum_{\substack{l_j=1 \\ (l_j\neq k_j)}}^{N_j} u_{k_j l_j}. \tag{3.46}$$

If particle-particle interaction energies are significantly large (attractive), then the system approaches an equilibrium state in which the energy $U(\{N_i\})$ is minimized, that is, one big cluster alone is formed.

From the above-mentioned discussion, it is physically reasonable to think that the cluster distributions are determined by the following quantity $F(\{N_i\})$:

$$\begin{aligned} F(\{N_i\}) &= U(\{N_i\}) - TS(\{N_i\}) = \sum_j U_j - kT\ln W(\{N_i\}) \\ &= \frac{1}{2}\sum_j \sum_{k_j=1}^{N_j}\sum_{\substack{l_j=1 \\ (l_j\neq k_j)}}^{N_j} u_{k_j l_j} - kT(\ln N! - \sum_j \ln N_j!). \end{aligned} \tag{3.47}$$

This quantity is very similar to the Helmholtz free energy, but it is not exactly equivalent to the free energy since the first term on the right-hand side is not the internal energy of the system. In thermodynamic equilibrium, the cluster distribution which gives a minimum value of F in Eq. (3.47) should appear. The method of defining the cluster formation by means of the function F is called the cluster analysis method. Since the analytical solution of the cluster distribution which gives a minimum of F is highly difficult, cluster formation based on F is evaluated by the Metropolis method in simulations. It is noted that there is some freedom concerning the definition of F [20-22].

The probability density $p(\{N_i\})$ for the cluster distribution $\{N_i\}$ can be written as

$$p(\{N_i\}) = \frac{1}{Q}\exp\left[-\left(\frac{1}{kT}\sum_j U_j - \ln W(\{N_i\})\right)\right]. \tag{3.48}$$

The procedure of assigning the cluster distribution by the Metropolis method based on this probability density is combined into the usual procedures such as the particle movement in Monte Carlo simulations. We now show an algorithm concerning the analysis of the cluster configuration.

We consider the transition from the cluster distribution $\{N_i\}$ to the new cluster distribution $\{N_i'\}$ by changing the belonging cluster of particle l from cluster i to cluster j. In this case, the transition probability $p_{i(l)j}$ can be written as

$$p_{i(l)j} = \exp\left(-\frac{1}{kT}\Delta U + \Delta\{\ln W\} \right), \tag{3.49}$$

in which

$$\begin{aligned} \Delta U &= \left(U_j(N_j') + U_i(N_i')\right) - \left(U_j(N_j) + U_i(N_i)\right) \\ &= \left(U_j(N_j') - U_j(N_j)\right) + \left(U_i(N_i') - U_i(N_i)\right) = \sum_{k=1}^{N_j} u_{k,N_j+1} - \sum_{\substack{k=1 \\ (k\neq l)}}^{N_i} u_{kl}, \end{aligned} \tag{3.50}$$

$$\begin{aligned} \Delta\{\ln W\} &= \{\ln W(\{N_i'\}) - \ln W(\{N_i\})\} = -(\ln N_j'! + \ln N_i'!) + (\ln N_j! + \ln N_i!) \\ &= -(\ln(N_j+1)! + \ln(N_i-1)!) + (\ln N_j! + \ln N_i!) = -\ln(N_j+1) + \ln N_i. \end{aligned} \tag{3.51}$$

In Eq. (3.49), we set $p_{i(l)j}$ =1 if $p_{i(l)j}$ >1.

The procedure of separating a particle from its assigned cluster is also necessary. If the particle which belongs to cluster i separates from the cluster and becomes a single-moving particle, the transition probability $p_{i(l)}$ can be written, from a similar procedure to the preceding discussion, as

$$p_{i(l)} = \exp\left[\frac{1}{kT} \sum_{\substack{k=1 \\ (k\neq l)}}^{N_i} u_{kl} + \ln N_i \right], \tag{3.52}$$

in which the constitution energy of a single particle, U_l, is zero.

Finally we show the main part of the algorithm of the cluster analysis method below.

1. Choose a particle randomly, and consider the particle and its cluster as particle l and cluster i, respectively
2. Choose a cluster randomly, and consider the cluster as cluster j
3. If $j=i$, go to step 5
4. Test for acceptance of move to new cluster:
   4.1 Compute ΔU and $\Delta\{\ln W\}$
   4.2 If $\Delta F = \Delta U - kT\,\Delta\{\ln W\} \leq 0$, change the assigned cluster of particle l from cluster i to cluster j, and return to step 1

4.3 If $\Delta F>0$, take a random number R from a uniform random number sequence ranging from 0 to 1

4.3.1 If $\exp\{-\Delta F/kT\}\geq R$, change the assigned cluster of particle l from cluster i to cluster j, and return to step 1

4.3.2 If $\exp\{-\Delta F/kT\}<R$, go to step 5

5. Move of particle to self-cluster:

5.1 Compute ΔU and $\Delta\{\ln W\}$

5.2 If $\Delta F=\Delta U-kT\,\Delta\{\ln W\}\leq 0$, separate particle l from cluster i, and return to step 1

5.3 If $\Delta F>0$, take another random number R from the uniform random number sequence

5.3.1 If $\exp\{-\Delta F/kT\}\geq R$, separate particle l from cluster i, and return to step 1

5.3.2 If $\exp\{-\Delta F/kT\}<R$, don't separate, retain particle l in cluster i, and return to step 1

These procedures have to be repeated until the distribution becomes stable or stationary. The cluster analysis procedure is combined into the ordinary particle-moving procedures in Monte Carlo simulations. However, since it is computationally expensive to conduct the cluster analysis procedure every time step, such procedure is carried out every certain time steps. The cluster analysis procedure starts under the assumption that a system is composed of single-moving particles alone without any clusters.

3.7 Some Examples of Monte Carlo Simulations

In this section, we concentrate our attention on the results which were obtained by the author to understand the role of molecular-microsimulations in investigating physical phenomena for colloidal dispersions.

3.7.1 Aggregation phenomena in non-magnetic colloidal dispersions

As already pointed out, for a strongly-interacting system, the ordinary Metropolis MC method shows a very slow convergence rate and, therefore, cannot capture a physically reasonable aggregate structure of colloidal particles. To circumvent this difficulty, the cluster-moving MC method has been developed. We here show the results [19] to verify the validity of the cluster-moving MC method for an ordinary colloidal dispersion which does not exhibit magnetic or electric properties.

the cluster-moving procedure is finished. Large aggregates with no scattered small clusters are seen to be formed in this figure. As already mentioned in Sec. 3.5, the present method requires the further attainment to equilibrium by the procedure without cluster movements, after the system almost converges to equilibrium by the procedure with cluster movements. Since the procedure with cluster movements determines the main structure of aggregates (the procedure without cluster movements merely refines the structure), physically reasonable aggregate structure cannot be obtained unless this procedure is sufficient. This insufficiency of the procedure with cluster movements is the reason the aggregate structure in Fig. 3.7(d) broke down to that in Fig. 3.7(c) after another 10,000 procedure without cluster movements.

Figure 3.6(b) shows the evolution of the interaction energy for η=70, where the influence of the maximum displacement of particles or clusters is investigated. Generally, the particles cannot aggregate with each other over a potential barrier with a height of more than $15kT$. Figure 3.5 shows the height of the potential barrier to be about $30kT$ for η =70, so that the particle aggregation cannot occur from a theoretical point of view. In simulations, we have to make the particles recognize this potential barrier and this may be accomplished by using an appropriate value of δr_{max} which is much smaller than the representative value of the potential barrier range. If this condition is not imposed on δr_{max}, the particles will be able to step over the barrier without recognizing the existence of the potential barrier, and thus false aggregate structures will be obtained. There are jumps in the energy around 10,000 and 15,000 MC steps for δr_{max}=0.2 in Fig. 3.6(b), which result from certain particles having jumped over the potential barrier due to the use of a large value of δr_{max}. It should, therefore, be noted that the use of large values of δr_{max} is generally restricted by the shape of the potential curve.

3.7.2 Aggregation phenomena in ferromagnetic colloidal dispersions

Ferrofluids are colloidal dispersions composed of nanometer-scale ferromagnetic particles dispersed in a carrier liquid; they are maintained in a stable state by a coating of surfactant molecules to prevent the aggregation of particles [23]. Since the diameter of the particles is typically about 10 nm, it is clear that the thick chainlike clusters [28,29], which are observable even with a microscope, are much thicker than the clusters which are obtained by the particles combining with each other to produce linear chains in a field direction. The formation of thick chainlike clusters can be explained very well by the concept of secondary particles [18]. That is, primary particles aggregate to form secondary particles due to isotropic attractive forces between primary particles. These secondary particles exhibit a magnetic moment in an applied magnetic field and so aggregate to form chainlike clusters along the field direction. Finally these clusters further aggregate to form thick chainlike cultures [18,30,31]. The investigation of the potential curves of linear chainlike clusters, which will be discussed in the next section, has shown that large attractive forces arise between chainlike clusters.

We here discuss the thick chainlike cluster formation for a three-dimensional system which was obtained by the cluster-moving Monte Carlo simulation [31].

A. The model of particles

Ferromagnetic particles are idealized as a spherical particle with a central magnetic dipole. If the magnitude of the magnetic moment is denoted by m and the magnetic field strength is denoted by $\mathbf{H}$ ($H=|\mathbf{H}|$), then the interactions between particle i and the magnetic field, and between particles i and j, are expressed, respectively, as

$$\left.\begin{aligned} u_i &= -kT\xi\,\mathbf{n}_i \cdot \mathbf{H}/H, \\ u_{ij} &= kT\lambda\frac{d^3}{r_{ij}^3}\left\{\mathbf{n}_i \cdot \mathbf{n}_j - 3(\mathbf{n}_i \cdot \mathbf{t}_{ji})(\mathbf{n}_j \cdot \mathbf{t}_{ji})\right\}, \end{aligned}\right\} \tag{3.57}$$

in which ξ and λ are the dimensionless parameters representing the strengths of particle-field and particle-particle interactions relative to the thermal energy kT, respectively. These parameters are written as

$$\xi = \mu_0 mH/kT, \qquad \lambda = \mu_0 m^2/4\pi d^3 kT, \tag{3.58}$$

in which k is Boltzmann's constant, μ_0 the permeability of free space, T the liquid temperature, $\mathbf{m}_i$ the magnetic moment ($m=|\mathbf{m}_i|$), and r_{ji} the magnitude of the vector $\mathbf{r}_{ji}$ drawn from particles i to j. Also $\mathbf{n}_i$ and $\mathbf{t}_{ji}$ are the unit vectors given by $\mathbf{n}_i=\mathbf{m}_i/m$ and $\mathbf{t}_{ji}=\mathbf{r}_{ji}/r_{ji}$, and d is the diameter of particles.

B. Potential curves of linear chainlike clusters

We will discuss in detail the results of the potential curves of linear chainlike clusters composed of secondary particles in the next section, so here we just make a short comment concerning the potential between clusters. For the parallel arrangement, only repulsive forces act between clusters and the attractive forces do not arise for any of the cluster-cluster separations. In contrast, for the staggered cluster arrangement, attractive forces are seen when two clusters are sufficiently close to each other.

C. Parameters for simulations

Three-dimensional Monte Carlo simulations were carried out under the following conditions. The system is assumed to be composed of N_{sys} particles, but the solvent molecules are not taken into account. The simulation region is a cube. The particle number N_{sys} and the dimensionless number density $n^*(=nd^3)$ were taken as $N_{sys}=1000$ and $n^*=0.1$. The effective distance for the cluster formation, r_c, is $1.2d$, where a cluster is defined as a group of particles in which adjacent particles are within the maximum particle-particle separation r_c. The maximum displacement of particles and clusters in one trial move is $0.5d$ and the maximum change in the particle orientation is 10°. It is desirable to take into account long-range magnetic interactions and there are some techniques of handling these long-range interactions. However, three-dimensional simulations require considerable computation time, even with a large value of the cutoff distance. Fortunately, the minimum image convention is a reasonable technique as a first approximation, even for this type of a dipolar system. Hence the introduction of the cutoff distance is allowed as a first approximation, unless it is too small. In consideration of the results in Sec. 3.7.3, the cutoff radius r_{coff} for particle-particle interactions was taken as $r_{coff}=8d$. We adopted the cluster-moving procedure every 70 MC steps and this procedure was continued until the end of each run.

We evaluate two types of the pair correlation function, $g_{\|}^{(2)}(\mathbf{r})$ and $g_{\perp}^{(2)}(\mathbf{r})$, which are for the directions parallel and normal to the magnetic field, respectively. In these cases, the small volume element ΔV in Eq. (11.7) is $\pi r^2(\Delta\theta)^2\Delta r/4$ and $2\pi r^2\Delta\theta\Delta r$ for $g_{\|}^{(2)}(\mathbf{r})$ and $g_{\perp}^{(2)}(\mathbf{r})$, respectively. The parameters Δr and $\Delta\theta$ were taken as $\Delta r=d/20$ and $\Delta\theta=\pi/18$. Each simulation was carried out to about 150,000 MC steps to obtain snapshots of the aggregate structures in equilibrium. Further 270,000 MC steps were continued for evaluating the mean values of the pair correlation function.

Ferromagnetic particles have a magnetic dipole moment, so change in the moment direction is necessary in conducting simulations. We here briefly explain the method of changing the direction of the magnetic moment. As shown in Fig. 3.8, the unit vector is assumed to point in the direction (θ,φ). The new coordinate system, XYZ, is made from the original one, xyz, by rotating the original coordinate system around the z-axis by the angle φ, and then rotating it around the y-axis by the angle θ. If the fundamental unit vectors for the two coordinate systems are denoted by $(\boldsymbol{\delta}_x,\boldsymbol{\delta}_y,\boldsymbol{\delta}_z)$ and $(\boldsymbol{\delta}_X,\boldsymbol{\delta}_Y,\boldsymbol{\delta}_Z)$, the xyz-components in the original coordinate system are related to the XYZ-component in the new coordinate system as

$$\begin{pmatrix} x \\ y \\ z \end{pmatrix} = \mathbf{T}_1^{-1}\mathbf{T}_2^{-1}\begin{pmatrix} X \\ Y \\ Z \end{pmatrix}, \tag{3.59}$$

in which $\mathbf{T}_1^{-1}$ and $\mathbf{T}_2^{-1}$ are the inverse matrices of $\mathbf{T}_1$ and $\mathbf{T}_2$, respectively, which are for the above-mentioned coordinate transformation. The expressions for $\mathbf{T}_1$ and $\mathbf{T}_2$ are written as

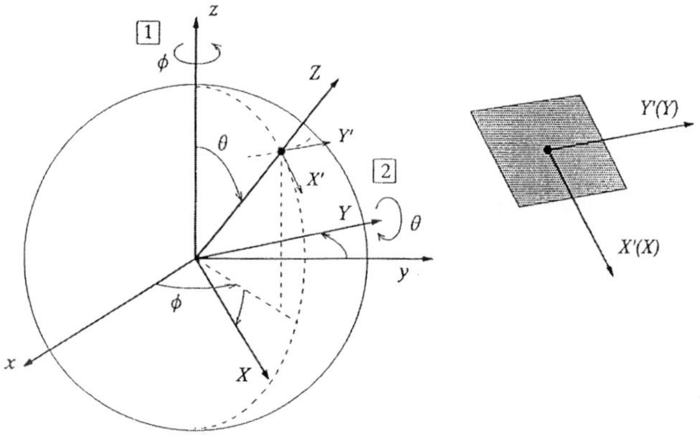

Figure 3.8 Method of changing the direction of the unit vector by a small angle.

$$\mathbf{T}_1 = \begin{pmatrix} \cos\varphi & \sin\varphi & 0 \\ -\sin\varphi & \cos\varphi & 0 \\ 0 & 0 & 1 \end{pmatrix}, \quad \mathbf{T}_2 = \begin{pmatrix} \cos\theta & 0 & -\sin\theta \\ 0 & 1 & 0 \\ \sin\theta & 0 & \cos\theta \end{pmatrix}. \tag{3.60}$$

From these expressions, $\mathbf{T}_1^{-1}\mathbf{T}_2^{-1}$ can be written as

$$\mathbf{T}_1^{-1}\mathbf{T}_2^{-1} = \begin{pmatrix} \cos\theta\cos\varphi & -\sin\varphi & \sin\theta\cos\varphi \\ \cos\theta\sin\varphi & \cos\varphi & \sin\theta\sin\varphi \\ -\sin\theta & 0 & \cos\theta \end{pmatrix}. \tag{3.61}$$

It is noted that the direction of the Z-axis is the same as the unit vector $\mathbf{n}$; that is, $\mathbf{n}=\boldsymbol{\delta}_Z$. Hence, in order to change the direction from $\mathbf{n}$ to $\mathbf{n}'$, the following method, using uniform random numbers R_1 and R_2, can be used:

$$\begin{pmatrix} n'_X \\ n'_Y \end{pmatrix} = \delta r \begin{pmatrix} 2R_1 - 1 \\ 2R_2 - 1 \end{pmatrix}, \tag{3.62}$$

in which δr is a constant specifying the maximum displacement. The value of $\mathbf{n}_Z'$ can be determined from the fact that $\mathbf{n}'$ is the unit vector, that is, $n_Z'=(1-n_X'^2-n_Y'^2)^{1/2}$. Thus the components (n_x', n_y', n_z') in the xyz-coordinate system can be obtained from Eq. (3.59). It is noted that this method is an approximate technique since a spherical surface is replaced with a tangential surface in the above procedure. In the present case the maximum angle displacement δr was taken as $\delta r=10°$.

D. Results and discussion

Figure 3.9 shows the dependence of the aggregate structures on the interactions between particles for ξ=30, in which snapshots in equilibrium are indicated for three cases of λ. The figures on the left-hand side are the views observed from an oblique angle, and those on the right-hand side are the views from the field direction. Figure 3.9(b) for λ=4 clearly shows that some thick chainlike clusters are formed along the applied field direction. For this case, it is seen from Fig. 3.12 that the height of the potential barrier is about $4kT$ (note that Λ=λ for the present case). This energy barrier is low enough for the thin chainlike clusters to overcome and form thick chainlike clusters. For strong particle-particle interactions such as λ=15, the aggregate structures are qualitatively different from those for λ=4. That is, although long chainlike clusters are observed along the field direction, they are much thinner than those shown in Fig.3.9(b). Figure 3.12 shows that it is impossible for these thin clusters to overcome the high energy barrier to form thick chainlike clusters, since the potential curve has an energy barrier with a height of about $15kT$ for λ=15. The thin chainlike clusters shown in Fig. 3.9(c), therefore, do not coalesce to form thick chainlike clusters as shown in Fig. 3.9(b). For a weak particle-particle interaction such as λ=3, although many short clusters are formed, long chainlike clusters are not observed. These aggregate structures are in significant contrast with those shown in Figs. 3.9(b) and (c).

Figure 3.10 shows the dependence of the aggregate structures on the magnetic field strength for λ=4, in which snapshots of the aggregate structures are given for two cases of ξ=1 and 5. The following understanding may be derived from comparing the results shown in Figs. 3.10 (a) and (b) with Fig. 3.9(b). For ξ=5, chainlike clusters are observed along the field direction, but they are loose structures compared with those for ξ=30. Since the strength of particle-particle interaction is nearly equal to that of particle-field interaction for ξ=5, the clusters do not necessarily have a closely linear form along the field direction. The chainlike clusters in Fig. 3.10(b) deviate considerably from the perfect linear form. We, therefore, may understand that sufficiently attractive forces do not arise between these winding clusters for the formation of dense thick chainlike clusters. For the case of ξ=1, thick chainlike clusters are not observed in Fig. 3.10(a), since the nonlinear cluster formation is preferred for the case where the strength of particle-field interaction is much smaller than that of particle-particle interaction [32,33].

Figure 3.11 shows the dependence of the pair correlation function on the interactions between particles for ξ=30, in which the pair correlation functions parallel and normal to the magnetic field are indicated in Figs. 3.11(a) and (b), respectively. These results were obtained

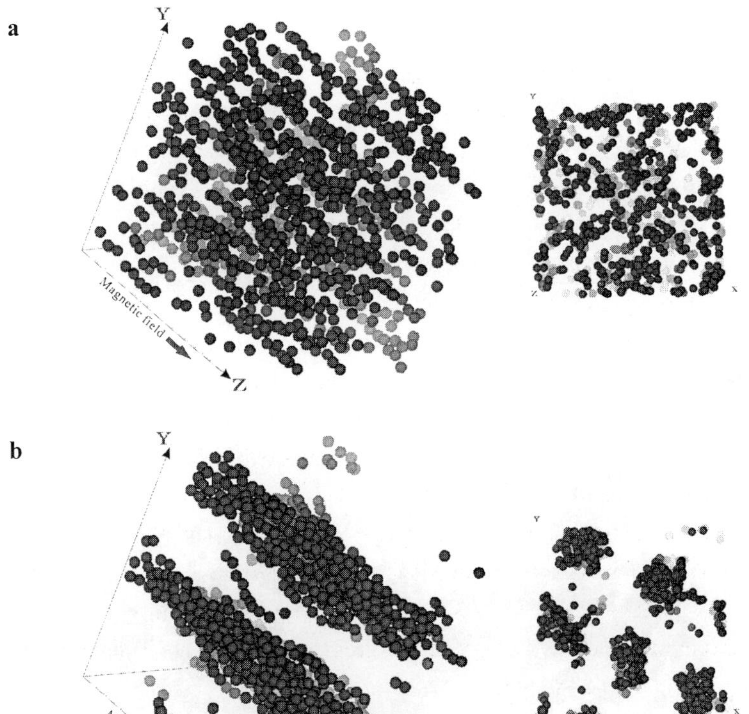

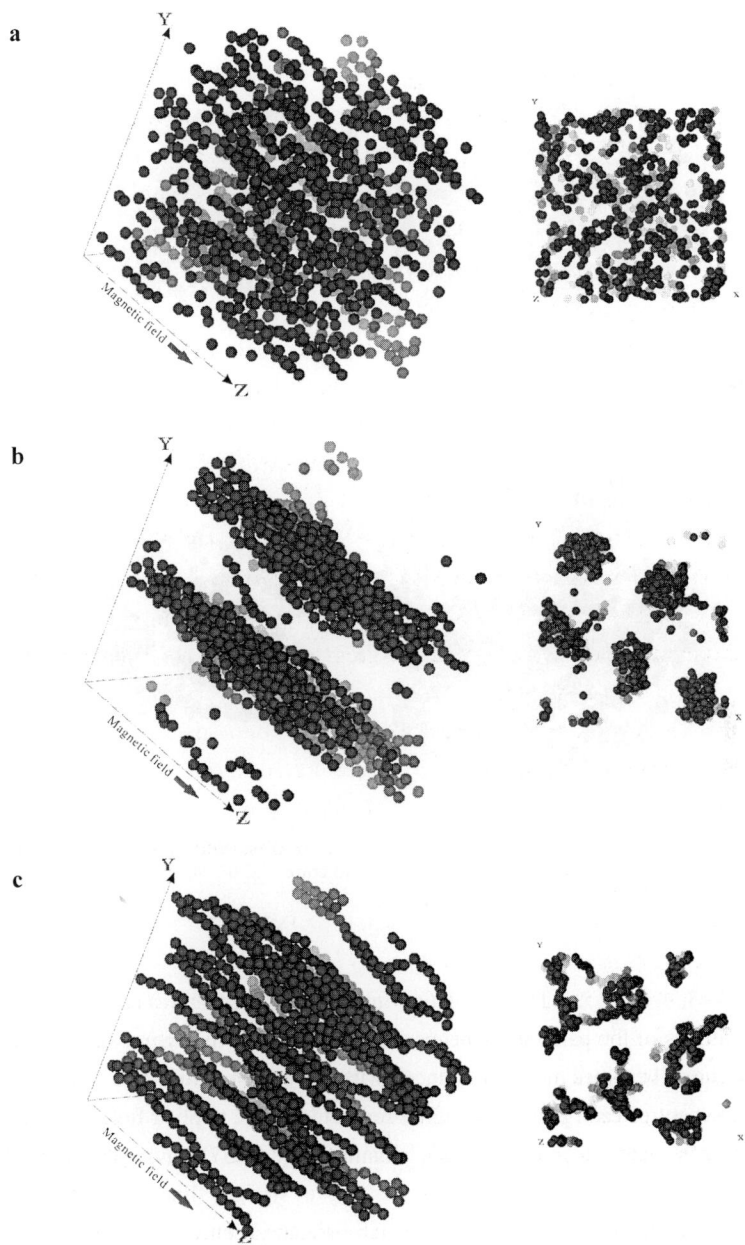

Figure 3.9 Influence of magnetic particle-particle interactions on aggregate structures for dimensionless magnetic field strength ξ=30: (a) for λ=3, (b) for λ=4, and (c) for λ=15.

52

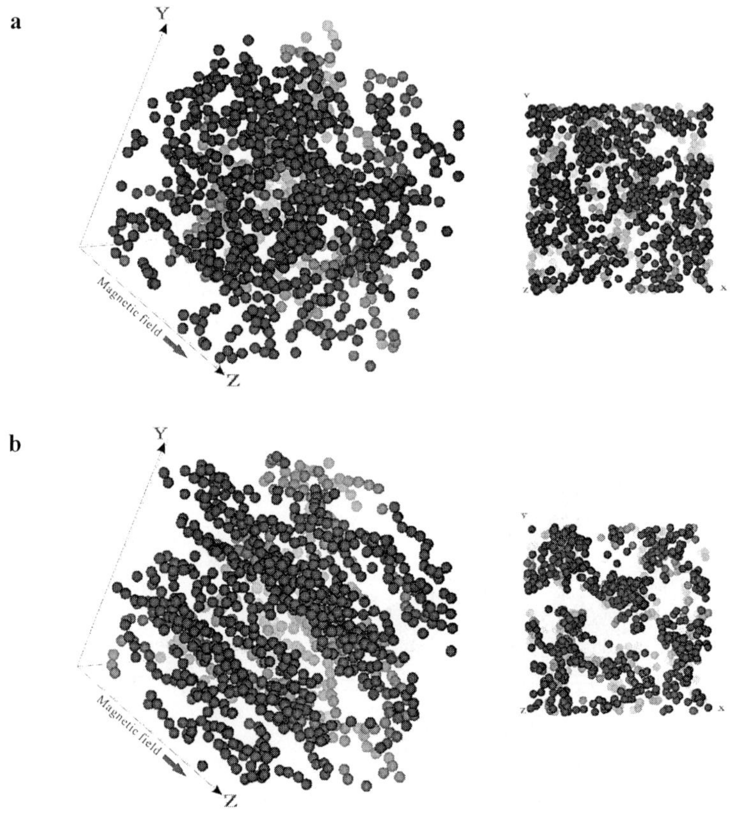

Figure 3.10 Influence of magnetic field strength on aggregate structures for the strength of magnetic particle-particle interactions λ=4: (a) for ξ=1, and (b) for ξ=5.

for three cases of λ=3, 4, and 15. The correlation functions in Fig. 3.11(a) have sharp peaks at nearly integer multiples of the particle diameter, which means that the aggregate structures are solidlike. This feature becomes more pronounced with increasing values of λ. The direction of the magnetic moment of each particle is highly restricted in the applied field direction for a strong magnetic field such as ξ=30. In this situation Eq. (3.57) says that such particle configurations where each magnetic moment lies on a straight line are preferred with increasing particle-particle interactions. This is why the sharper peaks in the correlation functions arise for the larger values of λ. Figure 3.11(b) shows that the pair correlation function for λ=3 has a gaslike feature. This agrees well with the aggregate structures shown in Fig. 3.9(a), where significant clusters are not formed. In contrast with this result, the

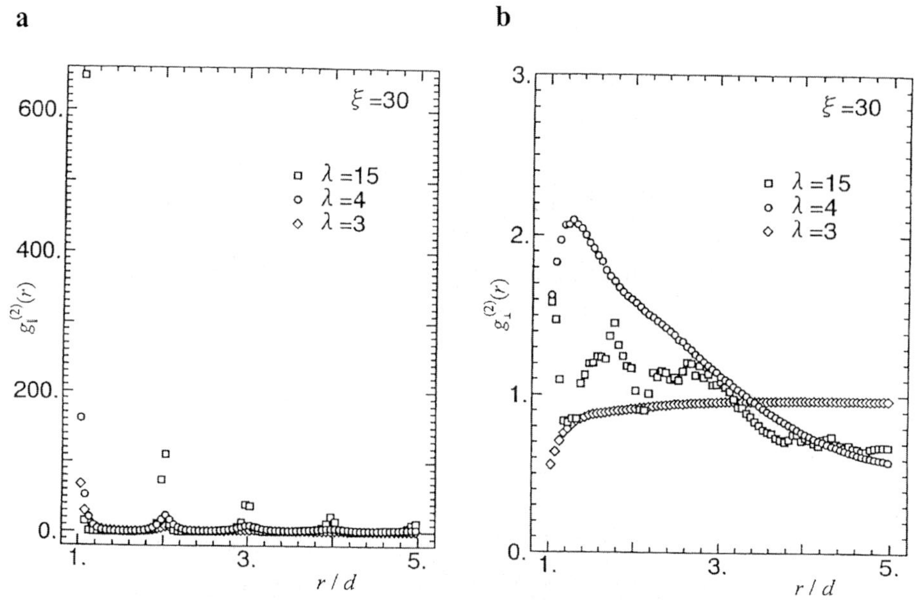

Figure 3.11 Dependence of pair correlation function on interactions between particles for ξ=30: (a) Correlations parallel to the field direction; (b) Correlations normal to the field direction.

correlation functions are liquidlike for λ=4 and 15. In particular, the result for λ=4 shows a strong correlation within a near range of clusters. This is highly reasonable since thick chainlike clusters are observed for λ= 4. The correlation for λ=15 is not so strong as that for λ=4, but suggests the existence of thicker chainlike cluster formation than that which is formed by the particles just aggregating along the field direction. This is in agreement with the result shown in Fig. 3.9(c). It is interesting that the correlation function for λ=15 has a periodic feature at short range.

3.7.3 Orientational distribution of magnetic moments of ferromagnetic particles

As already pointed out, the formation of thick chainlike clusters can be explained very well by the concept of secondary particles. Hence we here show how the potential curves of linear chainlike clusters, which are composed of magnetic secondary particles, change according to their arrangement [34]. Also we show results concerning how the orientational distribution of primary particles has influence on the interaction between clusters.

A. The model of secondary particles

The primary particles, forming the secondary particles, are idealized as spherical with a central point magnetic dipole. If the magnitude of the magnetic moment is constant and denoted by *m*, and the magnetic field strength is *H* (=|**H**|), then the interactions between particle *i* and magnetic field, and between particles *i* and *j*, are expressed in Eq. (3.57), in which ξ and λ are the dimensionless parameters representing the strengths of particle-field and particle-particle interactions relative to the thermal energy, respectively. We now consider magnetic interactions between secondary particles composed of *N* primary particles; *N*=19 for Fig. 3.15(a). If the diameter of the circumscribed circle is regarded as that of secondary particles and denoted by *D*, then the interaction between secondary particles *a* and *b* is written as

$$u_{ab}=\sum_{i_a=1}^{N}\sum_{j_b=1}^{N}u_{i_a j_b}=kT\lambda\sum_{i_a=1}^{N}\sum_{j_b=1}^{N}\frac{d^3}{r_{j_b i_a}^3}\left\{\mathbf{n}_{i_a}\cdot\mathbf{n}_{j_b}-3(\mathbf{n}_{i_a}\cdot\mathbf{t}_{j_b i_a})(\mathbf{n}_{j_b}\cdot\mathbf{t}_{j_b i_a})\right\}$$
$$=kT\Lambda\frac{1}{N^2}\sum_{i_a=1}^{N}\sum_{j_b=1}^{N}\frac{D^3}{r_{j_b i_a}^3}\left\{\mathbf{n}_{i_a}\cdot\mathbf{n}_{j_b}-3(\mathbf{n}_{i_a}\cdot\mathbf{t}_{j_b i_a})(\mathbf{n}_{j_b}\cdot\mathbf{t}_{j_b i_a})\right\}, \tag{3.63}$$

in which

$$\Lambda=\lambda(d/D)^3N^2=\frac{\mu_0(mN)^2}{4\pi D^3kT}. \tag{3.64}$$

By comparing Eq. (3.63) with Eq. (3.57), we see that Λ is similar to λ in the physical role, that is, the dimensionless parameter representing the strength of magnetic interaction between secondary particles. Equation (3.63) may, therefore, be regarded as the interaction between single particles with the magnetic moment *mN*. Numerical relationships between Λ and λ are as follows: Λ/λ=1.8, 2.9, 4.0, and 5.1 for *N*=7, 19, 37, and 61, for example. This data may show that it is possible for secondary particles to aggregate to form thick chainlike clusters for the case of larger secondary particles, even if the interactions between primary particles are weak as λ=1, which will be clarified later.

From Eq. (3.63), the interaction energy between two clusters, where one cluster is composed of N_{sp} secondary particles, is written as

$$u=\sum_{a=1}^{N_{sp}}\sum_{b=1}^{N_{sp}}u_{ab}=\sum_{a=1}^{N_{sp}}\sum_{b=1}^{N_{sp}}\sum_{i_a=1}^{N}\sum_{j_b=1}^{N}u_{j_b i_a}. \tag{3.65}$$

As shown in Fig. 3.15(d), a linear chainlike cluster, composed of secondary particles pointing in the magnetic field direction, is adopted.

B. Monte Carlo calculations of potential curves and the orientational distribution of magnetic moments

The orientational distribution of the magnetic moments of primary particles, and the magnetic interaction energies between the above-mentioned cluster models, were evaluated by means of the usual Monte Carlo method. The calculations were carried out for various values of the cluster-cluster separation and the cluster length, under the conditions of λ=1 and ξ=1 and 5. The sampling is about 30,000 MC steps for evaluating potential curves and about 300,000 MC steps for the orientational distribution of the magnetic moments.

C. Results and discussion

Figures 3.12 and 3.13 show the potential curves for two linear chainlike clusters, both containing N_{sp} secondary particles whose axes point in the magnetic field direction. Figure 3.13 shows the results for a parallel arrangement, in which one cluster is just beside the other. Figure 3.12 shows the results for a staggered arrangement, in which one cluster is shifted in the saturating field direction by the radius of the secondary particle relative to the other, e.g. Fig. 3.17(d). In the figures, the thick solid lines are for the special case of primary particles only (N=1), and the other lines are for secondary particles composed of N primary particles. In both cases, the results were obtained for the case of $\xi=\infty$, i.e., under the condition that the magnetic moment of each particle points in the field direction. The results obtained by the Monte Carlo method for finite ξ are also shown in Fig. 3.12, using symbols such as circles, triangles, etc., where r is the distance between the two cluster axes.

It is seen from Fig. 3.13 that, for the parallel arrangement, repulsive forces act between clusters and attractive forces do not occur for any cluster-cluster separations. These repulsive forces become strong as the values of N_{sp} increase, i.e., as the clusters increase in length. It is seen from Fig. 3.12 that, for the staggered arrangement, attractive forces come into play between clusters at small separations for all cases of N_{sp}=1,5,10, and 15 (the results for N_{sp}=1 and 10 are not shown here). These attractive forces increase with increasing length of clusters and the diameter D of the secondary particles. Also, they act over a longer distance with increasing cluster length. Another characteristic is that each potential curve has an energy barrier. The height of the barrier is almost independent of the cluster length, except for the case of short clusters (N_{sp}=1) and is about ΛkT. This means that larger magnetic moments of primary particles and larger secondary particles lead to a higher energy barrier, independent of the cluster length. In this situation, therefore, it is impossible for the chainlike clusters to overcome the high energy barrier and form thick chainlike clusters.

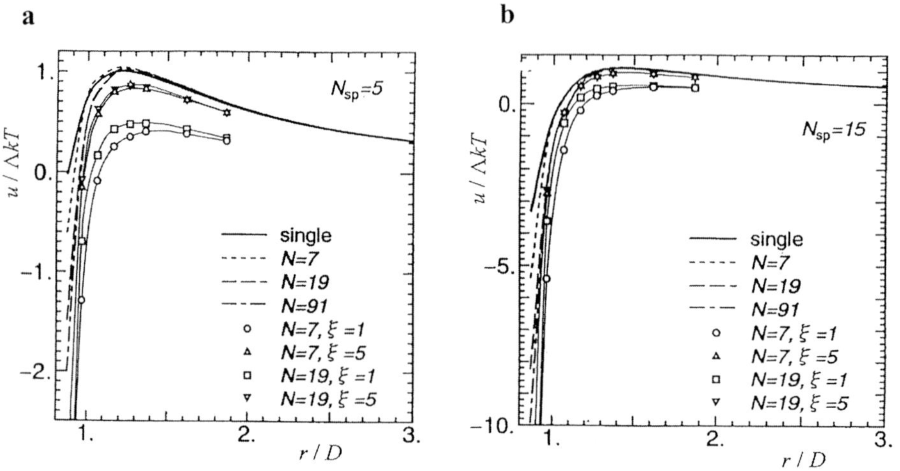

Figure 3.12 Potential curves for the linear chainlike clusters composed of secondary articles for the case of the staggered arrangement: (a) for N_{sp}=5, and (b) for N_{sp}=15.

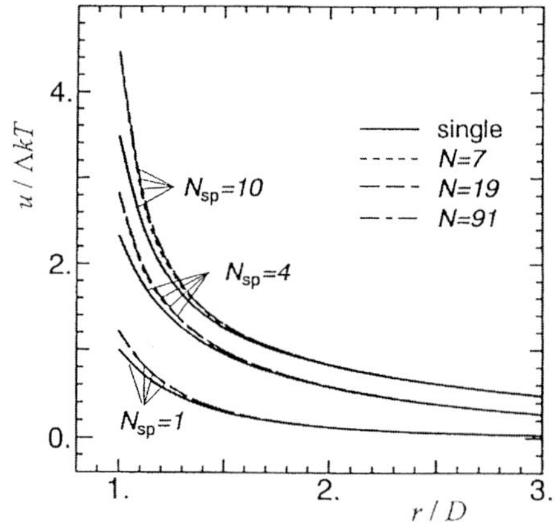

Figure 3.13 Potential curves of the chainlike clusters composed of secondary particles for the case of the parallel arrangement.

Figure 3.14 shows the influence of the cutoff radius on potential curves for N_{sp}=15. In Monte Carlo simulations, a potential cutoff radius is usually used to reduce the large amount of computation time. From a Monte Carlo simulation point of view, therefore, it is very important to clarify the dependence of the potential curves on the cutoff radius. Figures 3.14(a) and (b) show the results for N=1 and 19, respectively, for the staggered arrangement, in which r_{coff} represents the cutoff radius normalized by the secondary particle diameter D. It is seen from Fig. 3.14 that a higher energy barrier is predicted by a shorter cutoff radius. The height of the energy barrier for r_{coff}=3 is about three times that for $r_{coff}=\infty$, and also the use of r_{coff}=5 induces a large deviation from the other curves. Hence it is seen that we should use a cutoff radius of more than about r_{coff}=8 to simulate a realistic cluster formation via Monte Carlo simulations.

Finally, we make the following observation. The results obtained by using primary particles themselves as a secondary particle model (N=1) capture the essential characteristics of the potential curves for a real secondary particle model, composed of many primary particles, to first approximation. The reason for this is that the thick chain formation occurs in a relatively strong field which is sufficient to saturate the clusters and thereby minimize the effects of non-uniform magnetization. This means that it may be appropriate to adopt this simple model for the secondary particles in Monte Carlo simulations.

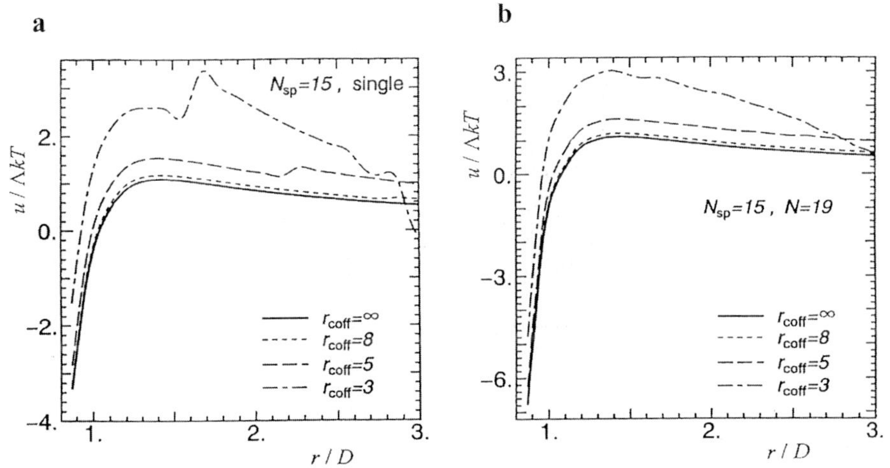

Figure 3.14 Influence of cutoff radius r_{coff} on the potential curves for the case of N_{sp}=15: (a) for N=1, and (b) N=19.

Figures 3.15 to 3.18 show the orientational distributions of ensemble-averaged magnetic moments of the primary particles. Figures 3.15 and 3.16 are for an isolated single cluster and Figs. 3.17 and 3.18 are for two similar clusters arranged in a staggered position. The magnetic field strength is taken as ξ=1 in Figs. 3.15 and 3.17, and ξ=5 in Figs. 3.16 and 3.18. It is noted that the results on the left- and right-hand sides in Figs. 3.17 and 3.18 were obtained for two cases of the cluster-cluster distance, namely, $r^*(=r/D)$=0.866 and 1.066, respectively. It should be noted that the magnetic moments are subject to thermal perturbations affecting the orientational distributions. The thermally averaged magnetic moments depend on the thermal energy, the magnetic particle-field, and particle-particle interactions, although the magnitude of each moment (not averaged) is constant. Consequently, the thermally averaged magnetic moments are drawn as vectors such that the difference in their magnitude is represented by the length of the vectors.

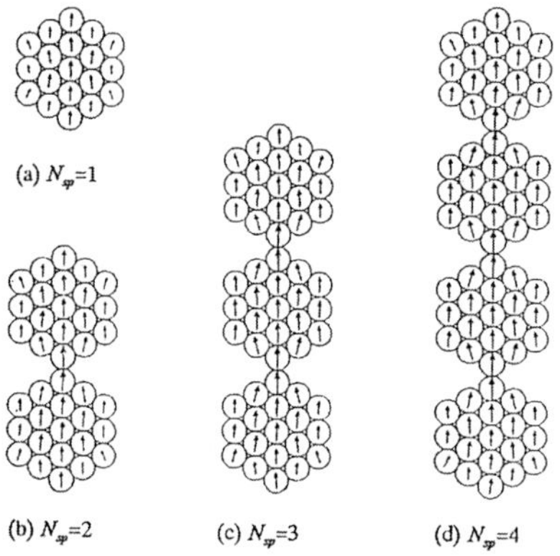

Figure 3.15 Orientational distributions of magnetic moments of primary particles in a single cluster for ξ =1.

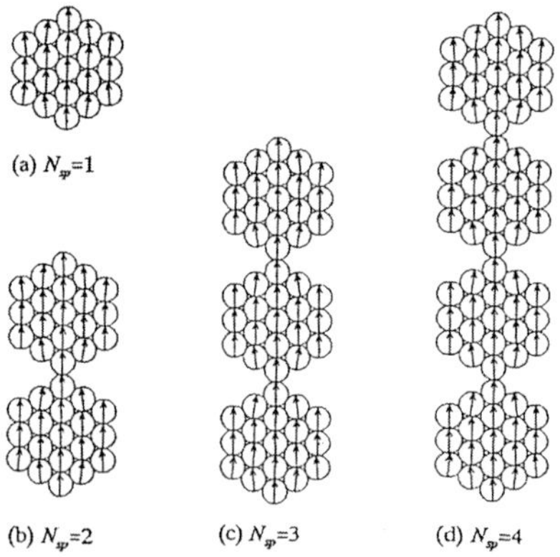

Figure 3.16 Orientational distributions of magnetic moments of primary particles in a single cluster for ξ =5.

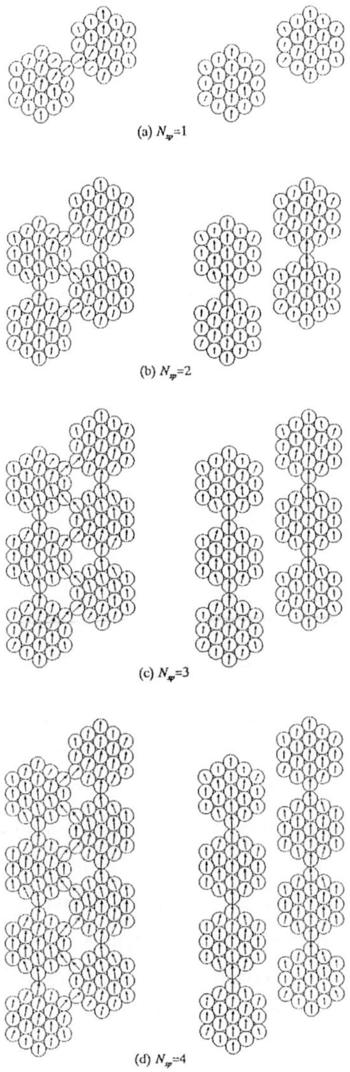

Figure 3.17 Influence of cluster-cluster interactions on the orientational distributions of magnetic moments for ξ=1. The results on the left-hand and right-hand sides are obtained for r^*=0.866 and 1.066, respectively.

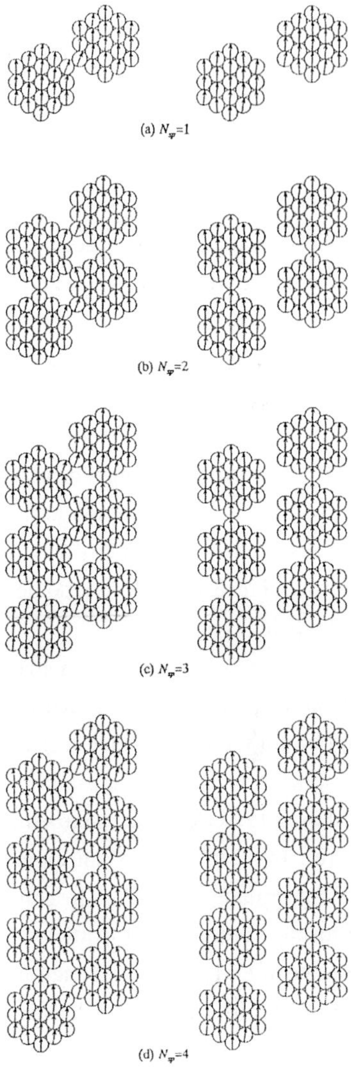

Figure 3.18 Influence of cluster-cluster interactions on the orientational distributions of magnetic moments for ξ=5. The results on the left-hand and right-hand sides are obtained for r^*=0.866 and 1.066, respectively.

Figures 3.15(a) and 3.16(a) clearly show that the orientational distribution of magnetic moments is not uniform for the case of a single secondary particle, and this nonuniformity becomes more pronounced for a weaker magnetic field, such as ξ=1. The magnitude of the averaged values of magnetic moment for ξ=5 are almost equal to m, which means that the direction of each magnetic moment is strongly constrained to the field direction for a strong magnetic field. On the other hand, since the magnetic moments are fluctuating around the applied field direction for the case of ξ=1, the magnitude of the magnetic moments shows much smaller values than m. It is seen from comparing Figs. 3.15(a), (b), (c), and (d) that the interaction between primary particles belonging to different secondary particles tend to suppress the fluctuations in the magnetic moments around the average direction. Also, the directions of primary particles lying at the edge of the secondary particles vary significantly from those of single secondary particles. The deviations from uniformity are in such a sense as to reduce the surface charge density of an individual secondary particle. It is also interesting to note that the reduction in the thermally averaged value of the magnetization, for those particles close to the edge of a secondary particle, would have a similar effect. The results shown in Fig. 3.16, however, do not clearly show the above-mentioned features because the magnetic field is too strong.

Figures 3.17 and 3.18 clearly show that, if the two clusters are in contact each other in the staggered arrangement, the orientational distributions of magnetic moments change significantly from those of isolated clusters. In particular, for a weak magnetic field such as ξ=1, this influence extends to almost all primary particles, which is clear from comparing Fig. 3.17 with Fig. 3.15. However, the orientational distributions for the longer chains shown in Figs. 3.17(b), (c), and (d) are not significantly different from each other concerning the corresponding primary particles. In contrast, for ξ=5, the interactions between the clusters influence mainly those primary particles close to the edge of the clusters. For the cluster-cluster distance of r^*=1.066, the orientational distributions of magnetic moments shown in Figs. 3.17 and 3.18 are not significantly different from those of a single cluster in Figs. 3.15 and 3.16.

References

1. M.P. Allen and D.J. Tildesley, "Computer Simulation of Liquids," Clarendon Press, Oxford, 1987
2. D. W. Heermann, "Computer Simulation Methods in Theoretical Physics," 2nd Ed., Springer-Verlag, Berlin, 1990
3. K. Binder and D. W. Heermann, "Monte Carlo Simulation in Statistical Physics:

An Introduction," Springer-Verlag, Berlin, 1988
4. K. Binder (Ed.), "Monte Carlo Methods in Statistical Physics," Springer-Verlag, Berlin, 1979
5. W. W. Woods, Chapter 5: Monte Carlo Studies of Simple Liquid Models, in Physics of Simple Liquids, H. N. V. Temperley, J. S. Rowlinson, and G.S. Rushbrooke (Eds.), North-Holland, Amsterdam, 1968
6. D. Frenkel and B. Smit, "Understanding Molecular Simulation," Academic Press, San Diego, 1996
7. M. M. Woolfson and G. J. Pert, "An Introduction to Computer Simulation," Oxford University Press, Oxford, 1999
8. M. Tanaka and R. Yamamoto, Eds., "Computational Physics and Computational Chemistry," Kaibundou, Tokyo, 1988 (in Japanese)
9. I. Okada and E. Ohsawa, "An Introduction to Molecular Simulation," Kaibundou, Tokyo, 1989 (in Japanese)
10. N. Metropolis, A.W. Rosenbluth, M.N. Rosenbluth, and A. Teller, Equation of State Calculations by Fast Computing Machines, J. Chem. Phys., 21(1953)1087
11. L. Breiman, "Probability," Addison-Wesley, Massachusetts, 1968
12. J.R. Norris, "Markov Chains," Cambridge University Press, Cambridge, 1997
13. M. Kijima, "Markov Processes for Stochastic Modeling," Chapman & Hall, Cambridge, 1997
14. E. Behrends, "Introduction to Markov Chains," Vieweg & Sohn, Wiesbaden, 2000
15. G. E. Norman and V. S. Filinov, Investigations of Phase Transitions by a Monte Carlo Method, High Temp., 7 (1969) 216
16. L. A. Rowley, D. Nicholson, and N.G. Parsonage, Monte Carlo Grand Canonical Ensemble Calculation in a Gas-Liquid Transition Region for 12-6 Argon, J. Comput. Phys., 17 (1975) 401
17. J. Yao, R.A. Greenkorn, and K.C. Chao, Monte Carlo Simulation of the Grand Canonical Ensemble, Molec. Phys., 46 (1982) 587
18. A. Satoh and S. Kamiyama, Analysis of Particles' Aggregation Phenomena in Magnetic Fluids (Consideration of Hydrophobic Bonds), in Continuum Mechanics and Its Applications, 731-739, Hemisphere Publishers, 1989
19. A. Satoh, A New Technique for Metropolis Monte Carlo Simulation to Capture Aggregate Structures of Fine Particles (Cluster-Moving Monte Carlo Algorithm), J. Coll. Interface Sci., 150(1992) 461
20. G.N. Coverdale, R.W. Chantrell, A. Hart, and D. Parker, A 3-D Simulation of a Particulate Dispersion, J. Magnet. Magnet. Mater., 120(1993) 210

21. G.N. Coverdale, R.W. Chantrell, A. Hart, and D. Parker, A Computer Simulation of the Microstructure of a Particulate Dispersion, J. Appl. Phys., 75(1994) 5574
22. G.N. Coverdale, R.W. Chantrell, G.A.R. Martin, A. Bradbury, A. Hart, and D. Parker, Cluster Analysis of the Microstructure of Colloidal Dispersions using the Maximum Entropy Technique, J. Magn. Magn. Mater., 188 (1998) 41
23. R.E. Rosensweig, "Ferrohydrodynamics," Cambridge University Press, Cambridge, 1985
24. R.J. Hunter, "Foundations of Colloid Science," Vol. 1, Clarendon Press, Oxford, 1989
25. B.V. Derjaguin and L. Landau, Acta. Physicochim., USSR, 14(195 4) 633
26. E.J. Verwey and J.T.G. Overbeek, "Theory of Stability of Lyophobic Colloids," Elsevier, Amsterdam, 1948
27. H.C. Hamaker, The London-Van der Waals Attraction between Spherical Particles, Phyisica, 4(1937)1058
28. C.F. Hayes, Observation of Association in a Ferromagnetic Colloid, J. Coll. Interface Sci., 52(1975)239.
29. S. Wells, K.J. Davies, S.W. Charles, and P.C. Fannin, Characterization of Magnetic Fluids in which Aggregation of Particles is Extensive, in Proceedings of the International Symposium on Aerospace and Fluid Science, 621, Tohoku University (1993)
30. A. Satoh, R.W. Chantrell, S. Kamiyama, and G.N. Coverdale, Two-Dimensional Monte Carlo Simulations to Capture Thick Chainlike Clusters of Ferromagnetic Particles in Colloidal Dispersions, J. Colloid Inter. Sci., 178 (1996) 620
31. A. Satoh, R.W. Chantrell, S. Kamiyama, and G.N. Coverdale, Three-Dimensional Monte Carlo Simulations to Thick Chainlike Clusters Composed of Ferromagnetic Fine Particles, J. Colloid Inter. Sci., 181 (1996) 422
32. R.W. Chantrell, A. Bradbury, J. Popplewell, and S.W. Charles, Particle Cluster Configuration in Magnetic Fluids, J. Phys. D: Appl. Phys., 13(1980) L119
33. R.W. Chantrell, A. Bradbury, J. Popplewell, and S.W. Charles, Agglomerate Formation in a Magnetic Fluid, J. Appl. Phys., 53(1982), 2742.
34. A. Satoh, R.W. Chantrell, S. Kamiyama, and G.N. Coverdale, Potential Curves and Orientational Distributions of Magnetic Moments of Chainlike Clusters Composed of Secondary Particles, J. Magn. Magn. Mater., 154 (1996)183

Exercises

3.1 By summing both sides of Eq. (3.19) with respect to *j*, and taking account of the condition (2) in Eq. (3.18), derive the condition (3) in Eq. (3.18)

3.2 In the grand canonical MC method, the probability density function $\rho(\underline{\mathbf{r}}^N)$ can be expressed, for a N particle system, as

$$\rho(\underline{\mathbf{r}}^N) = \frac{\dfrac{1}{N!\Lambda^{3N}} \exp(\mu N/kT) \exp\left\{-U(\underline{\mathbf{r}}^N)/kT\right\}}{\Xi}. \tag{3.66}$$

By taking account of this expression, derive Eqs. (3.25) and (3.26).

3.3 In the isothermal–isobaric MC method, show that Y_U defined by Eq. (3.34) can be written as the following equation by transforming the coordinate system according to Eq. (3.36) and using the dimensionless value V^* ($=V/V_0$):

$$Y_U = V_0^{N+1} \iint V^{*N} \exp(-PV_0V^*/kT) \exp\left\{-U(L\underline{\mathbf{s}};V)/kT\right\} d\underline{\mathbf{s}} dV^*. \tag{3.67}$$

With this equation, derive Eqs. (3.37) and (3.38).

3.4 We consider the situation in which a ferromagnetic particle i with a magnetic dipole moment $\mathbf{m}_i$ is placed in a uniform applied magnetic field $\mathbf{H}$. In this case, the magnetic particle-field and particle-particle interaction energies can be expressed as

$$\left.\begin{aligned} u_i &= -\mu_0 \mathbf{m}_i \cdot \mathbf{H}, \\ u_{ij} &= \frac{\mu_0}{4\pi r_{ij}^3} \left\{\mathbf{m}_i \cdot \mathbf{m}_j - 3(\mathbf{m}_i \cdot \mathbf{r}_{ji})(\mathbf{m}_j \cdot \mathbf{r}_{ji})\right\}. \end{aligned}\right\} \tag{3.68}$$

If the energies are nodimensionalized by the thermal energy kT, and distances by the particle diameter d, then show that Eq. (3.68) can be nondimensionalized as

$$\left.\begin{aligned} u_i^* &= -\xi \mathbf{n}_i \cdot \mathbf{H}/H, \\ u_{ij}^* &= \lambda \frac{1}{r_{ij}^{*3}} \left\{\mathbf{n}_i \cdot \mathbf{n}_j - 3(\mathbf{n}_i \cdot \mathbf{t}_{ji})(\mathbf{n}_j \cdot \mathbf{t}_{ji})\right\}. \end{aligned}\right\} \tag{3.69}$$

in which $m = |\mathbf{m}_i|$, $H = |\mathbf{H}|$, the quantities with the superscript * are dimensionless, and other symbols such as $\mathbf{n}_i$ and $\mathbf{t}_{ji}$ have already been defined in Eqs. (3.57) and (3.58).

CHAPTER 4

GOVERNING EQUATIONS OF THE FLOW FIELD

4.1 The Navier-Stokes Equation

The motion of a colloidal particle is determined by the forces and torques exerted on it by the other colloidal particles and the ambient fluid (dispersion medium). Since the particle motion has an influence on the other particles through the fluid, it is very important to understand the basic equations by which the motion of a fluid is governed. In this section we outline important equations such as the equation of continuity and the Navier-Stokes equation, which can be derived from the conservation equations of mass and momentum, respectively.

We consider first the mass conservation in an arbitrary fixed volume V, located within a fluid, shown in Fig. 4.1. If the local mass density at a position $\mathbf{r}$ at time t is denoted by $\rho(\mathbf{r},t)$, the total mass within this small volume V can be written as

$$M = \int_V \rho(\mathbf{r},t)d\mathbf{r} . \tag{4.1}$$

Since the volume V is fixed in space, the rate of change of mass within V is evaluated as

$$\frac{dM}{dt} = \frac{d}{dt}\int_V \rho(\mathbf{r},t)d\mathbf{r} = \int_V \frac{\partial \rho}{\partial t} d\mathbf{r} . \tag{4.2}$$

This rate of change of mass should be given by the mass flows through the surface of the volume, S. We focus our attention on the mass flow through the surface element dS in Fig. 4.1. If $\mathbf{n}$ is a unit vector directed normally outward from the enclosed volume, the rate of change of mass due to the fluid flowing in and out of the volume can be evaluated as

$$\frac{dM}{dt} = -\int_S \mathbf{n}\cdot(\rho\mathbf{u})dS = -\int_V \nabla\cdot(\rho\mathbf{u})d\mathbf{r} , \tag{4.3}$$

in which the velocity $\mathbf{u}$ is a function of the position $\mathbf{r}$ and time t, that is, $\mathbf{u}(\mathbf{r},t)$. To obtain the final expression on the right-hand side, the surface integral has been transformed to a volume integral by means of Gauss' divergence theorem, which is shown in Eq. (A1.22). Hence, since the rate of change of mass in Eq. (4.2) has to be equal to that in Eq. (4.3), the following equation can be obtained:

$$\int_V \left\{ \frac{\partial \rho}{\partial t} + \nabla\cdot(\rho\mathbf{u}) \right\} d\mathbf{r} = 0 . \tag{4.4}$$

Since the volume V can be taken arbitrarily, Eq. (4.4) finally reduces to the following equation:

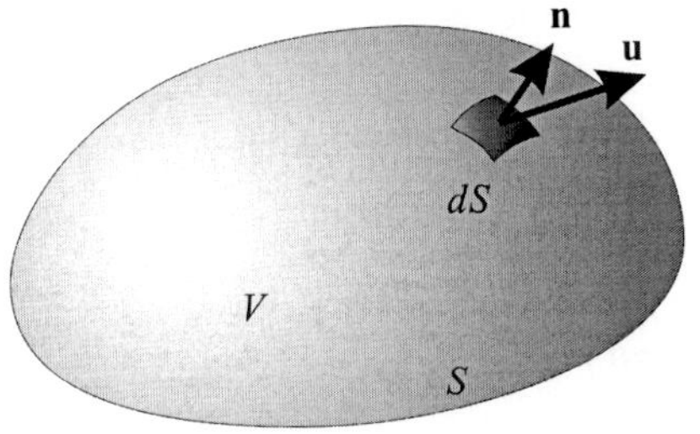

Figure 4.1 An arbitrary fixed volume in a fluid.

$$\frac{\partial \rho}{\partial t} + \nabla \cdot (\rho \mathbf{u}) = 0 . \tag{4.5}$$

This is known as the equation of continuity, which has been derived from the conservation law of mass. If the fluid is incompressible, the equation of continuity can be simplified as

$$\nabla \cdot \mathbf{u} = 0 . \tag{4.6}$$

Next we consider the conservation law of momentum. Since the momentum per unit volume located at $\mathbf{r}$ is given by $\rho(\mathbf{r},t)\mathbf{u}(\mathbf{r},t)$, the total momentum $\mathbf{L}$ within the volume V is written as

$$\mathbf{L} = \int_V \rho(\mathbf{r},t)\mathbf{u}(\mathbf{r},t) d\mathbf{r} . \tag{4.7}$$

Hence, the rate of change of momentum can be expressed as

$$\frac{d\mathbf{L}}{dt} = \frac{d}{dt} \int_V \rho(\mathbf{r},t)\mathbf{u}(\mathbf{r},t) d\mathbf{r} = \int_V \frac{\partial}{\partial t}(\rho \mathbf{u}) d\mathbf{r} . \tag{4.8}$$

This momentum change is given by three factors: (1) body forces, such as the gravitational force, which act on the volume, (2) forces exerted by the fluid, such as stress or pressure, which act through the surface of the volume, and (3) the flow convection, in which momentum flows in and out of the volume V. Firstly, we consider the third contribution. If $\Delta \mathbf{L}^{(flow)}$ denotes the rate of change of momentum due to the flow convection, this can be derived in a way similar to Eq. (4.3) as

$$\Delta \mathbf{L}^{(flow)} = -\int_S (\rho \mathbf{u})(\mathbf{n} \cdot \mathbf{u}) dS = -\int_V \nabla \cdot (\rho \mathbf{u}\mathbf{u}) d\mathbf{r} . \tag{4.9}$$

Next, if the body force per unit mass acting on the volume located at $\mathbf{r}$ is denoted by $\mathbf{K}(\mathbf{r})$, the force acting on the volume, $\mathbf{F}^{(body)}$, can be expressed as

$$\mathbf{F}^{(body)} = \int_V \rho \mathbf{K} d\mathbf{r} . \tag{4.10}$$

Finally we consider the contribution due to the stress. The stress is not a body force, but acts on the surface of the volume. This is a tensor quantity and usually denoted by $\boldsymbol{\tau}(\mathbf{r},t)$, which is a function of the location $\mathbf{r}$ and time t. The force acting on the unit area located at $\mathbf{r}$, at the surface of the volume, is $\mathbf{n}\cdot\boldsymbol{\tau}$. Hence, the force acting on the volume, $\mathbf{F}^{(stress)}$, can be written with the divergence theorem as

$$\mathbf{F}^{(stress)} = \int_S \mathbf{n}\cdot\boldsymbol{\tau}dS = \int_V \nabla\cdot\boldsymbol{\tau}d\mathbf{r}\,. \tag{4.11}$$

Since the change of the momentum in the volume is due to three contributions, the following equation has to be satisfied:

$$\frac{d\mathbf{L}}{dt} = \Delta\mathbf{L}^{(flow)} + \mathbf{F}^{(body)} + \mathbf{F}^{(stress)}\,. \tag{4.12}$$

By substituting Eqs. (4.8)~(4.11) into this equation, the following expression can be obtained:

$$\int_V \left\{\frac{\partial}{\partial t}(\rho\mathbf{u}) + \nabla\cdot(\rho\mathbf{u}\mathbf{u}) - \rho\mathbf{K} - \nabla\cdot\boldsymbol{\tau}\right\}d\mathbf{r} = 0\,. \tag{4.13}$$

Since the volume V can be taken arbitrarily, Eq. (4.13) reduces to

$$\frac{\partial}{\partial t}(\rho\mathbf{u}) + \nabla\cdot(\rho\mathbf{u}\mathbf{u}) = \rho\mathbf{K} + \nabla\cdot\boldsymbol{\tau}\,. \tag{4.14}$$

If we take into account the equation of continuity, Eq. (4.5), and Eq. (A1.19), this equation can be written in different form as

$$\rho\left(\frac{\partial\mathbf{u}}{\partial t} + (\mathbf{u}\cdot\nabla)\mathbf{u}\right) = \rho\mathbf{K} + \nabla\cdot\boldsymbol{\tau}\,. \tag{4.15}$$

In deriving this equation, we have not assumed that the fluid is incompressible and we have not used explicit form of the stress tensor. Hence Eqs. (4.14) and (4.15) are valid for general cases, that is, applicable to compressible or non-Newtonian fluid flows. Since a dispersion medium (base liquid) can be regarded as incompressible in almost all cases of physical phenomena in colloidal dispersions, from now on we will always assume that a dispersion medium is incompressible. In addition, since the body force can generally be combined into the pressure term in the stress tensor, the equation of motion will be simplified as

$$\rho\left(\frac{\partial\mathbf{u}}{\partial t} + (\mathbf{u}\cdot\nabla)\mathbf{u}\right) = \nabla\cdot\boldsymbol{\tau}\,. \tag{4.16}$$

The flow field can then be solved using this equation and the equation of continuity in Eq. (4.6).

Since the fluid of interest is assumed to be incompressible and Newtonian, the stress tensor $\boldsymbol{\tau}$ is expressed as [1]

$$\boldsymbol{\tau} = -p\mathbf{I} + \eta\left\{\nabla\mathbf{u} + (\nabla\mathbf{u})'\right\}, \tag{4.17}$$

in which p is the pressure, η is the viscosity of the fluid which, for Newtonian fluids, is constant under isothermal conditions. Also $\mathbf{I}$ is the unit tensor, and the superscript t means a transposed tensor. By substituting Eq. (4.17) into Eq. (4.16), the following Navier-Stokes equation can be obtained [2,3]:

$$\rho\left\{\frac{\partial \mathbf{u}}{\partial t}+(\mathbf{u}\cdot\nabla)\mathbf{u}\right\}=-\nabla p+\eta\nabla^2\mathbf{u}. \tag{4.18}$$

The flow field can be completely described by solving Eqs. (4.18) and (4.6) with appropriate initial and boundary conditions. From the viewpoint of numerical analysis approach, the dimensional governing equations are seldom solved, rather more generally the nondimensionalized equations are used. For example, consider a uniform flow around a sphere where the distances and velocities may be nondimensionalized by the diameter of a sphere and the velocity of a uniform flow, respectively. If the representative values of distance and velocity are denoted by L and U, respectively, Eq. (4.18) is nondimensionalized as

$$\frac{\partial \mathbf{u}^*}{\partial t^*}+(\mathbf{u}^*\cdot\nabla^*)\mathbf{u}^*=\frac{1}{Re}\left(-\nabla^* p^*+\nabla^{*2}\mathbf{u}^*\right), \tag{4.19}$$

in which the superscript * means dimensionless quantities. Time is nondimensionalized by (L/U), and pressure by $(\eta U/L)$. The nondimensional number $Re(=\rho UL/\eta)$, which has appeared in Eq. (4.19), is called the Reynolds number and plays a very important role in the field of fluid dynamics. Even for different values of (η, U, L, ρ), the solution for the flow field is the same if the Reynolds number is the same for each case in the nondimensional system, providing the initial and boundary conditions are the same.

Since Eq. (4.19) is a nonlinear partial differential equation, it is extraordinarily difficult to solve this equation except for some simple flow problems. Hence, numerical analysis methods, such as finite difference and finite element methods, are usually used to obtain a solution for flow fields. This approach is common in the mechanical and aeronautical engineering fields.

4.2 The Stokes Equation

In almost all cases of flow problems for colloidal dispersions, the inertia term on the left-hand side of Eq. (4.18) can be neglected, and the Navier-Stokes equation can be simplified to the Stokes equation:

$$\nabla p=\eta\nabla^2\mathbf{u}. \tag{4.20}$$

This equation and the equation of continuity in Eq. (4.6) must be solved to obtain the flow field around colloidal particles, then the forces acting on particles from the ambient fluid can be evaluated. It should be noted that an explicit time dependent term is not included in Eqs.

(4.6) and (4.20). The dependence of the solution on time appears through the boundary conditions at particle surfaces (due to the particle motion). A point to be noted is that the Stokes equation is a linear partial differential equation, so that it is essentially possible to obtain analytical solutions. Equation (4.20) is written in the Laplace form, using Eq. (4.6), as

$$\nabla^2 p = 0 . \tag{4.21}$$

We now consider a uniform flow directed in the x-direction past a sphere with a radius a as a sample of an analytical solution for the Stokes equation. Since it is not so difficult to derive the analytical solution, we show only the final expressions for the velocity and pressure:

$$\frac{\mathbf{u}}{U} = \boldsymbol{\delta}_x - \frac{3a}{4}\left(\frac{1}{r}\boldsymbol{\delta}_x + \frac{x}{r^3}\mathbf{r}\right) - \frac{a^3}{4}\left(\frac{1}{r^3}\boldsymbol{\delta}_x - \frac{3x}{r^5}\mathbf{r}\right), \tag{4.22}$$

$$p = p_\infty - \frac{3}{2}\eta U a \frac{x}{r^3}, \tag{4.23}$$

in which p_∞ is the pressure at the infinity position, $\mathbf{r}$ is the position vector as $\mathbf{r}=(x,y,z)$, $\boldsymbol{\delta}_x$ is the unit vector in the x-direction, and r is the magnitude of the vector $\mathbf{r}$. Since Eq. (4.22) is unchanged by the replacement of $\mathbf{r}$ with $-\mathbf{r}$, the flow line, (a line connected tangentially to the velocity vectors), is symmetric with respect to the y-axis. This property of the solution for the Stokes equation is generally in disagreement with that for the Navier-Stokes equation, but the solution of the Stokes equation approaches that of the Navier-Stokes equation in the limit of $Re \to 0$. In contrast, the pressure is not symmetric with respect to the y-axis, and the maximum pressure appears at the front stagnation point at $\mathbf{r}=-a\boldsymbol{\delta}_x$. By substituting Eqs. (4.22) and (4.23) into Eq. (4.17) and carrying out the integration in Eq. (4.11) at the particle surface, the force exerted on the sphere by the fluid, F, can be obtained as

$$F = 6\pi\eta a U . \tag{4.24}$$

This relation is called Stokes' drag formula.

Finally we show the flow fields, for a flow around a circular cylinder, that were obtained by solving numerically the nondimensionalized Navier-Stokes equation. Figure 4.2 shows the flow lines for different cases of the Reynolds number. It is clearly seen in Fig. 4.2(a) that, for $Re \ll 1$, the stream lines are symmetric in the upstream and downstream with respect to the y-axis (ordinate), which has already been pointed out. As the Reynolds number increases, a pair of vortices comes to appear behind the cylinder and this area expands in the backward direction. This kind of flow field cannot be expressed by the Stokes equation. It is noted, however, that the condition of $Re \ll 1$ can be satisfied for almost all problems of colloidal hydrodynamics.

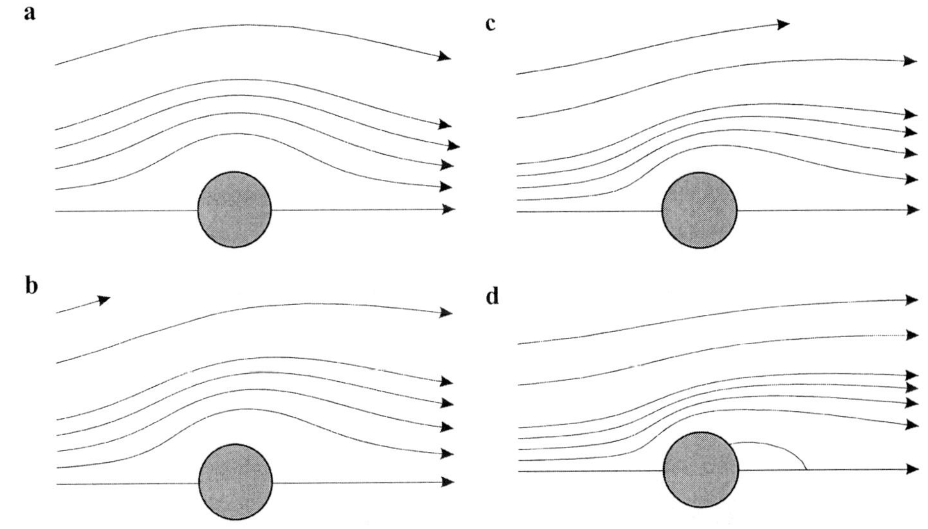

Figure 4.2 Numerical solutions of the Navier-Stokes equation for a uniform flow around a cylinder: (a) for Re=0.1, (b) for Re=1, (c) for Re=5, and (d) for Re=20.

For microsimulations of colloidal dispersions, the linear flow field is generally used to investigate particle behavior in a flow field and rheological properties. If an arbitrary point is defined by the vector **r**, the linear velocity field can be expressed in general form as

$$\mathbf{u}(\mathbf{r}) = \mathbf{u}_0 + \mathbf{r} \cdot \mathbf{\Gamma}, \tag{4.25}$$

in which $\mathbf{u}_0$ is the uniform flow velocity and $\mathbf{\Gamma}$ is the velocity gradient tensor with constants as its components. $\mathbf{\Gamma}$ is expressed as

$$\mathbf{\Gamma} = \nabla \mathbf{u} = \begin{bmatrix} \partial u_x / \partial x & \partial u_y / \partial x & \partial u_z / \partial x \\ \partial u_x / \partial y & \partial u_y / \partial y & \partial u_z / \partial y \\ \partial u_x / \partial z & \partial u_y / \partial z & \partial u_z / \partial z \end{bmatrix}, \tag{4.26}$$

in which $\mathbf{r}=(x,y,z)$ and $\mathbf{u}=(u_x,u_y,u_z)$.

Equation (4.25) can be derived straightforwardly. If the velocity vector $\mathbf{u}(\mathbf{r})$ is expanded in a Taylor series about any point $\mathbf{r}_0$, this is expressed as

$$\mathbf{u}(\mathbf{r}) = \mathbf{u}(\mathbf{r}_0) + (\mathbf{r} - \mathbf{r}_0) \cdot \nabla \mathbf{u}(\mathbf{r}_0) + \cdots, \tag{4.27}$$

in which $\nabla \mathbf{u}(\mathbf{r}_0)$ means that $\nabla \mathbf{u}(\mathbf{r})$ is performed and then **r** is taken as $\mathbf{r}=\mathbf{r}_0$. If only the linear

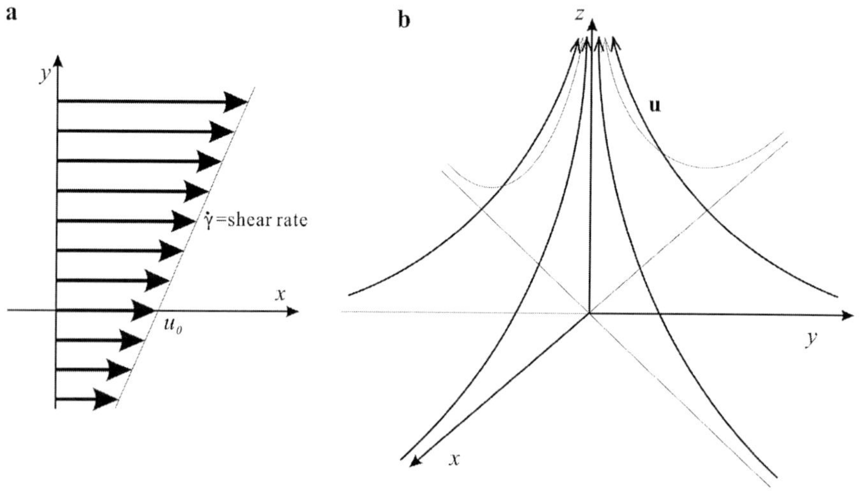

Figure 4.3 A simple shear flow and an extensional flow: (a) simple shear flow in x-direction and (b) extensional flow in z-direction.

4.3 The Linear Velocity Field

term with respect to **r** is retained and the higher terms are neglected, Eq. (4.25) can be obtained.

Now we use the angular velocity vector $\mathbf{\Omega}$ and the rate-of-strain tensor **E** defined, respectively, as

$$\mathbf{\Omega} = \frac{1}{2}\nabla \times \mathbf{u}, \qquad \mathbf{E} = \frac{1}{2}(\mathbf{\Gamma} + \mathbf{\Gamma}^t) = \frac{1}{2}\left\{\nabla\mathbf{u} + (\nabla\mathbf{u})^t\right\}. \tag{4.28}$$

With these notations, Eq. (4.25) can be rewritten as

$$\mathbf{u}(\mathbf{r}) = \mathbf{u}_0 + \mathbf{\Omega} \times \mathbf{r} + \mathbf{E} \cdot \mathbf{r}. \tag{4.29}$$

The superscript t in Eq. (4.28) means a transposed tensor. In some books, the rate-of-strain tensor **E** is defined without the constant 1/2, but we use the present definition in Eq. (4.28) in this book.

For the reader, we consider some examples to show explicit expressions for the linear field $\mathbf{u}(\mathbf{r})$, $\mathbf{\Omega}$, and **E**. If a simple shear flow directed in the x-direction with a shear rate $\dot{\gamma}$, as shown in Fig. 4.3(a), is considered, the flow field $\mathbf{u}(\mathbf{r})$ can be expressed as

$$\mathbf{u}(\mathbf{r}) = u_0 \begin{bmatrix} 1 \\ 0 \\ 0 \end{bmatrix} + \dot{\gamma} \begin{bmatrix} 0 & 1 & 0 \\ 0 & 0 & 0 \\ 0 & 0 & 0 \end{bmatrix} \begin{bmatrix} x \\ y \\ z \end{bmatrix}. \tag{4.30}$$

From this flow field, the expressions for $\mathbf{\Omega}$ and **E** can be derived straightforwardly, respectively, as

$$\mathbf{\Omega} = -\frac{\dot{\gamma}}{2}\begin{bmatrix}0\\0\\1\end{bmatrix}, \quad \mathbf{E} = \frac{\dot{\gamma}}{2}\begin{bmatrix}0 & 1 & 0\\1 & 0 & 0\\0 & 0 & 0\end{bmatrix}. \tag{4.31}$$

Hence, from Eq. (4.29), $\mathbf{u}(\mathbf{r})$ can be rewritten as

$$\mathbf{u}(\mathbf{r}) = u_0\begin{bmatrix}1\\0\\0\end{bmatrix} - \frac{\dot{\gamma}}{2}\begin{bmatrix}0\\0\\1\end{bmatrix}\times\begin{bmatrix}x\\y\\z\end{bmatrix} + \frac{\dot{\gamma}}{2}\begin{bmatrix}0 & 1 & 0\\1 & 0 & 0\\0 & 0 & 0\end{bmatrix}\begin{bmatrix}x\\y\\z\end{bmatrix}. \tag{4.32}$$

Another example is an extensional flow directed in the z-direction. If the strain rate is denoted by $\dot{\varepsilon}$, the flow field is expressed as

$$\mathbf{u}(\mathbf{r}) = \dot{\varepsilon}\begin{bmatrix}-1/2 & 0 & 0\\0 & -1/2 & 0\\0 & 0 & 1\end{bmatrix}\begin{bmatrix}x\\y\\z\end{bmatrix}. \tag{4.33}$$

It is straightforward, from this equation, to show that $\mathbf{\Omega}$ =0 and $\mathbf{E}$= $\mathbf{\Gamma}$. This means that there are no vortices in all the flow area for the extensional flow case.

4.4 Forces, Torques, and Stresslets

As already explained, the stress tensor $\boldsymbol{\tau}$ is written in Eq. (4.17) for Newtonian fluids. If a particle is dispersed in a liquid, the force and torque exerted on the particle by the fluid can be evaluated using the stress tensor. We now consider a simple particle model such as a spherical or prolate spheroid. In this case, the motion of a particle can be decomposed into the translational and rotational motion around its center of mass. If the center of mass of the particle is denoted by the position vector $\mathbf{r}_c$, then the force $\mathbf{F}$ and torque $\mathbf{T}$ exerted on the ambient fluid by the particle can be expressed, respectively, as

$$\mathbf{F} = -\int_{S_p}(\mathbf{n}\cdot\boldsymbol{\tau})dA, \tag{4.34}$$

$$\mathbf{T} = -\int_{S_p}(\mathbf{r}-\mathbf{r}_c)\times(\mathbf{n}\cdot\boldsymbol{\tau})dA, \tag{4.35}$$

in which the integration is carried out at the particle surface S_p, $\boldsymbol{\tau}$ is the stress tensor at $\mathbf{r}$, and $\mathbf{n}$ is the unit vector directed normally outward from the particle surface.

We will see in the following chapter that, if a spherical particle moves with the local velocity of a flow field and rotates with the local angular velocity, the particle exerts no forces and torques on the fluid. In this case, can we say that the particle motion makes no influence on the flow field of the ambient fluid? Einstein derived the following famous expression for the apparent viscosity η_{eff} of a suspension in which particles are dispersed [4]

$$\eta^{eff} = \eta_s (1 + \frac{5}{2}\varphi_V), \tag{4.36}$$

in which η_s is the viscosity of the base liquid in which the particles are dispersed, and φ_V is the volumetric fraction of particles. Equation (4.36) clearly shows that forces and torques are insufficient to describe the influence of the particle motion on the fluid, and therefore the stresslet **S**, which is a second-rank tensor, has to be evaluated:

$$\mathbf{S} = -\frac{1}{2}\int_{S_p}\left\{(\mathbf{r}-\mathbf{r}_c)(\mathbf{n}\cdot\boldsymbol{\tau}) + (\mathbf{n}\cdot\boldsymbol{\tau})(\mathbf{r}-\mathbf{r}_c) - \frac{2}{3}(\mathbf{n}\cdot\boldsymbol{\tau})\cdot(\mathbf{r}-\mathbf{r}_c)\mathbf{I}\right\}dA. \tag{4.37}$$

It will be clarified in Chap.5 that the torque arises due to the difference between the local angular velocity of the fluid and the angular velocity of the particle. However, even if these two angular velocities are the same, it is never ensured that the fluid velocity over the whole surface enclosing the particle, before and after the particle is added, coincides. This aspect of a particle is described by the stresslet. We discuss, therefore, the stresslet in more detail.

From an extension of the expression in Eq. (4.34), a second-rank tensor **G** is defined as

$$\mathbf{G} = -\int_{S_p}(\mathbf{n}\cdot\boldsymbol{\tau})(\mathbf{r}-\mathbf{r}_c)dA. \tag{4.38}$$

Since the diagonal components of **G** are of no dynamic significance, this contribution is subtracted from **G**. The result is decomposed into a symmetric tensor $\mathbf{H}^s$ and anti-symmetric tensor $\mathbf{H}^a$ as

$$\mathbf{G} - \frac{1}{3}(G_{11} + G_{22} + G_{33})\mathbf{I} = \mathbf{H}^s + \mathbf{H}^a, \tag{4.39}$$

in which

$$\mathbf{H}^s = -\frac{1}{2}\int_{S_p}\{(\mathbf{n}\cdot\boldsymbol{\tau})(\mathbf{r}-\mathbf{r}_c) + (\mathbf{r}-\mathbf{r}_c)(\mathbf{n}\cdot\boldsymbol{\tau})\}dA + \frac{1}{3}\int_{S_p}(\mathbf{n}\cdot\boldsymbol{\tau})\cdot(\mathbf{r}-\mathbf{r}_c)\mathbf{I}dA, \tag{4.40}$$

$$\mathbf{H}^a = -\frac{1}{2}\int_{S_p}\{(\mathbf{n}\cdot\boldsymbol{\tau})(\mathbf{r}-\mathbf{r}_c) - (\mathbf{r}-\mathbf{r}_c)(\mathbf{n}\cdot\boldsymbol{\tau})\}dA. \tag{4.41}$$

In this equation, $\mathbf{H}^s$ is clearly equal to the stresslet **S**, Eq. (4.37). On the other hand, from Appendix A1, $\mathbf{H}^a$ can be reformed as

$$\boldsymbol{\varepsilon}:\mathbf{H}^{a}=-\frac{1}{2}\int_{S_p}\{\boldsymbol{\varepsilon}:(\mathbf{n}\cdot\boldsymbol{\tau})(\mathbf{r}-\mathbf{r}_c)-\boldsymbol{\varepsilon}:(\mathbf{r}-\mathbf{r}_c)(\mathbf{n}\cdot\boldsymbol{\tau})\}dA$$
$$=-\frac{1}{2}\int_{S_p}\{(\mathbf{r}-\mathbf{r}_c)\times(\mathbf{n}\cdot\boldsymbol{\tau})-(\mathbf{n}\cdot\boldsymbol{\tau})\times(\mathbf{r}-\mathbf{r}_c)\}dA=-\int_{S_p}(\mathbf{r}-\mathbf{r}_c)\times(\mathbf{n}\cdot\boldsymbol{\tau})dA, \tag{4.42}$$

and this leads to the relation $\mathbf{T}=\boldsymbol{\varepsilon}:\mathbf{H}^{a}$. For two arbitrary vectors **a** and **b**, the following equation is satisfied:

$$\boldsymbol{\varepsilon}\cdot(\mathbf{a}\times\mathbf{b})=\boldsymbol{\varepsilon}\cdot(\boldsymbol{\varepsilon}:\mathbf{ba})=\mathbf{ab}-\mathbf{ba}. \tag{4.43}$$

$\mathbf{H}^{a}$ is, therefore, expressed using the torque **T** as

$$\mathbf{H}^{a}=-\frac{1}{2}\int_{S_p}\boldsymbol{\varepsilon}\cdot\{(\mathbf{n}\cdot\boldsymbol{\tau})\times(\mathbf{r}-\mathbf{r}_c)\}dA=-\frac{1}{2}\boldsymbol{\varepsilon}\cdot\mathbf{T}. \tag{4.44}$$

It is clear from these discussions that the second-rank tensor **G**, defined as the first moment of ($\mathbf{r}$-$\mathbf{r}_c$) in Eq. (4.38), gives the stresslet and a quantity related to the torque.

References

1. R.B. Bird, R.C. Armstrong, and O. Hassager, "Dynamics of Polymeric Liquids, Vol.1, Fluid Mechanics," John Wiley & Sons, New York, 1977
2. G.K. Batchelor, "An Introduction to Fluid Dynamics," Cambridge University Press, Cambridge, 1981
3. H. Lamb, "Hydrodynamics," 6th Ed., Cambridge University Press, Cambridge, 1932
4. A. Einstein, A New Determination of Molecular Dimensions, Ann. Phys., 19 (1906) 289

Exercises

4.1 Show that the components of the stress tensor $\boldsymbol{\tau}$ in Eq. (4.17) can be written in the rectangular coordinate system as

$$\boldsymbol{\tau}=\begin{bmatrix}-p+2\eta\partial u_x/\partial x & \eta(\partial u_y/\partial x+\partial u_x/\partial y) & \eta(\partial u_z/\partial x+\partial u_x/\partial z)\\ \eta(\partial u_x/\partial y+\partial u_y/\partial x) & -p+2\eta\partial u_y/\partial y & \eta(\partial u_z/\partial y+\partial u_y/\partial z)\\ \eta(\partial u_x/\partial z+\partial u_z/\partial x) & \eta(\partial u_y/\partial z+\partial u_z/\partial y) & -p+2\eta\partial u_z/\partial z\end{bmatrix}. \tag{4.45}$$

4.2 From the Navier-Stokes equation in Eq. (4.18) and the result in Eq. (4.45), derive the following equation which is satisfied by the x-component of the velocity, u_x:

$$\rho\left(\frac{\partial u_x}{\partial t}+u_x\frac{\partial u_x}{\partial x}+u_y\frac{\partial u_x}{\partial y}+u_z\frac{\partial u_x}{\partial z}\right)=-\frac{\partial p}{\partial x}+\eta\left(\frac{\partial^2 u_x}{\partial x^2}+\frac{\partial^2 u_x}{\partial y^2}+\frac{\partial^2 u_x}{\partial z^2}\right). \tag{4.46}$$

The equation for the y-component u_y can be obtained by replacing u_x, which is the object of the partial differential, with u_y and similarly the equation for u_z can be obtained.

4.3 The nabla operator ∇ in the cylindrical coordinate (r,θ, z) is written as

$$\nabla = \boldsymbol{\delta}_r \frac{\partial}{\partial r} + \boldsymbol{\delta}_\theta \frac{\partial}{r\partial \theta} + \boldsymbol{\delta}_z \frac{\partial}{\partial z}. \tag{4.47}$$

If this operator is applied to the fluid velocity $\mathbf{u}=u_r\boldsymbol{\delta}_r + u_\theta\boldsymbol{\delta}_\theta + u_z\boldsymbol{\delta}_z$, the following relations can be obtained:

$$\begin{aligned}\nabla\cdot\mathbf{u} &= \left\{\boldsymbol{\delta}_r \frac{\partial}{\partial r} + \boldsymbol{\delta}_\theta \frac{\partial}{r\partial \theta} + \boldsymbol{\delta}_z \frac{\partial}{\partial z}\right\}\cdot\{\boldsymbol{\delta}_r u_r + \boldsymbol{\delta}_\theta u_\theta + \boldsymbol{\delta}_z u_z\} \\ &= \frac{u_r}{r} + \frac{\partial u_r}{\partial r} + \frac{\partial u_\theta}{r\partial\theta} + \frac{\partial u_z}{\partial z},\end{aligned} \tag{4.48}$$

$$\begin{aligned}\nabla\mathbf{u} &= \left\{\boldsymbol{\delta}_r \frac{\partial}{\partial r} + \boldsymbol{\delta}_\theta \frac{\partial}{r\partial \theta} + \boldsymbol{\delta}_z \frac{\partial}{\partial z}\right\}\{\boldsymbol{\delta}_r u_r + \boldsymbol{\delta}_\theta u_\theta + \boldsymbol{\delta}_z u_z\} \\ &= \boldsymbol{\delta}_r\boldsymbol{\delta}_r \frac{\partial u_r}{\partial r} + \boldsymbol{\delta}_r\boldsymbol{\delta}_\theta \frac{\partial u_\theta}{\partial r} + \boldsymbol{\delta}_r\boldsymbol{\delta}_z \frac{\partial u_z}{\partial r} + \boldsymbol{\delta}_\theta\boldsymbol{\delta}_r\left(\frac{\partial u_r}{r\partial\theta} - \frac{u_\theta}{r}\right) \\ &+ \boldsymbol{\delta}_\theta\boldsymbol{\delta}_\theta\left(\frac{\partial u_\theta}{r\partial\theta} + \frac{u_r}{r}\right) + \boldsymbol{\delta}_\theta\boldsymbol{\delta}_z \frac{\partial u_z}{r\partial\theta} + \boldsymbol{\delta}_z\boldsymbol{\delta}_r \frac{\partial u_r}{\partial z} + \boldsymbol{\delta}_z\boldsymbol{\delta}_\theta \frac{\partial u_\theta}{\partial z} + \boldsymbol{\delta}_z\boldsymbol{\delta}_z \frac{\partial u_z}{\partial z}.\end{aligned} \tag{4.49}$$

Also the divergence of the stress tensor $\boldsymbol{\tau}$ is expressed as

$$\begin{aligned}\nabla\cdot\boldsymbol{\tau} &= \left\{\boldsymbol{\delta}_r \frac{\partial}{\partial r} + \boldsymbol{\delta}_\theta \frac{\partial}{r\partial \theta} + \boldsymbol{\delta}_z \frac{\partial}{\partial z}\right\}\cdot\{\boldsymbol{\delta}_r\boldsymbol{\delta}_r\tau_{rr} + \boldsymbol{\delta}_r\boldsymbol{\delta}_\theta\tau_{r\theta} + \boldsymbol{\delta}_r\boldsymbol{\delta}_z\tau_{rz} + \boldsymbol{\delta}_\theta\boldsymbol{\delta}_r\tau_{\theta r} + \boldsymbol{\delta}_\theta\boldsymbol{\delta}_\theta\tau_{\theta\theta} \\ &\quad + \boldsymbol{\delta}_\theta\boldsymbol{\delta}_z\tau_{\theta z} + \boldsymbol{\delta}_z\boldsymbol{\delta}_r\tau_{zr} + \boldsymbol{\delta}_z\boldsymbol{\delta}_\theta\tau_{z\theta} + \boldsymbol{\delta}_z\boldsymbol{\delta}_z\tau_{zz}\} \\ &= \boldsymbol{\delta}_r\left\{\frac{\partial}{r\partial r}(r\tau_{rr}) + \frac{\partial}{r\partial\theta}(\tau_{\theta r}) + \frac{\partial}{\partial z}(\tau_{zr}) - \frac{\tau_{\theta\theta}}{r}\right\} + \boldsymbol{\delta}_\theta\left\{\frac{\partial}{r^2\partial r}(r^2\tau_{r\theta}) + \frac{\partial}{r\partial\theta}(\tau_{\theta\theta})\right. \\ &\quad \left. + \frac{\partial}{\partial z}(\tau_{z\theta}) + \frac{\tau_{\theta r} - \tau_{r\theta}}{r}\right\} + \boldsymbol{\delta}_z\left\{\frac{\partial}{r\partial r}(r\tau_{rz}) + \frac{\partial}{r\partial\theta}(\tau_{\theta z}) + \frac{\partial}{\partial z}(\tau_{zz})\right\}.\end{aligned} \tag{4.50}$$

With these relations, show that the Navier-Stokes equation in Eq. (4.18) can be expressed in the cylindrical coordinate system as

$$\rho\left(\frac{\partial u_r}{\partial t} + u_r\frac{\partial u_r}{\partial r} + \frac{u_\theta}{r}\frac{\partial u_r}{\partial\theta} + u_z\frac{\partial u_r}{\partial z} - \frac{u_\theta^2}{r}\right) = -\frac{\partial p}{\partial r} + \eta\left(\nabla^2 u_r - \frac{u_r}{r^2} - \frac{2}{r^2}\frac{\partial u_\theta}{\partial\theta}\right), \tag{4.51}$$

$$\rho\left(\frac{\partial u_\theta}{\partial t} + u_r\frac{\partial u_\theta}{\partial r} + \frac{u_\theta}{r}\frac{\partial u_\theta}{\partial\theta} + u_z\frac{\partial u_\theta}{\partial z} + \frac{u_r u_\theta}{r}\right) = -\frac{\partial p}{r\partial\theta} + \eta\left(\nabla^2 u_\theta + \frac{2}{r^2}\frac{\partial u_r}{\partial\theta} - \frac{u_\theta}{r^2}\right), \tag{4.52}$$

$$\rho\left(\frac{\partial u_z}{\partial t} + u_r\frac{\partial u_z}{\partial r} + \frac{u_\theta}{r}\frac{\partial u_z}{\partial\theta} + u_z\frac{\partial u_z}{\partial z}\right) = -\frac{\partial p}{\partial z} + \eta\nabla^2 u_z, \tag{4.53}$$

in which the operator ∇^2 is written as

$$\nabla^2 = \frac{\partial^2}{\partial r^2} + \frac{\partial}{r\partial r} + \frac{\partial^2}{r^2\partial\theta^2} + \frac{\partial^2}{\partial z^2}\,. \tag{4.54}$$

4.4 By transforming the velocity components from the rectangular to the spherical coordinate system, show that the solution in Eq. (4.22) satisfies the no-slip boundary condition, that is, **u**=0, over the whole surface of the sphere.

4.5 By substituting Eqs. (4.22) and (4.23) into Eq. (4.17), using this expression for $\boldsymbol{\tau}$, and carrying out the integration over the surface of the sphere in Eq. (4.11), show that the force acting on the sphere by the fluid can be expressed as

$$F = 6\pi\eta a U\,. \tag{4.55}$$

CHAPTER 5

THEORY FOR THE MOTION OF A SINGLE PARTICLE AND TWO PARTICLES IN A FLUID

When a particle moves in a flow field, the particle motion influences the flow field. That is, the flow field without a particle is different from that including it. The flow field for a system including a particle can be analytically solved for $Re \ll 1$ using the Stokes equation, the equation of continuity, and the boundary conditions at the surface of the particle and at infinity.

The multipole expansion method may be used to solve analytically the Stokes equation for the motion of two particles in a fluid. But, since this analytical procedure has a strong applied mathematical aspect, it is not appropriate to deal with them in this book. If the reader is interested in the mathematical manipulation, an appropriate textbook dealing with such procedures should be referred to [1]. Hence, we here just outline the useful expressions that are of interest to microsimulations for colloidal dispersions. Unless otherwise specified, we consider a linear flow field, shown in Eq. (4.25) or (4.29), as the flow field before dispersing particles.

5.1 Theory for Single Particle Motion

5.1.1 The spherical particle

The solution shown in Eq. (4.22) is for a uniform flow, directed in the x-direction, past a spherical particle. By extending this solution to the general direction case, the flow field can be obtained for the particle moving in an arbitrary direction in a quiescent field. With this solution, the relationship between the force $\mathbf{F}$ exerted on the fluid by the particle and the particle velocity $\mathbf{v}$, can be derived straightforwardly from Eq. (4.11). This relation is well known as Stokes' drag formula and is expressed in vector form as

$$\mathbf{F} = 6\pi\eta a\mathbf{v}, \tag{5.1}$$

in which η is the viscosity for the fluid and a is the particle radius.

We next consider the flow problem of a particle placed in a flow field which is rotated with angular velocity $\mathbf{\Omega}$ around the origin. If the torque acting on the fluid from the particle is denoted by $\mathbf{T}$, the relationship between the particle angular velocity $\boldsymbol{\omega}$ and the torque $\mathbf{T}$ can be expressed as [2,3]

$$\mathbf{T} = 8\pi\eta a^3(\boldsymbol{\omega} - \mathbf{\Omega}). \tag{5.2}$$

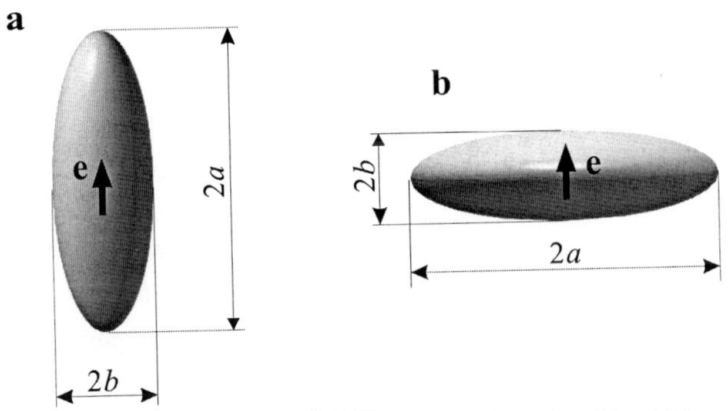

Figure 5.1 Spheroids with eccentricity $s=(a^2-b^2)^{1/2}/a$: (a) a prolate spheroid and (b) an oblate spheroid.

Hence, unless a torque acts on the particle, it will rotate with the angular velocity of the unperturbed velocity field. It is noted that Eqs. (5.1) and (5.2) are valid for the linear flow field shown in Eq. (4.25), although Eqs. (5.1) and (5.2) have been shown in separate discussions.

5.1.2 The spheroid

A spheroid is a very useful particle model in addition to the sphere. As a limiting case, a slender spheroid may be used as a model of a needlelike particle and a very flat spheroid as that of a disklike particle. In the following, we just outline the standard results [1] systematically without the derivation of each expression.

A. The prolate spheroid

As shown in Fig. 5.1(a), a prolate spheroid particle model is made from rotating an ellipse with the longest dimension (the major axis) $2a$ and shortest dimension (the minor axis) $2b$ around the major axis. For avoiding the reader's confusion, the notation **U** instead of **u** in Eq. (4.29) is used as the linear velocity field before dispersing particles. That is,

$$\mathbf{U}(\mathbf{r}) = \mathbf{U}_0 + \boldsymbol{\Omega} \times \mathbf{r} + \mathbf{E} \cdot \mathbf{r} . \tag{5.3}$$

If the orientation of the particle is denoted by the unit vector **e**, and the particle velocity and angular velocity are denoted by **v** and **ω**, respectively, then the force, torque and stresslet exerted on the ambient fluid by the particle are expressed as

$$\mathbf{F} = \eta \{ X^A \mathbf{ee} + Y^A (\mathbf{I} - \mathbf{ee}) \} \cdot (\mathbf{v} - \mathbf{U}) , \tag{5.4}$$

$$\mathbf{T} = \eta \{ X^C \mathbf{ee} + Y^C (\mathbf{I} - \mathbf{ee}) \} \cdot (\boldsymbol{\omega} - \boldsymbol{\Omega}) + \eta Y^H (\boldsymbol{\varepsilon} \cdot \mathbf{ee}) : \mathbf{E} , \tag{5.5}$$

$$\mathbf{S} = -\eta\{X^K \mathbf{e}^{(0)} + Y^K \mathbf{e}^{(1)} + Z^K \mathbf{e}^{(2)}\} : \mathbf{E} + \frac{1}{2}\eta Y^H [\{\boldsymbol{\varepsilon} : \mathbf{e}(\boldsymbol{\omega} - \boldsymbol{\Omega})\}\mathbf{e} + \mathbf{e}\{\boldsymbol{\varepsilon} : \mathbf{e}(\boldsymbol{\omega} - \boldsymbol{\Omega})\}], \tag{5.6}$$

in which $\boldsymbol{\varepsilon}$ is a third-rank tensor and its components ε_{ijk} (i,j,k=1,2,3) are expressed in Eq. (A1.10) in Appendix A1. The quantity **ee** is called the dyadic and is a second-rank tensor with the ij-component $e_i e_j$. Also, it is noted that the operation $\boldsymbol{\sigma}$:**A** of a third-rank tensor $\boldsymbol{\sigma}$ and a second-rank tensor **A**, becomes a first-rank tensor; many formulae concerning vectors and tensors are listed in Appendix A1, so the reader should refer to such expressions if necessary. Additionally, $\mathbf{e}^{(0)}$, $\mathbf{e}^{(1)}$, and $\mathbf{e}^{(2)}$ are fourth-rank tensors and expressed, respectively, as

$$\left.\begin{aligned} e^{(0)}_{ijkl} &= \frac{3}{2}(e_i e_j - \frac{1}{3}\delta_{ij})(e_k e_l - \frac{1}{3}\delta_{kl}), \\ e^{(1)}_{ijkl} &= \frac{1}{2}(e_i e_k \delta_{jl} + e_j e_k \delta_{il} + e_i e_l \delta_{jk} + e_j e_l \delta_{ik} - 4 e_i e_j e_k e_l), \\ e^{(2)}_{ijkl} &= \frac{1}{2}(\delta_{ik}\delta_{jl} + \delta_{jk}\delta_{il} - \delta_{ij}\delta_{kl} + e_i e_j \delta_{kl} + e_k e_l \delta_{ij} - e_i e_k \delta_{jl} \\ &\qquad - e_j e_k \delta_{il} - e_i e_l \delta_{jk} - e_j e_l \delta_{ik} + e_i e_j e_k e_l), \end{aligned}\right\} \tag{5.7}$$

in which δ_{ij} and similar symbols are Kronecker's delta. In Eq. (5.7) the symbols of subscripts i,j,k and l are used for expressing each component; for example, i=1 means the x-component.

X^A, Y^A, ..., Z^K, which have appeared in Eqs. (5.4) to (5.6), are called resistance functions and depend only on the geometric shape of a particle. If the eccentricity of the ellipse is denoted by $s = (a^2 - b^2)^{1/2}/a$, the resistance functions are written as

$$\left.\begin{aligned} \frac{X^A}{6\pi a} &= \frac{8}{3}\cdot\frac{s^3}{-2s + (1+s^2)L}, \\ \frac{Y^A}{6\pi a} &= \frac{16}{3}\cdot\frac{s^3}{2s + (3s^2 - 1)L}, \end{aligned}\right\} \tag{5.8}$$

$$\left.\begin{aligned} \frac{X^C}{8\pi a^3} &= \frac{4}{3}\cdot\frac{s^3(1-s^2)}{2s - (1-s^2)L}, \\ \frac{Y^C}{8\pi a^3} &= \frac{4}{3}\cdot\frac{s^3(2-s^2)}{-2s + (1+s^2)L}, \end{aligned}\right\} \tag{5.9}$$

$$\frac{Y^H}{8\pi a^3} = \frac{4}{3}\cdot\frac{s^5}{-2s + (1+s^2)L}, \tag{5.10}$$

$$\left.\begin{aligned}
\frac{X^K}{20\pi a^3/3} &= \frac{8}{15}\cdot\frac{s^5}{(3-s^2)L-6s}, \\
\frac{Y^K}{20\pi a^3/3} &= \frac{4}{5}\cdot\frac{s^5\{2s(1-2s^2)-(1-s^2)L\}}{\{2s(2s^2-3)+3(1-s^2)L\}\{-2s+(1+s^2)L\}}, \\
\frac{Z^K}{20\pi a^3/3} &= \frac{16}{5}\cdot\frac{s^5(1-s^2)}{3(1-s^2)^2L-2s(3-5s^2)},
\end{aligned}\right\} \tag{5.11}$$

in which L is a function of s and expressed as

$$L = L(s) = \ln\{(1+s)/(1-s)\}. \tag{5.12}$$

If **e**=(1,0,0), Eqs. (5.4) and (5.5) reduce to, respectively,

$$\begin{bmatrix} F_x \\ F_y \\ F_z \end{bmatrix} = \eta \begin{bmatrix} X^A(v_x - U_x) \\ Y^A(v_y - U_y) \\ Y^A(v_z - U_z) \end{bmatrix}, \tag{5.13}$$

$$\begin{bmatrix} T_x \\ T_y \\ T_z \end{bmatrix} = \eta \begin{bmatrix} X^C(\omega_x - \Omega_x) \\ Y^C(\omega_y - \Omega_y) \\ Y^C(\omega_z - \Omega_z) \end{bmatrix} + \eta Y^H \begin{bmatrix} 0 \\ E_{13} \\ -E_{12} \end{bmatrix}. \tag{5.14}$$

It is clearly seen from Eq. (5.13) that the motion of a spheroid can be decomposed into the motion in the major axis direction and the motion in the two minor axis directions. Also, the rotational motion can be decomposed in a similar way.

Next we show the asymptotic expressions for the resistance functions. A spheroid approaches a sphere as $s \to 0$, and in the case of $s \ll 1$ the resistance functions are expressed as

$$\left.\begin{aligned}
\frac{X^A}{6\pi a} &= 1 - \frac{2}{5}s^2 - \frac{17}{175}s^4 + \cdots, \\
\frac{Y^A}{6\pi a} &= 1 - \frac{3}{10}s^2 - \frac{57}{700}s^4 + \cdots,
\end{aligned}\right\} \tag{5.15}$$

$$\left.\begin{aligned}
\frac{X^C}{8\pi a^3} &= 1 - \frac{6}{5}s^2 + \frac{27}{175}s^4 + \cdots, \\
\frac{Y^C}{8\pi a^3} &= 1 - \frac{9}{10}s^2 + \frac{18}{175}s^4 + \cdots,
\end{aligned}\right\} \tag{5.16}$$

$$\frac{Y^H}{8\pi a^3} = \frac{1}{2}s^2 - \frac{1}{5}s^4 + \cdots, \tag{5.17}$$

$$\left.\begin{aligned}\frac{X^K}{20\pi a^3/3}&=1-\frac{6}{7}s^2+\frac{1}{49}s^4+\cdots,\\ \frac{Y^K}{20\pi a^3/3}&=1-\frac{13}{14}s^2+\frac{44}{735}s^4+\cdots,\\ \frac{Z^K}{20\pi a^3/3}&=1-\frac{8}{7}s^2+\frac{17}{147}s^4+\cdots.\end{aligned}\right\}\tag{5.18}$$

By setting s=0 in these equations, Eqs.(5.4) and (5.5) become equal to Eqs. (5.1) and (5.2), respectively.

On the other hand, a spheroid approaches a needle-like particle as $s\to 1$. If the notation ξ and P are used for $\xi=(1-s^2)^{1/2}$ and $P=\{\ln(2/\xi)\}^{-1}$, the resistance functions are expressed, in the case of $\xi \ll 1$, as

$$\left.\begin{aligned}\frac{X^A}{6\pi a}&=\frac{4P}{6-3P}-\frac{(8-6P)P}{12-12P+3P^2}\xi^2+\cdots,\\ \frac{Y^A}{6\pi a}&=\frac{8P}{6+3P}-\frac{4P^2}{12+12P+3P^2}\xi^2+\cdots,\end{aligned}\right\}\tag{5.19}$$

$$\left.\begin{aligned}\frac{X^C}{8\pi a^3}&=\frac{2}{3}\xi^2+\frac{2-2P}{3P}\xi^4+\cdots,\\ \frac{Y^C}{8\pi a^3}&=\frac{2P}{6-3P}+\frac{P^2}{12-12P+3P^2}\xi^2+\cdots,\end{aligned}\right\}\tag{5.20}$$

$$\frac{Y^H}{8\pi a^3}=\frac{2P}{6-3P}-\frac{(8-5P)P}{12-12P+3P^2}\xi^2+\cdots,\tag{5.21}$$

$$\left.\begin{aligned}\frac{X^K}{20\pi a^3/3}&=\frac{4P}{30-45P}-\frac{(24-26P)P}{60-180P+135P^2}\xi^2+\cdots,\\ \frac{Y^K}{20\pi a^3/3}&=\frac{2P}{10-5P}+\frac{16-32P+13P^2}{20-20P+5P^2}\xi^2+\cdots,\\ \frac{Z^K}{20\pi a^3/3}&=\frac{4}{5}\xi^2+\frac{2}{5}\xi^4+\cdots.\end{aligned}\right\}\tag{5.22}$$

Hence, if these results are applied to Eq. (5.14), it can be seen that the rotational motion around the major axis has less influence on the ambient fluid as a particle approaches a needlelike shape.

Finally we show how (**v**-**U**) and (**ω**-**Ω**) can be expressed using the force **F** and torque **T**. This may straightforwardly be accomplished by transforming Eqs. (5.4) and (5.5). If Eq. (5.4) is multiplied by the following quantity:

$$\frac{1}{\eta}\left\{(X^A)^{-1}\mathbf{ee}+(Y^A)^{-1}(\mathbf{I}-\mathbf{ee})\right\},\tag{5.23}$$

we can obtain the velocity equation as

$$\mathbf{v}-\mathbf{U}=\frac{1}{\eta}\left\{(X^{A})^{-1}\mathbf{ee}+(Y^{A})^{-1}(\mathbf{I}-\mathbf{ee})\right\}\cdot\mathbf{F}. \tag{5.24}$$

From a similar procedure concerning Eq. (5.5), the angular velocity can be expressed as

$$\boldsymbol{\omega}-\boldsymbol{\Omega}=\frac{1}{\eta}\left\{(X^{C})^{-1}\mathbf{ee}+(Y^{C})^{-1}(\mathbf{I}-\mathbf{ee})\right\}\cdot\mathbf{T}-\frac{Y^{H}}{Y^{C}}(\boldsymbol{\varepsilon}\cdot\mathbf{ee}):\mathbf{E}. \tag{5.25}$$

B. The oblate spheroid

Equations (5.4) and (5.5) are still valid for the oblate spheroidal particle, which is shown in Fig. 5.1(b). In this case, the resistance functions become the following expressions:

$$\left.\begin{aligned}\frac{X^{A}}{6\pi a}&=\frac{4}{3}\cdot\frac{s^{3}}{(2s^{2}-1)Q+s(1-s^{2})^{1/2}},\\ \frac{Y^{A}}{6\pi a}&=\frac{8}{3}\cdot\frac{s^{3}}{(2s^{2}+1)Q-s(1-s^{2})^{1/2}},\end{aligned}\right\} \tag{5.26}$$

$$\left.\begin{aligned}\frac{X^{C}}{8\pi a^{3}}&=\frac{2}{3}\cdot\frac{s^{3}}{Q-s(1-s^{2})^{1/2}},\\ \frac{Y^{C}}{8\pi a^{3}}&=\frac{2}{3}\cdot\frac{s^{3}(2-s^{2})}{s(1-s^{2})^{1/2}-(1-2s^{2})Q},\end{aligned}\right\} \tag{5.27}$$

$$\frac{Y^{H}}{8\pi a^{3}}=-\frac{2}{3}\cdot\frac{s^{5}}{s(1-s^{2})^{1/2}-(1-2s^{2})Q}, \tag{5.28}$$

$$\left.\begin{aligned}\frac{X^{K}}{20\pi a^{3}/3}&=\frac{4}{15}\cdot\frac{s^{5}}{(3-2s^{2})Q-3s(1-s^{2})^{1/2}},\\ \frac{Y^{K}}{20\pi a^{3}/3}&=\frac{2}{5}\cdot\frac{s^{5}\left\{s(1+s^{2})-(1-s^{2})^{1/2}Q\right\}}{\left\{3s-s^{3}-3(1-s^{2})^{1/2}Q\right\}\left\{s(1-s^{2})^{1/2}-(1-2s^{2})Q\right\}},\\ \frac{Z^{K}}{20\pi a^{3}/3}&=\frac{8}{5}\cdot\frac{s^{5}}{3Q-(2s^{3}+3s)(1-s^{2})^{1/2}},\end{aligned}\right\} \tag{5.29}$$

in which $s=(a^2-b^2)^{1/2}/a$, as already shown, and Q is expressed as

$$Q=Q(s)=\cot^{-1}\left\{(1-s^{2})^{1/2}/s\right\}. \tag{5.30}$$

An oblate spheroid approaches a spherical shape as $s\to 0$, and in the case of $s\ll 1$, the resistance functions are written as

$$\left.\begin{aligned}\frac{X^A}{6\pi a}&=1-\frac{1}{10}s^2-\frac{31}{1400}s^4+\cdots,\\ \frac{Y^A}{6\pi a}&=1-\frac{1}{5}s^2-\frac{79}{1400}s^4+\cdots,\end{aligned}\right\}\tag{5.31}$$

$$\left.\begin{aligned}\frac{X^C}{8\pi a^3}&=1-\frac{3}{10}s^2-\frac{99}{1400}s^4+\cdots,\\ \frac{Y^C}{8\pi a^3}&=1-\frac{3}{5}s^2+\frac{39}{1400}s^4+\cdots,\end{aligned}\right\}\tag{5.32}$$

$$\frac{Y^H}{8\pi a^3}=-\frac{1}{2}s^2+\frac{1}{20}s^4+\cdots,\tag{5.33}$$

$$\left.\begin{aligned}\frac{X^K}{20\pi a^3/3}&=1-\frac{9}{14}s^2-\frac{13}{392}s^4+\cdots,\\ \frac{Y^K}{20\pi a^3/3}&=1-\frac{4}{7}s^2-\frac{173}{5880}s^4+\cdots,\\ \frac{Z^K}{20\pi a^3/3}&=1-\frac{5}{14}s^2-\frac{95}{1176}s^4+\cdots.\end{aligned}\right\}\tag{5.34}$$

On the other hand, a spheroid approaches a disklike shape as $s\to 1$, and, in the case of $\xi=(1-s^2)^{1/2} \ll 1$, Eqs. (5.26)~(5.29) can be written as

$$\left.\begin{aligned}\frac{X^A}{6\pi a}&=\frac{8}{3\pi}\left(1+\frac{1}{2}\xi^2+\cdots\right),\\ \frac{Y^A}{6\pi a}&=\frac{16}{9\pi}\left(1+\frac{8}{3\pi}\xi-\frac{15\pi^2-128}{18\pi^2}\xi^2+\cdots\right),\end{aligned}\right\}\tag{5.35}$$

$$\left.\begin{aligned}\frac{X^C}{8\pi a^3}&=\frac{4}{3\pi}\left(1+\frac{4}{\pi}\xi+\frac{32-3\pi^2}{2\pi^2}\xi^2+\cdots\right),\\ \frac{Y^C}{8\pi a^3}&=\frac{4}{3\pi}\left(1+\frac{3}{2}\xi^2+\cdots\right),\end{aligned}\right\}\tag{5.36}$$

$$\frac{Y^H}{8\pi a^3}=-\frac{4}{3\pi}\left(1-\frac{1}{2}\xi^2+\cdots\right),\tag{5.37}$$

$$\left.\begin{aligned}
\frac{X^K}{20\pi a^3/3} &= \frac{8}{15\pi}\left(1+\frac{8}{\pi}\xi+\frac{128-9\pi^2}{2\pi^2}\xi^2+\cdots\right),\\
\frac{Y^K}{20\pi a^3/3} &= \frac{4}{5\pi}\left(1+\frac{\pi}{2}\xi+\frac{3\pi^2-20}{8}\xi^2+\cdots\right),\\
\frac{Z^K}{20\pi a^3/3} &= \frac{16}{15\pi}\left(1+\frac{16}{3\pi}\xi+\frac{512-45\pi^2}{18\pi^2}\xi^2+\cdots\right).
\end{aligned}\right\} \tag{5.38}$$

If we set s=1 and $\mathbf{e}$=(1,0,0), the following relations are obtained from Eqs. (5.4) and (5.5):

$$\begin{bmatrix} F_x \\ F_y \\ F_z \end{bmatrix} = 6\pi\eta a \begin{bmatrix} (8/3\pi)(v_x - U_x) \\ (16/9\pi)(v_y - U_y) \\ (16/9\pi)(v_z - U_z) \end{bmatrix}, \tag{5.39}$$

$$\begin{bmatrix} T_x \\ T_y \\ T_z \end{bmatrix} = 8\pi\eta a^3 \begin{bmatrix} (4/3\pi)(\omega_x - \Omega_x) \\ (4/3\pi)(\omega_y - \Omega_y) \\ (4/3\pi)(\omega_z - \Omega_z) \end{bmatrix} - 8\pi\eta a^3 \frac{4}{3\pi} \begin{bmatrix} 0 \\ E_{13} \\ -E_{12} \end{bmatrix}. \tag{5.40}$$

It is seen from Eq. (5.39) that, in the case of a disklike particle, the particle motion normal to the disk surface feels larger resistance than the motion parallel to it.

5.1.3 The arbitrarily shaped particle

For an axisymmetric particle, such as a sphere and spheroid, the force exerted on the ambient fluid by a particle depends only on the particle velocity, but is independent of the angular velocity. Similarly, the torque depends only on the angular velocity. However, for a particle with a general shape, both the force and torque are dependent on the velocity and angular velocity of the particle. We now outline the relationship between ($\mathbf{F}$,$\mathbf{T}$) and ($\mathbf{v}$,$\boldsymbol{\omega}$) for such an arbitrarily shaped particle for a later discussion [1].

A. The resistance formulation

As already defined, the force and torque exerted on the fluid by a particle are denoted by ($\mathbf{F}$,$\mathbf{T}$), respectively, and the particle velocity and angular velocity are denoted by ($\mathbf{v}$,$\boldsymbol{\omega}$), respectively. The relationship between these quantities including the stresslet $\mathbf{S}$ is expressed as

$$\begin{bmatrix} \mathbf{F} \\ \mathbf{T} \\ \mathbf{S} \end{bmatrix} = \eta \mathbf{R} \begin{bmatrix} \mathbf{v} - \mathbf{U} \\ \boldsymbol{\omega} - \boldsymbol{\Omega} \\ -\mathbf{E} \end{bmatrix}, \tag{5.41}$$

in which $\mathbf{E}$ is the rate-of-strain tensor, $\mathbf{R}$ is called the resistance matrix (sometimes called the grand resistance matrix) and expressed as, using second-rank tensors $\mathbf{A}$, $\mathbf{B}$, and $\mathbf{C}$, third-rank tensors $\mathbf{G}$ and $\mathbf{H}$, and fourth-rank tensor $\mathbf{K}$,

$$\mathbf{R} = \begin{bmatrix} \mathbf{A} & \tilde{\mathbf{B}} & \tilde{\mathbf{G}} \\ \mathbf{B} & \mathbf{C} & \tilde{\mathbf{H}} \\ \mathbf{G} & \mathbf{H} & \mathbf{K} \end{bmatrix}, \tag{5.42}$$

in which $\tilde{\mathbf{B}}$ is equal to $\mathbf{B}'$, which is the transposed tensor of $\mathbf{B}$. If the *ijk*-components of $(\mathbf{G},\mathbf{H})$ are denoted by (G_{ijk}, H_{ijk}), respectively, $\tilde{\mathbf{G}}$ and $\tilde{\mathbf{H}}$ can be related to $\mathbf{G}$ and $\mathbf{H}$ as

$$\tilde{G}_{ijk} = G_{jki}, \qquad \tilde{H}_{ijk} = H_{jki}. \tag{5.43}$$

If the *ijkl*-component of $\mathbf{K}$ is denoted by K_{ijkl}, the following relation has to be satisfied:

$$K_{ijkl} = K_{klij}. \tag{5.44}$$

Also $\mathbf{A}$ and $\mathbf{C}$ are symmetric tensors. These tensors, constituting the small matrices of the resistance matrix, are called the resistance tensors.

B. The mobility formulation

The particle velocity, angular velocity, and stresslet are expressed in mobility formulation as a function of the force, torque, and rate-of-strain tensor:

$$\begin{bmatrix} \mathbf{v}-\mathbf{U} \\ \boldsymbol{\omega}-\boldsymbol{\Omega} \\ \mathbf{S}/\eta \end{bmatrix} = \mathbf{M}\begin{bmatrix} \mathbf{F}/\eta \\ \mathbf{T}/\eta \\ \mathbf{E} \end{bmatrix} = \begin{bmatrix} \mathbf{a} & \tilde{\mathbf{b}} & \tilde{\mathbf{g}} \\ \mathbf{b} & \mathbf{c} & \tilde{\mathbf{h}} \\ \mathbf{g} & \mathbf{h} & \mathbf{k} \end{bmatrix}\begin{bmatrix} \mathbf{F}/\eta \\ \mathbf{T}/\eta \\ \mathbf{E} \end{bmatrix}, \tag{5.45}$$

in which $\mathbf{M}$ is called the mobility matrix, and the small matrices constituting the mobility matrix are called the mobility tensors. The $\mathbf{a}$, $\mathbf{b}$, and $\mathbf{c}$ are second-rank tensors, $\mathbf{g}$ and $\mathbf{k}$ are third-rank tensors, and $\mathbf{k}$ is a fourth-rank tensor. The second-rank tensor $\tilde{\mathbf{b}}$ is equal to $\mathbf{b}'$, and the relationships similar to Eq. (5.43) are satisfied concerning $\tilde{\mathbf{g}}$ and $\tilde{\mathbf{h}}$:

$$\tilde{g}_{ijk} = g_{jki}, \qquad \tilde{h}_{ijk} = h_{jki}. \tag{5.46}$$

Concerning the fourth-rank tensor $\mathbf{k}$, the relationship similar to Eq. (5.44) is satisfied:

$$k_{ijkl} = k_{klij}. \tag{5.47}$$

Finally, it is noted that $\mathbf{a}$ and $\mathbf{c}$ are symmetric tensors.

C. The relationship between resistance and mobility tensors

Since Eqs. (5.41) and (5.45) are just different expressions of the same solution, there should be relationships between (**A**, **B**, **C**, **G**, **H**, **K**) and (**a**, **b**, **c**, **g**, **h**, **k**). We now derive the relationships between these tensors. By substituting Eq. (5.45) into Eq. (5.41), the following equation can be obtained:

$$\begin{bmatrix}\mathbf{F}\\\mathbf{T}\end{bmatrix}=\begin{bmatrix}\mathbf{A} & \widetilde{\mathbf{B}}\\\mathbf{B} & \mathbf{C}\end{bmatrix}\begin{bmatrix}\mathbf{a} & \widetilde{\mathbf{b}}\\\mathbf{b} & \mathbf{c}\end{bmatrix}\begin{bmatrix}\mathbf{F}\\\mathbf{T}\end{bmatrix}+\eta\begin{bmatrix}\mathbf{A} & \widetilde{\mathbf{B}}\\\mathbf{B} & \mathbf{C}\end{bmatrix}\begin{bmatrix}\widetilde{\mathbf{g}}\\\widetilde{\mathbf{h}}\end{bmatrix}\mathbf{E}-\eta\begin{bmatrix}\widetilde{\mathbf{G}}\\\widetilde{\mathbf{H}}\end{bmatrix}\mathbf{E}. \tag{5.48}$$

Since this equation is valid for arbitrary values of **F** and **T**, we obtain the following relations:

$$\begin{bmatrix}\mathbf{A} & \widetilde{\mathbf{B}}\\\mathbf{B} & \mathbf{C}\end{bmatrix}\begin{bmatrix}\mathbf{a} & \widetilde{\mathbf{b}}\\\mathbf{b} & \mathbf{c}\end{bmatrix}=\begin{bmatrix}\mathbf{I} & \mathbf{0}\\\mathbf{0} & \mathbf{I}\end{bmatrix}, \tag{5.49}$$

$$\begin{bmatrix}\mathbf{A} & \widetilde{\mathbf{B}}\\\mathbf{B} & \mathbf{C}\end{bmatrix}\begin{bmatrix}\widetilde{\mathbf{g}}\\\widetilde{\mathbf{h}}\end{bmatrix}-\begin{bmatrix}\widetilde{\mathbf{G}}\\\widetilde{\mathbf{H}}\end{bmatrix}=\begin{bmatrix}\mathbf{0}\\\mathbf{0}\end{bmatrix}, \tag{5.50}$$

in which **I** is the unit tensor. Hence, from Eqs. (5.49) and (5.50), (**a**, **b**, **c**, **g**, **h**) can be related to (**A**, **B**, **C**, **G**, **H**) as

$$\begin{bmatrix}\mathbf{a} & \widetilde{\mathbf{b}}\\\mathbf{b} & \mathbf{c}\end{bmatrix}=\begin{bmatrix}\mathbf{A} & \widetilde{\mathbf{B}}\\\mathbf{B} & \mathbf{C}\end{bmatrix}^{-1}, \tag{5.51}$$

$$\begin{bmatrix}\widetilde{\mathbf{g}}\\\widetilde{\mathbf{h}}\end{bmatrix}=\begin{bmatrix}\mathbf{A} & \widetilde{\mathbf{B}}\\\mathbf{B} & \mathbf{C}\end{bmatrix}^{-1}\begin{bmatrix}\widetilde{\mathbf{G}}\\\widetilde{\mathbf{H}}\end{bmatrix}. \tag{5.52}$$

From a similar procedure concerning the stresslet **S**, the following equations are obtained:

$$\begin{bmatrix}\mathbf{g} & \mathbf{h}\end{bmatrix}=\begin{bmatrix}\mathbf{G} & \mathbf{H}\end{bmatrix}\begin{bmatrix}\mathbf{A} & \widetilde{\mathbf{B}}\\\mathbf{B} & \mathbf{C}\end{bmatrix}^{-1}, \tag{5.53}$$

$$\mathbf{k}=-\mathbf{K}+\begin{bmatrix}\mathbf{G} & \mathbf{H}\end{bmatrix}\begin{bmatrix}\mathbf{A} & \widetilde{\mathbf{B}}\\\mathbf{B} & \mathbf{C}\end{bmatrix}^{-1}\begin{bmatrix}\widetilde{\mathbf{G}}\\\widetilde{\mathbf{H}}\end{bmatrix}. \tag{5.54}$$

Similarly, by substituting Eq. (5.41) into Eq. (5.45), (**A**, **B**, **C**, **G**, **H**, **K**) can be expressed using (**a**, **b**, **c**, **g**, **h**, **k**).

5.2 The Far-Field Theory for the Motion of Two Particles

In this section, we explain the Oseen tensor [4] and the Rotne-Prager tensor [5] which are frequently used for the translational motion of particles in a dilute dispersion. We concentrate our attention on the motion of spherical particles

5.2.1 The Oseen tensor

Generally, the flow field for a system of two particles moving in a base liquid is obtained by solving the Stokes equation. However, if the colloidal dispersion of interest is sufficiently dilute, we can approximately evaluate the flow field by applying the exact solution of the single particle motion to a multi-particle system.

We assume that a single spherical particle moves in a quiescent flow field with a velocity $\mathbf{v}$. In this case, the flow velocity at the distance $\mathbf{r}$ from the particle center, $\mathbf{u}(\mathbf{r})$, can be obtained from the Stokes equation in Eq. (4.20) and the equation of continuity in Eq. (4.6):

$$\mathbf{u}(\mathbf{r}) = \frac{3}{4}\left(\frac{a}{r}\right)\left[\mathbf{I}+\frac{\mathbf{rr}}{r^2}\right]\cdot\mathbf{v}+\frac{1}{4}\left(\frac{a}{r}\right)^3\left[\mathbf{I}-3\frac{\mathbf{rr}}{r^2}\right]\cdot\mathbf{v}\,. \tag{5.55}$$

If we assume that the force exerted on the fluid by the particle, $\mathbf{F}$, is equal to the friction force, that is, $\mathbf{F}=6\pi\eta a\mathbf{v}$, then Eq. (5.55) can be rewritten as

$$\mathbf{u}(\mathbf{r}) = \frac{1}{8\pi\eta r}\left[\mathbf{I}+\frac{\mathbf{rr}}{r^2}\right]\cdot\mathbf{F}+\frac{a^2}{24\pi\eta r^3}\left[\mathbf{I}-3\frac{\mathbf{rr}}{r^2}\right]\cdot\mathbf{F}\,. \tag{5.56}$$

If the higher-order terms are neglected, this equation reduces to

$$\mathbf{u}(\mathbf{r}) = \frac{1}{8\pi\eta r}\left[\mathbf{I}+\frac{\mathbf{rr}}{r^2}\right]\cdot\mathbf{F}\,, \tag{5.57}$$

in which $(\mathbf{I}+\mathbf{rr}/r^2)\,/r$ is called the Oseen tensor. If two particles move sufficiently apart from each other, the velocity $\mathbf{v}_j$ of particle j can be evaluated by adding the velocity induced from particle i to its own velocity, and expressed as

$$\mathbf{v}_j = \frac{1}{6\pi\eta a}\mathbf{F}_j+\frac{1}{8\pi\eta r_{ji}}\left[\mathbf{I}+\frac{\mathbf{r}_{ji}\mathbf{r}_{ji}}{r_{ji}^2}\right]\cdot\mathbf{F}_i\,, \tag{5.58}$$

in which $\mathbf{F}_i$ is the force exerted on the fluid by particle i ($\mathbf{F}_j$ is similar), $\mathbf{r}_{ji}=\mathbf{r}_j-\mathbf{r}_i$, and $r_{ji}=|\mathbf{r}_{ji}|$.

5.2.2 The Rotne-Prager tensor

The velocity $\mathbf{u}(\mathbf{r})$ in Eq. (5.56) is the induced velocity at an arbitrary position $\mathbf{r}$ which is due to the motion of a single particle in a quiescent flow field. Since two particles move in a liquid simultaneously, the motion of one particle will be influenced by the motion of the other particle through the ambient liquid. A more exact expression for the induced velocity, therefore, should be obtained by taking into account this effect. This was done by Rotne and Prager [5], and the induced velocity $\mathbf{v}_{j(i)}$ of particle j due to particle i is expressed as

$$\mathbf{v}_{j(i)} = \frac{1}{8\pi\eta r_{ji}}\left[\mathbf{I} + \frac{\mathbf{r}_{ji}\mathbf{r}_{ji}}{r_{ji}^2}\right]\cdot\mathbf{F}_i + \frac{a^2}{12\pi\eta r_{ji}^3}\left[\mathbf{I} - 3\frac{\mathbf{r}_{ji}\mathbf{r}_{ji}}{r_{ji}^2}\right]\cdot\mathbf{F}_i. \tag{5.59}$$

Hence the velocity $\mathbf{v}_j$ of particle j can be written as

$$\mathbf{v}_j = \frac{1}{6\pi\eta a}\mathbf{F}_j + \mathbf{v}_{j(i)}. \tag{5.60}$$

From similarity to the Oseen tensor, the following quantity is called the Rotne-Prager tensor:

$$\frac{1}{r_{ji}}\left[\left\{\mathbf{I} + \frac{\mathbf{r}_{ji}\mathbf{r}_{ji}}{r_{ji}^2}\right\} + \frac{2}{3}\left(\frac{a}{r_{ji}}\right)^2\left\{\mathbf{I} - 3\frac{\mathbf{r}_{ji}\mathbf{r}_{ji}}{r_{ji}^2}\right\}\right]. \tag{5.61}$$

Some researchers may prefer another definition of the Rotne-Prager tensor in which a constant ($1/8\,r_{ji}$) appears in the front of Eq. (5.61) instead of ($1/r_{ji}$).

5.3 General Theory for the Motion of Two Particles

In this section, we don't assume that two particles are sufficiently apart from each other, rather we show the theory [1] which is applicable to nearly-touching particles. We concentrate our attention on a spherical particle system. The theory of this system will be extended to a multi-particle system in the later sections.

5.3.1 The resistance matrix formulation

We consider the problem that two particles with an arbitrary geometrical shape move in a flow field. In this case, the forces and torques are related to the velocities and angular velocities by extending the theory for the single particle motion, which has been explained in Sec. 5.1.3:

$$\begin{bmatrix}\mathbf{F}_1\\ \mathbf{F}_2\\ \mathbf{T}_1\\ \mathbf{T}_2\\ \mathbf{S}_1\\ \mathbf{S}_2\end{bmatrix} = \eta\begin{bmatrix}\mathbf{A}_{11} & \mathbf{A}_{12} & \tilde{\mathbf{B}}_{11} & \tilde{\mathbf{B}}_{12} & \tilde{\mathbf{G}}_{11} & \tilde{\mathbf{G}}_{12}\\ \mathbf{A}_{21} & \mathbf{A}_{22} & \tilde{\mathbf{B}}_{21} & \tilde{\mathbf{B}}_{22} & \tilde{\mathbf{G}}_{21} & \tilde{\mathbf{G}}_{22}\\ \mathbf{B}_{11} & \mathbf{B}_{12} & \mathbf{C}_{11} & \mathbf{C}_{12} & \tilde{\mathbf{H}}_{11} & \tilde{\mathbf{H}}_{12}\\ \mathbf{B}_{21} & \mathbf{B}_{22} & \mathbf{C}_{21} & \mathbf{C}_{22} & \tilde{\mathbf{H}}_{21} & \tilde{\mathbf{H}}_{22}\\ \mathbf{G}_{11} & \mathbf{G}_{12} & \mathbf{H}_{11} & \mathbf{H}_{12} & \mathbf{K}_{11} & \mathbf{K}_{12}\\ \mathbf{G}_{21} & \mathbf{G}_{22} & \mathbf{H}_{21} & \mathbf{H}_{22} & \mathbf{K}_{21} & \mathbf{K}_{22}\end{bmatrix}\begin{bmatrix}\mathbf{v}_1 - \mathbf{U}(\mathbf{r}_1)\\ \mathbf{v}_2 - \mathbf{U}(\mathbf{r}_2)\\ \boldsymbol{\omega}_1 - \boldsymbol{\Omega}\\ \boldsymbol{\omega}_2 - \boldsymbol{\Omega}\\ -\mathbf{E}\\ -\mathbf{E}\end{bmatrix}, \tag{5.62}$$

in which $\mathbf{r}$ is the position vector, $\mathbf{v}$ and $\boldsymbol{\omega}$ are the particle velocity and angular velocity, $\mathbf{F}$, $\mathbf{T}$, and $\mathbf{S}$ are the force , torque, and stresslet, respectively, exerted on the fluid by a particle. The subscripts 1 and 2 mean quantities related to particles 1 and 2, respectively. Also, **A**, **B**, and **C** are second-rank tensors, **G** and **H** are third-rank tensors, and **K** is a fourth-rank tensor. These tensors have the following properties:

$$\mathbf{A}_{\alpha\beta} = \mathbf{A}'_{\beta\alpha}, \quad \mathbf{C}_{\alpha\beta} = \mathbf{C}'_{\beta\alpha}, \quad \widetilde{\mathbf{B}}_{\alpha\beta} = \mathbf{B}'_{\beta\alpha} \quad (\alpha, \beta = 1,2), \tag{5.63}$$

$$K^{\alpha\beta}_{ijkl} = K^{\beta\alpha}_{klij}, \quad G^{\alpha\beta}_{ijk} = \widetilde{G}^{\beta\alpha}_{kij}, \quad H^{\alpha\beta}_{ijk} = \widetilde{H}^{\beta\alpha}_{kij} \quad (\alpha, \beta = 1,2), \tag{5.64}$$

in which the *ijk*-component of $\mathbf{H}_{\alpha\beta}$ is denoted by $H_{ijk}^{\alpha\beta}$ (similarly for other quantities). The above properties show that the resistance matrix with the small matrices of $\mathbf{A}_{11}$ to $\mathbf{K}_{22}$ in Eq. (5.62) is symmetric.

5.3.2 The mobility matrix formulation

Equation (5.62) can be rewritten in mobility matrix form as

$$\begin{bmatrix} \mathbf{v}_1 - \mathbf{U}(\mathbf{r}_1) \\ \mathbf{v}_2 - \mathbf{U}(\mathbf{r}_2) \\ \boldsymbol{\omega}_1 - \boldsymbol{\Omega} \\ \boldsymbol{\omega}_2 - \boldsymbol{\Omega} \\ \mathbf{S}_1/\eta \\ \mathbf{S}_2/\eta \end{bmatrix} = \begin{bmatrix} \mathbf{a}_{11} & \mathbf{a}_{12} & \widetilde{\mathbf{b}}_{11} & \widetilde{\mathbf{b}}_{12} & \widetilde{\mathbf{g}}_1 \\ \mathbf{a}_{21} & \mathbf{a}_{22} & \widetilde{\mathbf{b}}_{21} & \widetilde{\mathbf{b}}_{22} & \widetilde{\mathbf{g}}_2 \\ \mathbf{b}_{11} & \mathbf{b}_{12} & \mathbf{c}_{11} & \mathbf{c}_{12} & \widetilde{\mathbf{h}}_1 \\ \mathbf{b}_{21} & \mathbf{b}_{22} & \mathbf{c}_{21} & \mathbf{c}_{22} & \widetilde{\mathbf{h}}_2 \\ \mathbf{g}_{11} & \mathbf{g}_{12} & \mathbf{h}_{11} & \mathbf{h}_{12} & \mathbf{k}_1 \\ \mathbf{g}_{21} & \mathbf{g}_{22} & \mathbf{h}_{21} & \mathbf{h}_{22} & \mathbf{k}_2 \end{bmatrix} \begin{bmatrix} \mathbf{F}_1/\eta \\ \mathbf{F}_2/\eta \\ \mathbf{T}_1/\eta \\ \mathbf{T}_2/\eta \\ \mathbf{E} \end{bmatrix}. \tag{5.65}$$

In this equation, the matrix with the small matrices $\mathbf{a}_{11}$ to $\mathbf{k}_2$ is called the mobility matrix. As before, **a**, **b**, and **c** are second-rank tensors, **g** and **h** are third-rank tensors, and **k** is a fourth-rank tensor. These tensors have the following properties:

$$\boldsymbol{\alpha}_{\alpha\beta} = \boldsymbol{\alpha}'_{\beta\alpha}, \quad \mathbf{c}_{\alpha\beta} = \mathbf{c}'_{\beta\alpha}, \quad \widetilde{\mathbf{b}}_{\alpha\beta} = \mathbf{b}'_{\beta\alpha} \quad (\alpha, \beta = 1,2), \tag{5.66}$$

$$\left.\begin{aligned} &k^1_{ijkl} + k^2_{ijkl} = k^1_{klij} + k^2_{klij}, \\ &g^{1\alpha}_{ijk} + g^{2\alpha}_{ijk} = \widetilde{g}^{\alpha}_{kij} \qquad (\alpha = 1,2), \\ &h^{1\alpha}_{ijk} + h^{2\alpha}_{ijk} = \widetilde{h}^{\alpha}_{kij} \qquad (\alpha = 1,2), \end{aligned}\right\} \tag{5.67}$$

in which the notation rule similar to that in Sec. 5.3.1 has been used for components of tensors.

5.3.3 The relationship between resistance and mobility tensors

Since Eqs. (5.62) and (5.65) are just different expressions for the same solution, the tensors constituting the small matrices should be related to each other. These relations can straightforwardly be derived from Eqs. (5.62) and (5.65), and are expressed as

$$\begin{bmatrix} \mathbf{a}_{11} & \mathbf{a}_{12} & \widetilde{\mathbf{b}}_{11} & \widetilde{\mathbf{b}}_{12} \\ \mathbf{a}_{21} & \mathbf{a}_{22} & \widetilde{\mathbf{b}}_{21} & \widetilde{\mathbf{b}}_{22} \\ \mathbf{b}_{11} & \mathbf{b}_{12} & \mathbf{c}_{11} & \mathbf{c}_{12} \\ \mathbf{b}_{21} & \mathbf{b}_{22} & \mathbf{c}_{21} & \mathbf{c}_{22} \end{bmatrix} = \begin{bmatrix} \mathbf{A}_{11} & \mathbf{A}_{12} & \widetilde{\mathbf{B}}_{11} & \widetilde{\mathbf{B}}_{12} \\ \mathbf{A}_{21} & \mathbf{A}_{22} & \widetilde{\mathbf{B}}_{21} & \widetilde{\mathbf{B}}_{22} \\ \mathbf{B}_{11} & \mathbf{B}_{12} & \mathbf{C}_{11} & \mathbf{C}_{12} \\ \mathbf{B}_{21} & \mathbf{B}_{22} & \mathbf{C}_{21} & \mathbf{C}_{22} \end{bmatrix}^{-1}, \tag{5.68}$$

$$\begin{bmatrix} \mathbf{g}_{11} & \mathbf{g}_{12} & \mathbf{h}_{11} & \mathbf{h}_{12} \\ \mathbf{g}_{21} & \mathbf{g}_{22} & \mathbf{h}_{21} & \mathbf{h}_{22} \end{bmatrix} = \begin{bmatrix} \mathbf{G}_{11} & \mathbf{G}_{12} & \mathbf{H}_{11} & \mathbf{H}_{12} \\ \mathbf{G}_{21} & \mathbf{G}_{22} & \mathbf{H}_{21} & \mathbf{H}_{22} \end{bmatrix} \begin{bmatrix} \mathbf{a}_{11} & \mathbf{a}_{12} & \widetilde{\mathbf{b}}_{11} & \widetilde{\mathbf{b}}_{12} \\ \mathbf{a}_{21} & \mathbf{a}_{22} & \widetilde{\mathbf{b}}_{21} & \widetilde{\mathbf{b}}_{22} \\ \mathbf{b}_{11} & \mathbf{b}_{12} & \mathbf{c}_{11} & \mathbf{c}_{12} \\ \mathbf{b}_{21} & \mathbf{b}_{22} & \mathbf{c}_{21} & \mathbf{c}_{22} \end{bmatrix}, \tag{5.69}$$

$$\begin{bmatrix} \mathbf{k}_1 \\ \mathbf{k}_2 \end{bmatrix} = -\begin{bmatrix} \mathbf{K}_{11} + \mathbf{K}_{12} \\ \mathbf{K}_{21} + \mathbf{K}_{22} \end{bmatrix} + \begin{bmatrix} \mathbf{G}_{11} & \mathbf{G}_{12} & \mathbf{H}_{11} & \mathbf{H}_{12} \\ \mathbf{G}_{21} & \mathbf{G}_{22} & \mathbf{H}_{21} & \mathbf{H}_{22} \end{bmatrix} \begin{bmatrix} \widetilde{\mathbf{g}}_1 \\ \widetilde{\mathbf{g}}_2 \\ \widetilde{\mathbf{h}}_1 \\ \widetilde{\mathbf{h}}_2 \end{bmatrix}. \tag{5.70}$$

5.3.4 Axisymmetric particles

For axisymmetric particles, which are very useful as a model particle, the tensors constituting the resistance matrix in Eq. (5.62) are expressed as

$$\left.\begin{aligned} \mathbf{A}_{\alpha\beta} &= X^A_{\alpha\beta}\mathbf{ee} + Y^A_{\alpha\beta}(\mathbf{I} - \mathbf{ee}), \\ \mathbf{B}_{\alpha\beta} &= Y^B_{\alpha\beta}(\boldsymbol{\varepsilon}\cdot\mathbf{e}), \\ \mathbf{C}_{\alpha\beta} &= X^C_{\alpha\beta}\mathbf{ee} + Y^C_{\alpha\beta}(\mathbf{I} - \mathbf{ee}), \end{aligned}\right\} \tag{5.71}$$

$$\left.\begin{aligned} G^{\alpha\beta}_{ijk} &= X^G_{\alpha\beta}(e_i e_j - \frac{1}{3}\delta_{ij})e_k + Y^G_{\alpha\beta}(e_i\delta_{jk} + e_j\delta_{ik} - 2e_ie_je_k), \\ H^{\alpha\beta}_{ijk} &= Y^H_{\alpha\beta}\sum_{l=1}^{3}(\varepsilon_{ikl}e_le_j + \varepsilon_{jkl}e_le_i), \\ K^{\alpha\beta}_{ijkl} &= X^K_{\alpha\beta}e^{(0)}_{ijkl} + Y^K_{\alpha\beta}e^{(1)}_{ijkl} + Z^K_{\alpha\beta}e^{(2)}_{ijkl}, \end{aligned}\right\} \tag{5.72}$$

in which $\mathbf{e}=\mathbf{r}_{21}/r_{21}$, $\mathbf{r}_{21}=\mathbf{r}_2-\mathbf{r}_1$, $r_{21}=|\mathbf{r}_{21}|$, $\boldsymbol{\varepsilon}$ is the third-rank tensor as expressed in Eq. (A1.10), $\mathbf{ee}$ is the dyadic product, δ_{ij} is Kronecker's delta, and $e_{ijkl}^{(0)}$ and other notations have already been shown in Eq. (5.7). The quantities $X_{\alpha\beta}^{A}$, $Y_{\alpha\beta}^{A}$, ..., $Z_{\alpha\beta}^{K}$ (α,β=1,2) which have appeared in Eqs. (5.71) and (5.72) are called the resistance functions.

Similarly, the tensors constituting the mobility matrix in Eq. (5.65) can be expressed in similar expressions to Eqs. (5.71) and (5.72):

If two particles are in a widely-separated, these resistance functions are written as

$$X_{11}^A = 6\pi a \sum_{k=0}^{\infty} \left(\frac{1}{2s}\right)^{2k} f_{2k}^X, \quad X_{12}^A = -6\pi a \sum_{k=0}^{\infty} \left(\frac{1}{2s}\right)^{2k+1} f_{2k+1}^X, \tag{5.85}$$

$$Y_{11}^A = 6\pi a \sum_{k=0}^{\infty} \left(\frac{1}{2s}\right)^{2k} f_{2k}^Y, \quad Y_{12}^A = -6\pi a \sum_{k=0}^{\infty} \left(\frac{1}{2s}\right)^{2k+1} f_{2k+1}^Y, \tag{5.86}$$

in which

$$\left.\begin{aligned} &f_0^X = 1,\ f_1^X = 3,\ f_2^X = 9,\ f_3^X = 19,\ f_4^X = 93,\ f_5^X = 387,\ f_6^X = 1197, \\ &f_7^X = 5331,\ f_8^X = 19821,\ f_9^X = 76115,\ f_{10}^X = 320173,\ f_{11}^X = 1178451, \end{aligned}\right\} \tag{5.87}$$

$$\left.\begin{aligned} &f_0^Y = 1,\ f_1^Y = 3/2,\ f_2^Y = 9/4,\ f_3^Y = 59/8,\ f_4^Y = 465/16,\ f_5^Y = 2259/32, \\ &f_6^Y = 14745/64,\ f_7^Y = 89643/128,\ f_8^Y = 570017/256, \\ &f_9^Y = 4451395/512,\ f_{10}^Y = 33678825/1024,\ f_{11}^Y = 266862875/2048. \end{aligned}\right\} \tag{5.88}$$

B. Expressions for mobility functions $x_{\alpha\beta}{}^a$ and $y_{\alpha\beta}{}^a$

For the nearly-touching case,

$$\left.\begin{aligned} (6\pi a)x_{11}^a &= 0.7750 + 0.930\xi + 0.900\xi^2 \ln\xi - 2.685\xi^2 + O(\xi^3(\ln\xi)^2), \\ (6\pi a)x_{12}^a &= 0.7750 - 1.070\xi - 0.900\xi^2 \ln\xi + 2.697\xi^2 + O(\xi^3(\ln\xi)^2), \end{aligned}\right\} \tag{5.89}$$

$$\left.\begin{aligned} (6\pi a)y_{11}^a &= \frac{0.89056(\ln\xi^{-1})^2 + 5.77196\ln\xi^{-1} + 7.06897}{(\ln\xi^{-1})^2 + 6.04250\ln\xi^{-1} + 6.32549} + O(\xi(\ln\xi)^3), \\ (6\pi a)y_{12}^a &= \frac{0.48951(\ln\xi^{-1})^2 + 2.80545\ln\xi^{-1} + 1.98174}{(\ln\xi^{-1})^2 + 6.04250\ln\xi^{-1} + 6.32549} + O(\xi(\ln\xi)^3). \end{aligned}\right\} \tag{5.90}$$

For the widely-separating case,

$$x_{11}^a = \frac{1}{6\pi a} \sum_{k=0}^{\infty} \left(\frac{1}{2s}\right)^{2k} f_{2k}^x, \quad x_{12}^a = -\frac{1}{6\pi a} \sum_{k=0}^{\infty} \left(\frac{1}{2s}\right)^{2k+1} f_{2k+1}^x, \tag{5.91}$$

$$y_{11}^a = \frac{1}{6\pi a} \sum_{k=0}^{\infty} \left(\frac{1}{2s}\right)^{2k} f_{2k}^y, \quad y_{12}^a = \frac{1}{6\pi a} \sum_{k=0}^{\infty} \left(\frac{1}{2s}\right)^{2k+1} f_{2k+1}^y, \tag{5.92}$$

in which

$$\left.\begin{array}{l} f_0^x = 1,\ f_1^x = -3,\ f_2^x = 0,\ f_3^x = 8,\ f_4^x = -60,\ f_5^x = 0,\ f_6^x = 352, \\ f_7^x = -2400,\ f_8^x = 2688,\ f_9^x = 3840,\ f_{10}^x = -85504,\ f_{11}^x = 201216, \end{array}\right\} \tag{5.93}$$

$$\left.\begin{array}{l} f_0^y = 1,\ f_1^y = 3/2,\ f_2^y = 0,\ f_3^y = 4,\ f_4^y = f_5^y = 0,\ f_6^y = -68, \\ f_7^y = 0,\ f_8^y = -320,\ f_9^y = 0,\ f_{10}^y = -4416,\ f_{11}^y = 9072. \end{array}\right\} \tag{5.94}$$

5.3.6 Numerical solutions of resistance and mobility functions for two equal diameter spherical particles

In the preceding section, we have showed the analytical expressions for the resistance and mobility functions in two asymptotic cases. But, if a numerical method is used, solutions which cover nearly all the range of particle-particle separations can numerically be obtained. Such a numerical approach has been developed by Kim and Mifflin [6]. In an actual simulation, the values of the resistance and mobility functions are not computed when required, rather the tabulated data of these functions are made before the simulations. It is generally used with an interpolation procedure, whereby appropriate values at the particle-particle separation of interest can be computed. Hence, Kim and Miflin's numerical method is very valuable for simulation applications. We outline their method in Appendix A10, and a computational program using their method is shown in Appendix A11.7. We here show the results of resistance and mobility functions, which were obtained using this program, in Figs. 5.2 to 5.13.

Since particle overlap is not physically allowed for rigid spherical particles, the values of the resistance functions become significantly large as particles approach to nearly-touching. This feature is clearly seen in the curves of X_{11}^{A*} and X_{12}^{A*} in Fig. 5.2 and X_{11}^{G*} and X_{12}^{G*} in Fig. 5.5. In contrast, the results of the mobility functions do not exhibit such a property.

Figures 5.12 and 5.13 show the results of the comparison between the analytical solutions in the preceding section and the numerical solutions, in which X_{11}^{A*} and x_{11}^{g*} are treated as typical examples. Figures 5.12(a) and 5.13(a) are for comparing the differences in the nearly-touching case, and Figs. 5.12(b) and 5.13(b) are for the widely-separating case. It should be noted, therefore, that the scale is different between Figs. 5.12(a) and 5.12(b), and between Figs. 5.13(a) and 5.13(b). It is clearly seen from these figures that the analytical solutions agree well with the numerical results under the conditions that the analytical solutions were derived. That is, the errors in the analytical solutions for the nearly-touching case significantly increase in the range above r^*=2.02. In contrast, the difference between the numerical results and the analytical solutions for the widely-separating case becomes significant in the range below

r^*=2.5

References

1. S. Kim and S.J. Karrila, "Microhydrodynamics: Principles and Selected Applications," Butterworth-Heinemann, Stoneham, 1991
2. G.K. Batchelor, "An Introduction to Fluid Dynamics," Cambridge University Press, Cambridge, 1981
3. H. Brenner, Rheology of a Dilute Suspension of Axisymmetric Brownian Particles, Int. J. Multiphase Flow, 1 (1974) 195
4. W.B. Russel, D.A. Saville, and W.R. Schowalter, "Colloidal Dispersions," Cambridge University Press, Cambridge, 1989
5. J. Rotne and S. Prager, Variational Treatment of Hydrodynamic Interaction in Polymers, J. Chem. Phys., 50 (1969) 4831
6. S. Kim and R. T. Mifflin, The Resistance and Mobility Functions of Two Equal Spheres in Low-Reynolds-Number Flow, Phys. Fluids, 28 (1985) 2034

Exercises

5.1 Derive Eqs. (5.13) and (5.14) by taking account of the following equations:

$$\mathbf{ee} = \begin{bmatrix} 1 & 0 & 0 \\ 0 & 0 & 0 \\ 0 & 0 & 0 \end{bmatrix}, \tag{5.95}$$

$$\boldsymbol{\varepsilon} \cdot \mathbf{ee} = \sum_{i=1}^{3}\sum_{j=1}^{3}\sum_{k=1}^{3}(\varepsilon_{ijk}\boldsymbol{\delta}_i\boldsymbol{\delta}_j\boldsymbol{\delta}_k)\cdot(\boldsymbol{\delta}_1\boldsymbol{\delta}_1) = \sum_{i=1}^{3}\sum_{j=1}^{3}\varepsilon_{ij1}\boldsymbol{\delta}_i\boldsymbol{\delta}_j\boldsymbol{\delta}_1, \tag{5.96}$$

$$(\boldsymbol{\varepsilon}\cdot\mathbf{ee}):\mathbf{E} = \sum_{i=1}^{3}\boldsymbol{\delta}_i(\sum_{j=1}^{3}\varepsilon_{ij1}E_{1j}) = \sum_{i=1}^{3}\boldsymbol{\delta}_i(\varepsilon_{i21}E_{12}+\varepsilon_{i31}E_{13}) = -\boldsymbol{\delta}_3E_{12}+\boldsymbol{\delta}_2E_{13}. \tag{5.97}$$

In a similar way, derive Eqs. (5.39) and (5.40).

5.2 By substituting Eq. (5.41) into (5.45), show that **A**, **B**, **C**, **G**, **H**, and **K** are related to **a**, **b**, **c**, **g**, **h**, and **k** as

$$\begin{bmatrix} \mathbf{A} & \tilde{\mathbf{B}} \\ \mathbf{B} & \mathbf{C} \end{bmatrix} = \begin{bmatrix} \mathbf{a} & \tilde{\mathbf{b}} \\ \mathbf{b} & \mathbf{c} \end{bmatrix}^{-1}, \tag{5.98}$$

$$\begin{bmatrix} \tilde{\mathbf{G}} \\ \tilde{\mathbf{H}} \end{bmatrix} = \begin{bmatrix} \mathbf{a} & \tilde{\mathbf{b}} \\ \mathbf{b} & \mathbf{c} \end{bmatrix}^{-1} \begin{bmatrix} \tilde{\mathbf{g}} \\ \tilde{\mathbf{h}} \end{bmatrix}, \tag{5.99}$$

$$[\mathbf{G} \quad \mathbf{H}] = [\mathbf{g} \quad \mathbf{h}] \begin{bmatrix} \mathbf{a} & \tilde{\mathbf{b}} \\ \mathbf{b} & \mathbf{c} \end{bmatrix}^{-1}, \tag{5.100}$$

$$\mathbf{K} = -\mathbf{k} + [\mathbf{g} \quad \mathbf{h}] \begin{bmatrix} \mathbf{a} & \tilde{\mathbf{b}} \\ \mathbf{b} & \mathbf{c} \end{bmatrix}^{-1} \begin{bmatrix} \tilde{\mathbf{g}} \\ \tilde{\mathbf{h}} \end{bmatrix}. \tag{5.101}$$

5.3 When a spherical particle with a radius a moves with velocity $\mathbf{v}$ in a quiescent flow field, the flow velocity at the position $\mathbf{r}$ from the particle center, $\mathbf{u}(\mathbf{r})$, can be written in Eq. (5.55). Verify the no-slip condition at the particle surface, that is, verify that the flow velocity $\mathbf{u}(\mathbf{r})$ at $\mathbf{r}=a\mathbf{e}$ ($\mathbf{e}$ is an arbitrarily oriented unit vector) is equal to the particle velocity $\mathbf{v}$.

5.4 Derive Eqs (5.68) to (5.70) in a similar way that Eqs. (5.51) to (5.54) were derived. The starting equations for such a derivation are Eqs. (5.62) and (5.65).

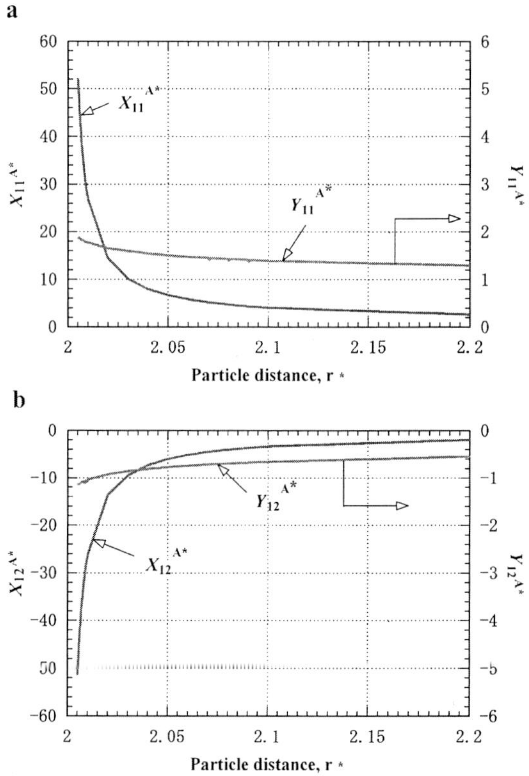

Figure 5.2 Resistance functions for spherical particles (X_{11}^{A*}, Y_{11}^{A*}, X_{12}^{A*}, Y_{12}^{A*}): (a) X_{11}^{A*} and Y_{11}^{A*}, and (b) X_{12}^{A*} and Y_{12}^{A*}.

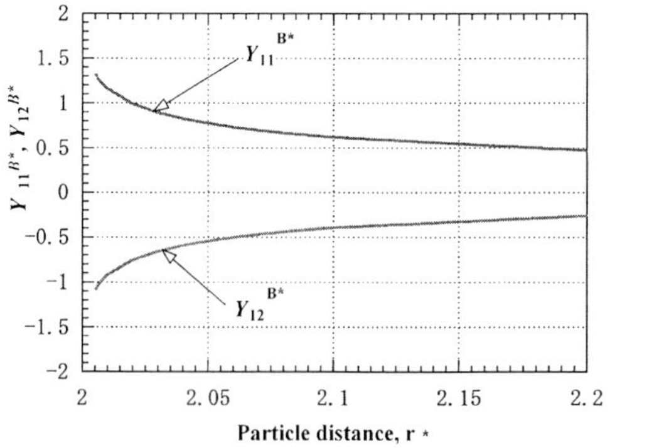

Figure 5.3 Resistance functions for spherical particles (Y_{11}^{B*}, Y_{12}^{B*}).

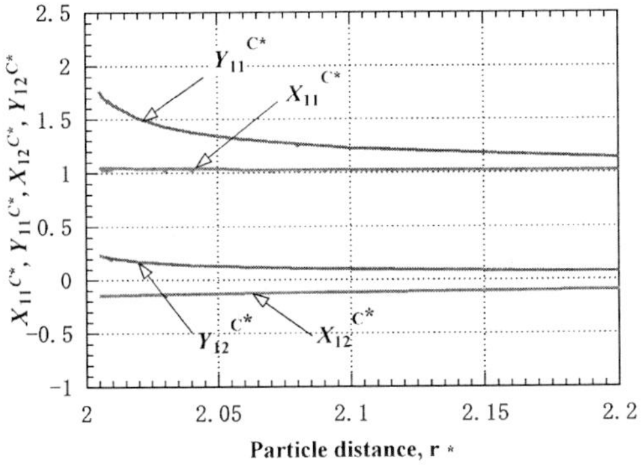

Figure 5.4 Resistance functions for spherical particles (X_{11}^{C*}, Y_{11}^{C*}, X_{12}^{C*}, Y_{12}^{C*}).

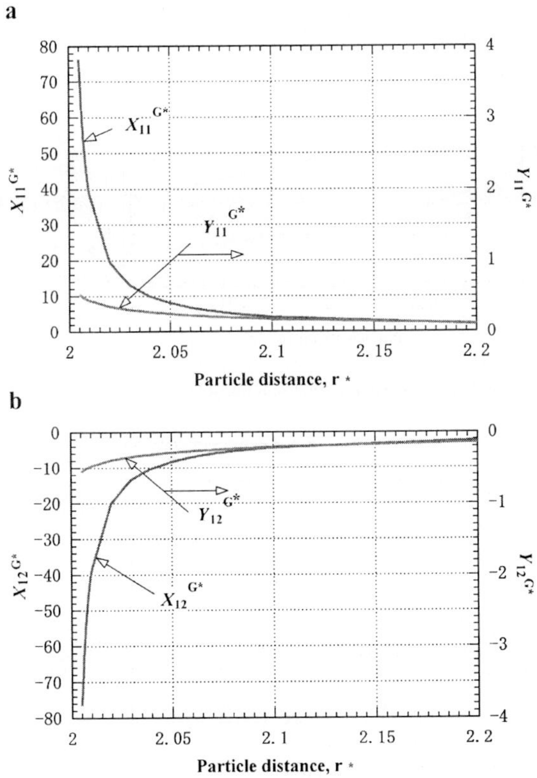

Figure 5.5 Resistance functions for spherical particles (X_{11}^{G*}, Y_{11}^{G*}, X_{12}^{G*}, Y_{12}^{G*}): (a) X_{11}^{G*} and Y_{11}^{G*}, and (b) X_{12}^{G*} and Y_{12}^{G*}.

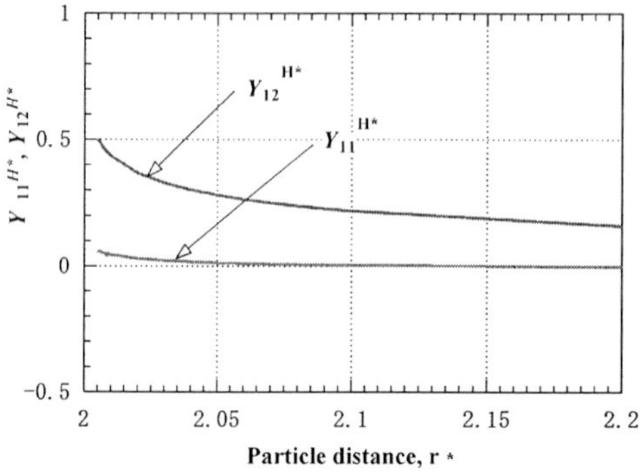

Figure 5.6 Resistance functions for spherical particles (Y_{11}^{H*}, Y_{12}^{H*}).

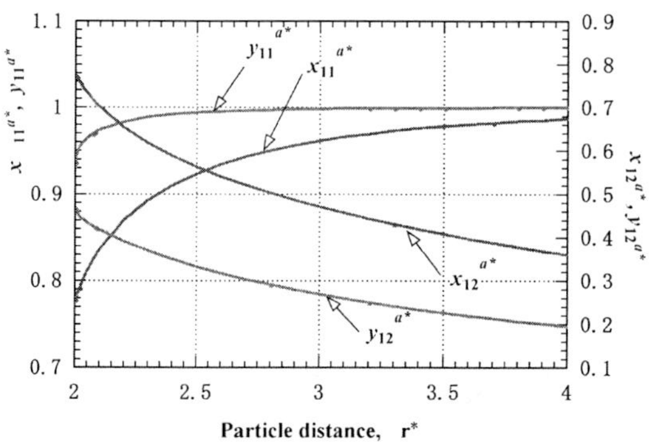

Figure 5.7 Mobility functions for spherical particles (x_{11}^{a*}, y_{11}^{a*}, x_{12}^{a*}, y_{12}^{a*}).

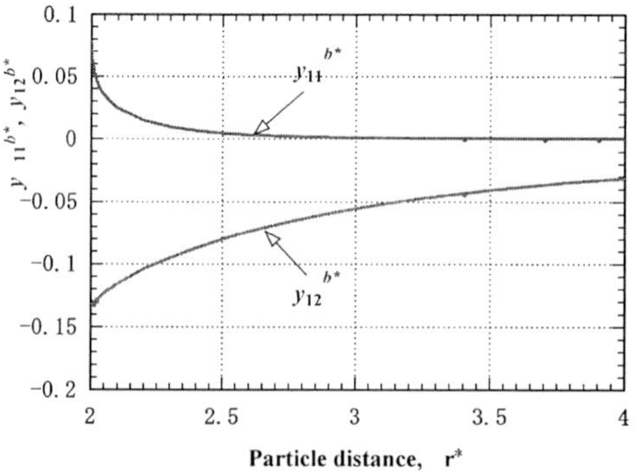

Figure 5.8 Mobility functions for spherical particles (y_{11}^{b*}, y_{12}^{b*}).

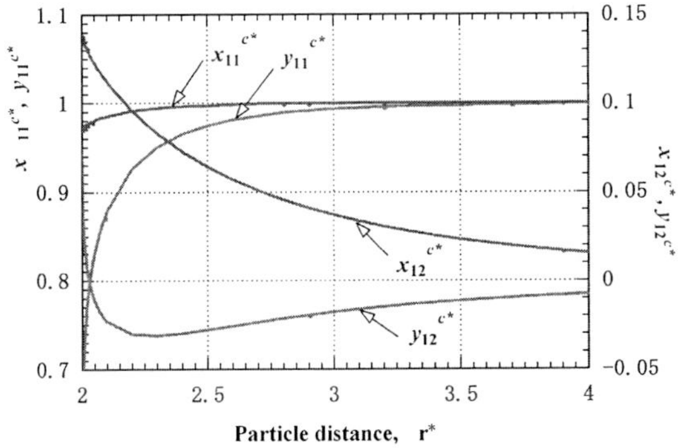

Figure 5.9 Mobility functions for spherical particles ($x_{11}^{c*}, y_{11}^{c*}, x_{12}^{c*}, y_{12}^{c*}$).

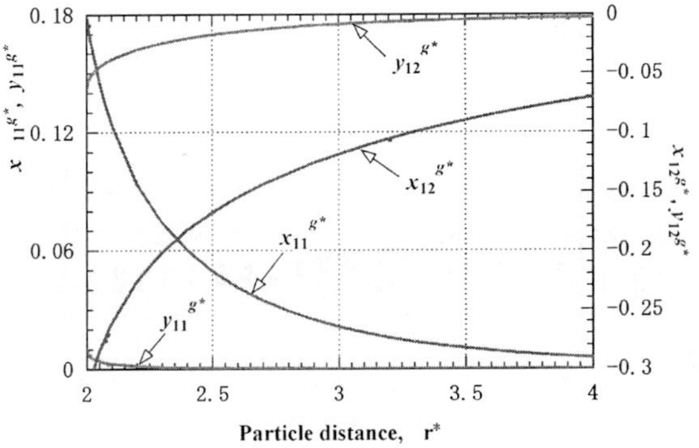

Figure 5.10 Mobility functions for spherical particles ($x_{11}^{g*}, y_{11}^{g*}, x_{12}^{g*}, y_{12}^{g*}$).

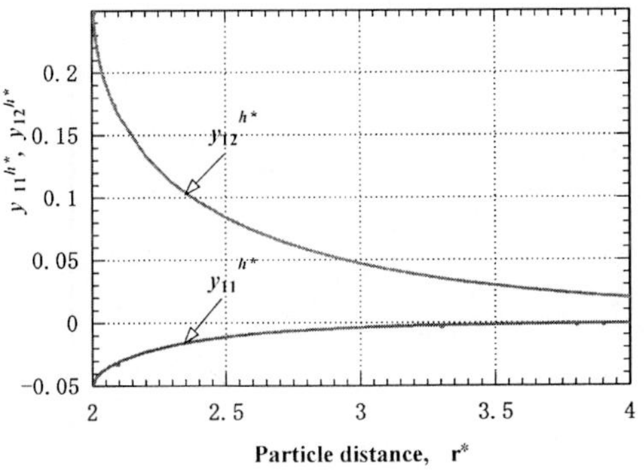

Figure 5.11 Mobility functions for spherical particles (y_{11}^{h*}, y_{12}^{h*}).

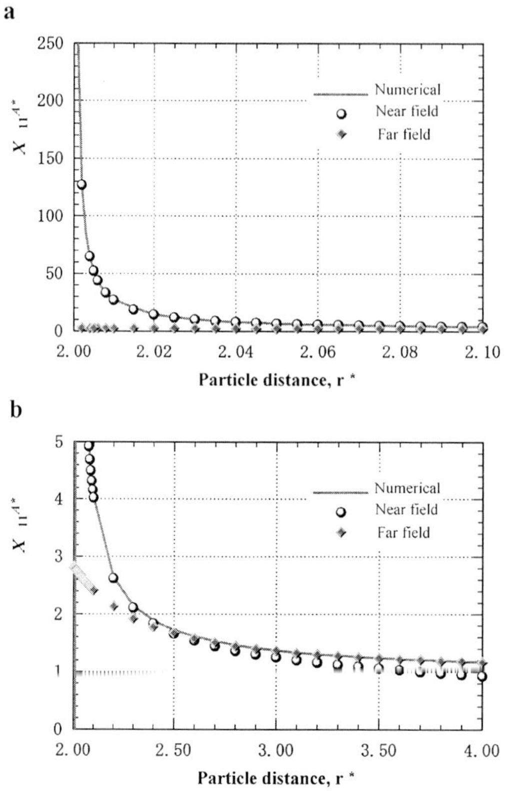

Figure 5.12 Comparison between analytical and numerical solutions for X_{11}^{A*}: (a) for a nearly-touching situation and (b) for a widely-separating situation.

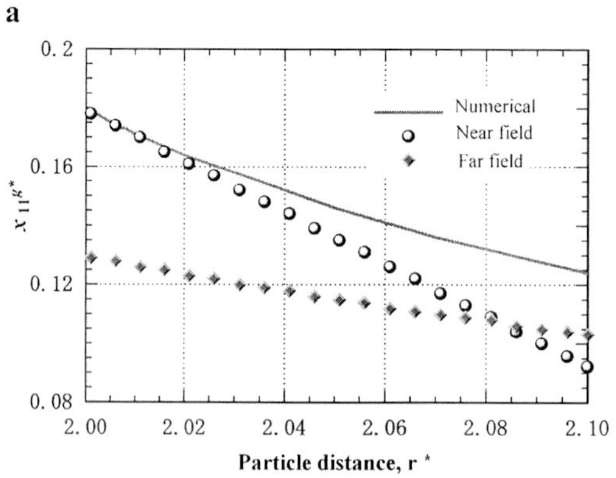

Figure 5.13 Comparison between analytical and numerical solutions for x_{11}^{g*}: (a) for a nearly-touching situation and (b) for a widely-separating situation.

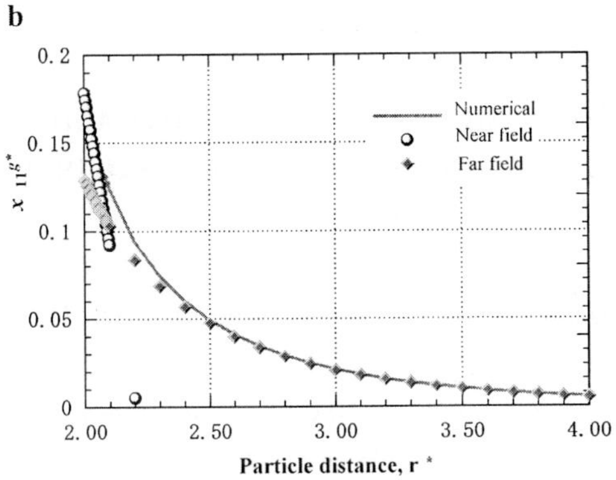

Figure 5.13 (continued).

CHAPTER 6

THE APPROXIMATION OF MULT-BODY HYDRODYNAMIC INTERACTIONS AMONG PARTICLES IN A DENSE COLLOIDAL DISPERSION

In the preceding chapter, we have explained the theory for two particles moving in a fluid. In a real colloidal dispersion, the motion of the particles is determined by multi-body hydrodynamic interactions among particles. However, a solution for the multi-body problem cannot be found in the case of a multi-particle system of the size required for simulations. It is very difficult to obtain exact solutions, even for a three-particle system. To circumvent this difficulty in dealing with multi-body hydrodynamic interactions, it is possible to apply the theory for a two-particle system to a multi-particle system, for which there are two useful approximations, known as the additivity of forces and the additivity of velocities. The former approximation can reproduce more accurately the lubrication effect, which arises when particles nearly touch. However, an inverse procedure for the resistance matrix is necessary in the method based on this approximation, so that the simulations are usually restricted to a very small system. In contrast, although the additivity of velocities is inferior regarding the accuracy of the lubrication effect, the calculation of the inverse matrix is unnecessary in simulations. One can, therefore, expand the simulations based on this approximation to a much larger system. Due to the above-mentioned drawback for simulations, the method based on the additivity of velocities has not been applied generally; the application to ferromagnetic colloidal dispersions is one of its successful applications. We explain these approximations below and more accurate treatments of multi-body hydrodynamic interactions will be discussed in Chapter 13.

6.1 The Additivity of Forces

When the internal structure and dynamic properties of a pure liquid are investigated by means of statistical-mechanical approaches, multi-body interactions among molecules may be approximated as the sum of pair interactions between molecules; three-body interactions are usually neglected. Even with this approximation, the essential characteristics of a liquid can be evaluated to a considerable degree. A similar approximation may be applicable to a colloidal dispersion composed of many particles. It is noted, however, that hydrodynamic interactions between particles are essentially different from those for a pure liquid system, since they act through the medium of the ambient fluid around the particles.

If we apply Eq. (5.62) to two particles α and β, the force exerted on the ambient fluid by particle α, $\mathbf{F}_\alpha$, is written as

$$\begin{aligned}\mathbf{F}_\alpha/\eta &= \mathbf{A}_{\alpha\alpha}\cdot(\mathbf{v}_\alpha-\mathbf{U}(\mathbf{r}_\alpha))+\mathbf{A}_{\alpha\beta}\cdot(\mathbf{v}_\beta-\mathbf{U}(\mathbf{r}_\beta))+\widetilde{\mathbf{B}}_{\alpha\alpha}\cdot(\boldsymbol{\omega}_\alpha-\boldsymbol{\Omega})\\&\quad+\widetilde{\mathbf{B}}_{\alpha\beta}\cdot(\boldsymbol{\omega}_\beta-\boldsymbol{\Omega})-(\widetilde{\mathbf{G}}_{\alpha\alpha}+\widetilde{\mathbf{G}}_{\alpha\beta}):\mathbf{E}.\end{aligned}\tag{6.1}$$

This force $\mathbf{F}_\alpha$ can be expressed as the sum of the force $\mathbf{F}_\alpha^\infty$ in the absence of particle β and the force $\Delta\mathbf{F}_{\alpha\beta}$ due to particle β :

$$\mathbf{F}_\alpha=\mathbf{F}_\alpha^\infty+\Delta\mathbf{F}_{\alpha\beta},\tag{6.2}$$

in which $\mathbf{F}_\alpha^\infty$ can be obtained from $\mathbf{F}_\alpha$ in the limiting case where the particle-particle separation is taken as infinity, and expressed as

$$\mathbf{F}_\alpha^\infty/\eta=\mathbf{A}_{\alpha\alpha}^\infty\cdot(\mathbf{v}_\alpha-\mathbf{U}(\mathbf{r}_\alpha))+\widetilde{\mathbf{B}}_{\alpha\alpha}^\infty\cdot(\boldsymbol{\omega}_\alpha-\boldsymbol{\Omega})-\widetilde{\mathbf{G}}_{\alpha\alpha}^\infty:\mathbf{E}.\tag{6.3}$$

In deriving this equation, we have taken into account the property that the tensors $\mathbf{A}_{\alpha\beta}^\infty$, $\widetilde{\mathbf{B}}_{\alpha\beta}^\infty$, and $\widetilde{\mathbf{G}}_{\alpha\beta}^\infty$ approach zero with increasing separation between particles. Hence the correction term $\Delta\mathbf{F}_{\alpha\beta}$ due to the influence of particle β can be written as

$$\begin{aligned}\Delta\mathbf{F}_{\alpha\beta}/\eta &= (\mathbf{A}_{\alpha\alpha}-\mathbf{A}_{\alpha\alpha}^\infty)\cdot(\mathbf{v}_\alpha-\mathbf{U}(\mathbf{r}_\alpha))+\mathbf{A}_{\alpha\beta}\cdot(\mathbf{v}_\beta-\mathbf{U}(\mathbf{r}_\beta))\\&\quad+(\widetilde{\mathbf{B}}_{\alpha\alpha}-\widetilde{\mathbf{B}}_{\alpha\alpha}^\infty)\cdot(\boldsymbol{\omega}_\alpha-\boldsymbol{\Omega})+\widetilde{\mathbf{B}}_{\alpha\beta}\cdot(\boldsymbol{\omega}_\beta-\boldsymbol{\Omega})-\left\{(\widetilde{\mathbf{G}}_{\alpha\alpha}-\widetilde{\mathbf{G}}_{\alpha\alpha}^\infty)+\widetilde{\mathbf{G}}_{\alpha\beta}\right\}:\mathbf{E}.\end{aligned}\tag{6.4}$$

It should be noted that, in this equation, the tensors $\mathbf{A}_{\alpha\alpha}$, $\widetilde{\mathbf{B}}_{\alpha\alpha}$, and $\widetilde{\mathbf{G}}_{\alpha\alpha}$ have no subscript β, but they are dependent on the position of particle β.

Now we explain the additivity of forces using the above-mentioned equations. In this approximation, the contributions from the other particles are summed, but the friction term of the particle itself has to be taken into account just once, i.e. not summed together with the contributions of the other particles. The force acting on the ambient fluid by particle α, therefore, can be expressed, in the approximation of the additivity of forces, as

$$\mathbf{F}_\alpha=\mathbf{F}_\alpha^\infty+\sum_{\beta=1(\neq\alpha)}^{N}\Delta\mathbf{F}_{\alpha\beta}.\tag{6.5}$$

The torque acting on the ambient fluid from particle α, $\mathbf{T}_\alpha$, can be expressed in a similar way, and finally, the forces and torques can be expressed in one matrix form as

$$\begin{bmatrix} \mathbf{F}_1 \\ \mathbf{F}_2 \\ \vdots \\ \mathbf{F}_N \\ \mathbf{T}_1 \\ \mathbf{T}_2 \\ \vdots \\ \mathbf{T}_N \end{bmatrix} = \eta \begin{bmatrix} \mathbf{A}'_{11} & \mathbf{A}_{12} & \cdots & \mathbf{A}_{1N} & \widetilde{\mathbf{B}}'_{11} & \widetilde{\mathbf{B}}_{12} & \cdots & \widetilde{\mathbf{B}}_{1N} \\ \mathbf{A}_{21} & \mathbf{A}'_{22} & \cdots & \mathbf{A}_{2N} & \widetilde{\mathbf{B}}_{22} & \widetilde{\mathbf{B}}'_{22} & \cdots & \widetilde{\mathbf{B}}_{2N} \\ \vdots & \vdots & & \vdots & \vdots & \vdots & & \vdots \\ \mathbf{A}_{N1} & \mathbf{A}_{N2} & \cdots & \mathbf{A}'_{NN} & \widetilde{\mathbf{B}}_{N1} & \widetilde{\mathbf{B}}_{N2} & \cdots & \widetilde{\mathbf{B}}'_{NN} \\ \mathbf{B}'_{11} & \mathbf{B}_{12} & & \mathbf{B}_{1N} & \mathbf{C}'_{11} & \mathbf{C}_{12} & \cdots & \mathbf{C}_{1N} \\ \mathbf{B}_{21} & \mathbf{B}'_{22} & \cdots & \mathbf{B}_{2N} & \mathbf{C}_{21} & \mathbf{C}'_{22} & \cdots & \mathbf{C}_{2N} \\ \vdots & \vdots & & \vdots & \vdots & \vdots & & \vdots \\ \mathbf{B}_{N1} & \mathbf{B}_{N2} & \cdots & \mathbf{B}'_{NN} & \mathbf{C}_{N1} & \mathbf{C}_{N2} & \cdots & \mathbf{C}'_{NN} \end{bmatrix} \begin{bmatrix} \mathbf{v}_1 - \mathbf{U}(\mathbf{r}_1) \\ \mathbf{v}_2 - \mathbf{U}(\mathbf{r}_2) \\ \vdots \\ \mathbf{v}_N - \mathbf{U}(\mathbf{r}_N) \\ \boldsymbol{\omega}_1 - \boldsymbol{\Omega} \\ \boldsymbol{\omega}_2 - \boldsymbol{\Omega} \\ \vdots \\ \boldsymbol{\omega}_N - \boldsymbol{\Omega} \end{bmatrix} - \eta \begin{bmatrix} \widetilde{\mathbf{G}}'_1 \\ \widetilde{\mathbf{G}}'_2 \\ \vdots \\ \widetilde{\mathbf{G}}'_N \\ \widetilde{\mathbf{H}}'_1 \\ \widetilde{\mathbf{H}}'_2 \\ \vdots \\ \widetilde{\mathbf{H}}'_N \end{bmatrix} : \mathbf{E}, \tag{6.6}$$

in which the quantities with the superscript prime are written as

$$\mathbf{A}'_{\alpha\alpha} = \mathbf{A}^{\infty}_{\alpha\alpha} + \sum_{\beta=1(\neq\alpha)}^{N} (\mathbf{A}_{\alpha\alpha} - \mathbf{A}^{\infty}_{\alpha\alpha}) \quad (\alpha = 1,2,\cdots,N), \tag{6.7}$$

$$\widetilde{\mathbf{B}}'_{\alpha\alpha} = \widetilde{\mathbf{B}}^{\infty}_{\alpha\alpha} + \sum_{\beta=1(\neq\alpha)}^{N} (\widetilde{\mathbf{B}}_{\alpha\alpha} - \widetilde{\mathbf{B}}^{\infty}_{\alpha\alpha}) \quad (\alpha = 1,2,\cdots,N), \tag{6.8}$$

$$\mathbf{B}'_{\alpha\alpha} = \mathbf{B}^{\infty}_{\alpha\alpha} + \sum_{\beta=1(\neq\alpha)}^{N} (\mathbf{B}_{\alpha\alpha} - \mathbf{B}^{\infty}_{\alpha\alpha}) \quad (\alpha = 1,2,\cdots,N), \tag{6.9}$$

$$\mathbf{C}'_{\alpha\alpha} = \mathbf{C}^{\infty}_{\alpha\alpha} + \sum_{\beta=1(\neq\alpha)}^{N} (\mathbf{C}_{\alpha\alpha} - \mathbf{C}^{\infty}_{\alpha\alpha}) \quad (\alpha = 1,2,\cdots,N), \tag{6.10}$$

$$\widetilde{\mathbf{G}}'_{\alpha} = \widetilde{\mathbf{G}}^{\infty}_{\alpha\alpha} + \sum_{\beta=1(\neq\alpha)}^{N} \left\{ (\widetilde{\mathbf{G}}_{\alpha\alpha} - \widetilde{\mathbf{G}}^{\infty}_{\alpha\alpha}) + \widetilde{\mathbf{G}}_{\alpha\beta} \right\} \quad (\alpha = 1,2,\cdots,N), \tag{6.11}$$

$$\widetilde{\mathbf{H}}'_{\alpha} = \widetilde{\mathbf{H}}^{\infty}_{\alpha\alpha} + \sum_{\beta=1(\neq\alpha)}^{N} \left\{ (\widetilde{\mathbf{H}}_{\alpha\alpha} - \widetilde{\mathbf{H}}^{\infty}_{\alpha\alpha}) + \widetilde{\mathbf{H}}_{\alpha\beta} \right\} \quad (\alpha = 1,2,\cdots,N). \tag{6.12}$$

If the column vector containing the forces and torques, the column vector containing the translational and angular velocities of all N particles, and the column vector containing $\hat{\mathbf{G}}_1{}' \sim \hat{\mathbf{H}}_N{}'$ are denoted by $\hat{\mathbf{F}}$, $\hat{\mathbf{v}}$, and $\mathbf{\Phi}$, respectively, and with the resistance matrix denoted by $\mathbf{R}$, then Eq. (6.6) can be expressed in simple form:

$$\hat{\mathbf{F}} = \eta(\mathbf{R} \cdot \hat{\mathbf{v}} - \mathbf{\Phi} : \mathbf{E}). \tag{6.13}$$

From this equation, the vector $\hat{\mathbf{v}}$ can be obtained as

$$\hat{\mathbf{v}} = \mathbf{R}^{-1} \cdot (\hat{\mathbf{F}} / \eta + \mathbf{\Phi} : \mathbf{E}). \tag{6.14}$$

The expression for the stresslet can be derived from a similar procedure.

It is clear from Eq. (6.14) that the inverse of the resistance matrix has to be calculated at each time step in simulations to obtain the translational and angular velocities. If we consider a three-dimensional system of N particles, then $6N$ variables of the velocities and angular velocities have to be treated. That is, the $6N \times 6N$ resistance matrix has to be treated in simulations. Hence, the simulations based on this approximation are computationally very expensive and generally restricted to a very small system. On the other hand, the lubrication effect [1], which becomes an important factor in the nearly-touching case, is exactly taken into account on the level of pair interactions, so that particle velocities arise to prevent the particles from overlapping. Hence one may say that the approximation of the additivity of forces reproduces the lubrication effect more accurately.

Before proceeding to the next section, we show the expressions for the resistance tensors with the superscript ∞ in Eqs. (6.7) to (6.12) for the case of a system of rigid spherical particles with a radius a. From Eqs. (5.71), (5.72), Sec. 5.3.5, and Appendix A4, these expressions are written as

$$\mathbf{A}_{\alpha\alpha}^{\infty} = 6\pi a\mathbf{I},\ \ \widetilde{\mathbf{B}}_{\alpha\alpha}^{\infty} = 0,\ \ \mathbf{B}_{\alpha\alpha}^{\infty} = 0,\ \ \mathbf{C}_{\alpha\alpha}^{\infty} = 8\pi a^3\mathbf{I},\ \ \widetilde{\mathbf{G}}_{\alpha\alpha}^{\infty} = 0,\ \ \widetilde{\mathbf{H}}_{\alpha\alpha}^{\infty} = 0\,. \tag{6.15}$$

6.2 The Additivity of Velocities

In a similar manner to the additivity of forces, if we rewrite Eq. (5.65) for the relation between particles α and β, the velocity of particle α, $\mathbf{v}_\alpha$, can be written as

$$\mathbf{v}_{\alpha} - \mathbf{U}(\mathbf{r}_{\alpha}) = \frac{1}{\eta}\left\{\mathbf{a}_{\alpha\alpha}\cdot\mathbf{F}_{\alpha} + \mathbf{a}_{\alpha\beta}\cdot\mathbf{F}_{\beta} + \widetilde{\mathbf{b}}_{\alpha\alpha}\cdot\mathbf{T}_{\alpha} + \widetilde{\mathbf{b}}_{\alpha\beta}\cdot\mathbf{T}_{\beta}\right\} + \widetilde{\mathbf{g}}_{\alpha}:\mathbf{E}\,. \tag{6.16}$$

The term on the right-hand side can be expressed as the sum of the contribution in the absence of particle β and the contribution due to particle β:

$$\mathbf{v}_{\alpha} = \mathbf{v}_{\alpha}^{\infty} + \Delta\mathbf{v}_{\alpha\beta}\,. \tag{6.17}$$

In a similar manner to Eqs. (6.3) and (6.4), $\mathbf{v}_\alpha^{\infty}$ and $\Delta\mathbf{v}_{\alpha\beta}$ are written as

$$\mathbf{v}_{\alpha}^{\infty} - \mathbf{U}(\mathbf{r}_{\alpha}) = \frac{1}{\eta}\left\{\mathbf{a}_{\alpha\alpha}^{\infty}\cdot\mathbf{F}_{\alpha} + \widetilde{\mathbf{b}}_{\alpha\alpha}^{\infty}\cdot\mathbf{T}_{\alpha}\right\} + \widetilde{\mathbf{g}}_{\alpha}^{\infty}:\mathbf{E}\,, \tag{6.18}$$

$$\begin{aligned}\Delta\mathbf{v}_{\alpha\beta} = \frac{1}{\eta}\Big\{&(\mathbf{a}_{\alpha\alpha} - \mathbf{a}_{\alpha\alpha}^{\infty})\cdot\mathbf{F}_{\alpha} + \mathbf{a}_{\alpha\beta}\cdot\mathbf{F}_{\beta} + (\widetilde{\mathbf{b}}_{\alpha\alpha} - \widetilde{\mathbf{b}}_{\alpha\alpha}^{\infty})\cdot\mathbf{T}_{\alpha} + \widetilde{\mathbf{b}}_{\alpha\beta}\cdot\mathbf{T}_{\beta}\Big\} \\ &+ (\widetilde{\mathbf{g}}_{\alpha} - \widetilde{\mathbf{g}}_{\alpha}^{\infty}):\mathbf{E}\,.\end{aligned} \tag{6.19}$$

As in the preceding section, the tensors with the superscript ∞ are quantities which are obtained by taking the particle-particle separation at infinity. Also it has been taken into account that $\mathbf{a}_{\alpha\beta}^{\infty}$ and $\widetilde{\mathbf{b}}_{\alpha\beta}^{\infty}$ approach zero with increasing particle-particle separation. Furthermore, it is noted that $\mathbf{a}_{\alpha\alpha}$, $\widetilde{\mathbf{b}}_{\alpha\alpha}$, and $\widetilde{\mathbf{g}}_{\alpha}$ are dependent on the position of particle β, although the subscript β is not attached to these quantities. If the contributions from particle α itself and the other particles are summed, the velocity of particle α can be obtained as

$$\mathbf{v}_\alpha = \mathbf{v}_\alpha^\infty + \sum_{\beta=1(\neq\alpha)}^{N} \Delta\mathbf{v}_{\alpha\beta} \, . \tag{6.20}$$

The angular velocity $\boldsymbol{\omega}_\alpha$ can also be expressed in a similar way, and finally the velocities and angular velocities can be expressed in one compact matrix form as

$$\begin{bmatrix} \mathbf{v}_1 - \mathbf{U}(\mathbf{r}_1) \\ \mathbf{v}_2 - \mathbf{U}(\mathbf{r}_2) \\ \vdots \\ \mathbf{v}_N - \mathbf{U}(\mathbf{r}_N) \\ \boldsymbol{\omega}_1 - \boldsymbol{\Omega} \\ \boldsymbol{\omega}_2 - \boldsymbol{\Omega} \\ \vdots \\ \boldsymbol{\omega}_N - \boldsymbol{\Omega} \end{bmatrix} = \frac{1}{\eta} \begin{bmatrix} \mathbf{a}'_{11} & \mathbf{a}_{12} & \cdots & \mathbf{a}_{1N} & \tilde{\mathbf{b}}'_{11} & \tilde{\mathbf{b}}_{12} & \cdots & \tilde{\mathbf{b}}_{1N} \\ \mathbf{a}_{21} & \mathbf{a}'_{22} & \cdots & \mathbf{a}_{2N} & \tilde{\mathbf{b}}_{22} & \tilde{\mathbf{b}}'_{22} & \cdots & \tilde{\mathbf{b}}_{2N} \\ \vdots & \vdots & & \vdots & \vdots & \vdots & & \vdots \\ \mathbf{a}_{N1} & \mathbf{a}_{N2} & \cdots & \mathbf{a}'_{NN} & \tilde{\mathbf{b}}_{N1} & \tilde{\mathbf{b}}_{N2} & \cdots & \tilde{\mathbf{b}}'_{NN} \\ \mathbf{b}'_{11} & \mathbf{b}_{12} & & \mathbf{b}_{1N} & \mathbf{c}'_{11} & \mathbf{c}_{12} & \cdots & \mathbf{c}_{1N} \\ \mathbf{b}_{21} & \mathbf{b}'_{22} & \cdots & \mathbf{b}_{2N} & \mathbf{c}_{21} & \mathbf{c}'_{22} & \cdots & \mathbf{c}_{2N} \\ \vdots & \vdots & & \vdots & \vdots & \vdots & & \vdots \\ \mathbf{b}_{N1} & \mathbf{b}_{N2} & \cdots & \mathbf{b}'_{NN} & \mathbf{c}_{N1} & \mathbf{c}_{N2} & \cdots & \mathbf{c}'_{NN} \end{bmatrix} \begin{bmatrix} \mathbf{F}_1 \\ \mathbf{F}_2 \\ \vdots \\ \mathbf{F}_N \\ \mathbf{T}_1 \\ \mathbf{T}_2 \\ \vdots \\ \mathbf{T}_N \end{bmatrix} + \begin{bmatrix} \tilde{\mathbf{g}}'_1 \\ \tilde{\mathbf{g}}'_2 \\ \vdots \\ \tilde{\mathbf{g}}'_N \\ \tilde{\mathbf{h}}'_1 \\ \tilde{\mathbf{h}}'_2 \\ \vdots \\ \tilde{\mathbf{h}}'_N \end{bmatrix} : \mathbf{E} \, , \tag{6.21}$$

in which the quantities with the superscript prime are written as

$$\mathbf{a}'_{\alpha\alpha} = \mathbf{a}_{\alpha\alpha}^\infty + \sum_{\beta=1(\neq\alpha)}^{N} (\mathbf{a}_{\alpha\alpha} - \mathbf{a}_{\alpha\alpha}^\infty) \quad (\alpha = 1, 2, \cdots, N) \, , \tag{6.22}$$

$$\tilde{\mathbf{b}}'_{\alpha\alpha} = \tilde{\mathbf{b}}_{\alpha\alpha}^\infty + \sum_{\beta=1(\neq\alpha)}^{N} (\tilde{\mathbf{b}}_{\alpha\alpha} - \tilde{\mathbf{b}}_{\alpha\alpha}^\infty) \quad (\alpha = 1, 2, \cdots, N) \, , \tag{6.23}$$

$$\mathbf{b}'_{\alpha\alpha} = \mathbf{b}_{\alpha\alpha}^\infty + \sum_{\beta=1(\neq\alpha)}^{N} (\mathbf{b}_{\alpha\alpha} - \mathbf{b}_{\alpha\alpha}^\infty) \quad (\alpha = 1, 2, \cdots, N) \, , \tag{6.24}$$

$$\mathbf{c}'_{\alpha\alpha} = \mathbf{c}_{\alpha\alpha}^\infty + \sum_{\beta=1(\neq\alpha)}^{N} (\mathbf{c}_{\alpha\alpha} - \mathbf{c}_{\alpha\alpha}^\infty) \quad (\alpha = 1, 2, \cdots, N) \, , \tag{6.25}$$

$$\tilde{\mathbf{g}}'_{\alpha} = \tilde{\mathbf{g}}_{\alpha}^\infty + \sum_{\beta=1(\neq\alpha)}^{N} (\tilde{\mathbf{g}}_{\alpha} - \tilde{\mathbf{g}}_{\alpha}^\infty) \quad (\alpha = 1, 2, \cdots, N) \, , \tag{6.26}$$

$$\tilde{\mathbf{h}}'_{\alpha} = \tilde{\mathbf{h}}_{\alpha}^\infty + \sum_{\beta=1(\neq\alpha)}^{N} (\tilde{\mathbf{h}}_{\alpha} - \tilde{\mathbf{h}}_{\alpha}^\infty) \quad (\alpha = 1, 2, \cdots, N) \, . \tag{6.27}$$

In a similar way to Eq. (6.6), we use the notation $\hat{\mathbf{v}}$ for the column vector on the left-hand side in Eq. (6.21), $\hat{\mathbf{F}}$ for the forces and torques, $\mathbf{M}$ for the mobility matrix, and $\boldsymbol{\Psi}$ for the matrix including $\tilde{\mathbf{g}}_1{}' \sim \tilde{\mathbf{h}}_N{}'$ on the right-hand side in Eq. (6.21). With these notations, Eq. (6.21) can be rewritten in simple form as

$$\hat{\mathbf{v}} = \mathbf{M} \cdot \hat{\mathbf{F}} / \eta + \boldsymbol{\Psi} : \mathbf{E} \, . \tag{6.28}$$

This is the equation for the particle velocities and angular velocities based on the approximation of the additivity of velocities. It is clear from Eq. (6.28) that the calculation of an inverse matrix is unnecessary to evaluate the velocities and angular velocities of particles for this additivity approximation. This means, therefore, that simulations based on the additivity of velocities can be applied to a much larger system. However, the lubrication effect is taken into account only at the level of the additivity of velocities, so that a velocity does not arise to prevent particles from overlapping. This is quite understandable from comparing Eq. (6.28) with Eq. (6.14). Although the lubrication effect of a pair of particles is exactly taken into account in both approximations, the lubrication effect only appears through the sum of a pair-interaction of particles for the approximation of the additivity of velocities. In contrast, for the approximation of the additivity of forces, pair-interactions appear indirectly as multi-body hydrodynamic interactions through the inverse procedure of the resistance matrix. In other words, the additivity of velocities is inferior to the additivity of forces in regard to the accuracy of the lubrication effect, so that physically unreasonable overlaps of particles may occur in simulations based on the additivity of velocities. Hence, this inferior point may become a serious problem and result in the divergence of the system in simulations for a rigid particle system. Thus one has to be very careful when attempting to apply the approximation of the additivity of velocities to a rigid particle system. However, in many cases of colloidal problems, the above-mentioned drawback may not be serious, since the electrical double layers [2] or the steric repulsion due to surfactant layers [3] arises between particles in general colloidal dispersions and may help to prevent overlap.

Finally, we show the expressions for the mobility tensors with the superscript ∞ in Eqs. (6.22)~(6.27) for the case of the rigid spherical particles with a radius a. With Eqs. (5.73), (5.74), Sec. 5.3.5, and Appendix A4, such tensors can be expressed as

$$\mathbf{a}_{\alpha\alpha}^{\infty} = \frac{1}{6\pi a}\mathbf{I},\ \tilde{\mathbf{b}}_{\alpha\alpha}^{\infty} = 0,\ \mathbf{b}_{\alpha\alpha}^{\infty} = 0,\ \mathbf{c}_{\alpha\alpha}^{\infty} = \frac{1}{8\pi a^3}\mathbf{I},\ \tilde{\mathbf{g}}_{\alpha}^{\infty} = 0,\ \tilde{\mathbf{h}}_{\alpha}^{\infty} = 0 . \tag{6.29}$$

References

1. G.K. Batchelor, "An Introduction to Fluid Dynamics," Cambridge University Press, Cambridge, 1967
2. R.J. Hunter, "Foundations of Colloid Science," Vol.1, Clarendon Press, Oxford, 1986
3. R. E. Rosensweig, "Ferrohydrodynamics," Cambridge University Press, Cambridge, 1985

Exercises

6.1 Derive Eqs. (6.15) and (6.29) by taking the limit of $s\to\infty$ for the expressions in Appendix A4, using Eqs. (5.71) to (5.74), and referring to Sec. 5.3.5.

6.2 Show that Eq. (6.6) reduces to Eq. (5.62) for a two-particle system. Similarly, show that Eq. (6.21) reduces to Eq. (5.65) for a two-particle system.

CHAPTER 7

MOLECULAR DYNAMICS METHODS FOR A DILUTE COLLOIDAL DISPERSION

Molecular dynamics methods can be used for a dilute dispersion in which both hydrodynamic interactions among particles and particle Brownian motion are negligible. In this chapter we explain molecular dynamics methods for a dilute dispersion. We here concentrate our attention on a spherical or spheroidal particle system, which is a very important model dispersion from a simulation point of view. The methodology such as the specification of an initial configuration and the generation of a simple shear flow will be discussed in detail in Chapter 11.

7.1 Molecular Dynamics for a Spherical Particle System

If the rotational motion of spherical particles is negligible, a physical phenomenon is governed by the particle translational motion alone. If we use m_i for the mass of spherical particle i, $\mathbf{r}_i$ for the particle position vector, $\mathbf{v}_i$ for the particle velocity vector, and $\mathbf{F}_i$ for the force exerted on particle i by the other particles, then the equation of motion of particle i can be written as

$$m_i \frac{d^2\mathbf{r}_i}{dt^2} = \mathbf{F}_i - \xi_i \mathbf{v}_i , \tag{7.1}$$

in which ξ_i is the friction coefficient and expressed as $\xi_i=6\pi\eta a_i$ for a spherical particle with a radius a_i, which is just from Stokes' drag formula, and η is the viscosity of the base liquid. In almost all fluid problems of colloidal dispersions, the term on the left-hand side in Eq. (7.1) is negligible, and then Eq. (7.1) reduces to the following simple equation:

$$\mathbf{v}_i = \mathbf{F}_i / \xi_i . \tag{7.2}$$

This equation is valid for an arbitrary particle in a system.

We now show the condition under which the inertia term on the left-hand side in Eq. (7.1) can be neglected. A system of particles with the same diameter a and mass m is considered for simplification for the discussion. If the force is nondimensionalized by F_0, the distance by $2a$, the velocity by $(F_0/6\pi\eta a)$, and the time by $(2a/(F_0/6\pi\eta a))$, then Eq. (7.1) can be nondimensionalized as

$$\frac{mF_0}{(6\pi\eta a)^2 2a} \cdot \frac{d^2\mathbf{r}_i^*}{dt^{*2}} = \mathbf{F}_i^* - \mathbf{v}_i^* , \tag{7.3}$$

in which the quantities with the superscript * are dimensionless quantities. If the nondimensional number, appeared in Eq. (7.3), satisfies the following condition:

$$\frac{mF_0}{(6\pi\eta a)^2 2a} << 1, \tag{7.4}$$

then the inertia term in Eq. (7.1) can be neglected and Eq. (7.1) reduces to Eq. (7.2). The condition in Eq. (7.4) can be satisfied in almost all fluid problems of colloidal dispersions. The condition in Eq. (7.4) can also be derived by comparing the characteristic time t_f (=12 $\pi\eta a^2/F_0$) for the case of neglecting the inertia term and the characteristic time t_m (=$(2ma/F_0)^{1/2}$) for the case of neglecting the friction term. That is, for the inertial term to be negligible, these two characteristic times have to satisfy the following condition:

$$t_m << t_f . \tag{7.5}$$

It is straightforward to verify that the condition in Eq. (7.5) is equivalent to that in Eq. (7.4).

We now return to the previous main discussion. If particles have magnetic properties, the force exerted on the fluid by particle i is equal to the sum of the magnetic forces exerted on particle i by the other particles. Hence, if the particle configuration at a certain time t is known, $\mathbf{F}_i$ can be calculated and the velocities of all particles at time t can be evaluated from Eq. (7.2). With these solutions for the particle velocities, the particle position at the next time step ($t+\Delta t$) can be calculated from the following equation, which is obtained from the finite difference approximation of $\mathbf{v}_i(t)=d\mathbf{r}_i(t)/dt$:

$$\mathbf{r}_i(t+\Delta t) = \mathbf{r}_i(t) + \Delta t \mathbf{v}_i(t) . \tag{7.6}$$

It is clear from these discussions that, in molecular dynamics simulations of colloidal dispersions, the initial particle velocities can be calculated from a given initial configuration from Eq. (7.2), so that the initial velocities of particles do not need to be specified according to the Maxwellian distribution function. In other words, the temperature of a system only appears through the viscosity of the base liquid and the forces between particles. Hence, the system temperature cannot be defined by the thermal motion of particles. This feature is significantly different from that for a pure molecular system.

7.2 Molecular Dynamics for a Spheroidal Particle System

As shown in Sec. 5.1.2, for an axisymmetric particle, the particle motion can be decomposed into translational and rotational motion and these two motion can be treated separately. Furthermore, the translational motion can be decomposed into motion parallel and normal to the particle axis and each motion can be dealt with separately.

We use the notation $\mathbf{n}_i$ for the direction of the particle axis, and $\mathbf{F}_i$ for the force exerted on the ambient fluid by particle *i*. If $\mathbf{F}_i$ is expressed as the sum of the force $\mathbf{F}_i^{\parallel}$ parallel to the particle axis and the force $\mathbf{F}_i^{\perp}$ normal to it, then the respective particle velocities $\mathbf{v}_i^{\parallel}$ and $\mathbf{v}_i^{\perp}$ can be obtained from Eq. (5.24) or (5.13) as

$$\mathbf{v}_i^{\parallel} = \mathbf{U}^{\parallel}(\mathbf{r}_i) + \frac{1}{\eta X^A}\mathbf{F}_i^{\parallel}, \tag{7.7}$$

$$\mathbf{v}_i^{\perp} = \mathbf{U}^{\perp}(\mathbf{r}_i) + \frac{1}{\eta Y^A}\mathbf{F}_i^{\perp}, \tag{7.8}$$

in which $\mathbf{U}_i^{\parallel}$ and $\mathbf{U}_i^{\perp}$ are the imposed flow velocities parallel and normal to the particle axis at the position $\mathbf{r}_i$, respectively. X^A and Y^A have already been shown in Eqs. (5.8) and (5.26). The velocity of particle *i*, $\mathbf{v}_i$, can be obtained as the sum of velocities $\mathbf{v}_i^{\parallel}$ and $\mathbf{v}_i^{\perp}$, that is , $\mathbf{v}_i = \mathbf{v}_i^{\parallel} + \mathbf{v}_i^{\perp}$. Furthermore, the following relations should be noted:

$$\mathbf{F}_i^{\parallel} = (\mathbf{F}_i \cdot \mathbf{n}_i)\mathbf{n}_i, \quad \mathbf{F}_i^{\perp} = \mathbf{F}_i - (\mathbf{F}_i \cdot \mathbf{n}_i)\mathbf{n}_i, \tag{7.9}$$

$$\mathbf{U}_i^{\parallel} = (\mathbf{U}_i \cdot \mathbf{n}_i)\mathbf{n}_i, \quad \mathbf{U}_i^{\perp} = \mathbf{U}_i - (\mathbf{U}_i \cdot \mathbf{n}_i)\mathbf{n}_i. \tag{7.10}$$

If the particles have magnetic properties, the force exerted on the ambient fluid by the particle, $\mathbf{F}_i$, is equal to the sum of magnetic forces exerted on the particle by the other particles.

Next we consider the rotational motion of an axisymmetric particle. As before, we use the superscripts $\parallel$ and $\perp$ for the vectors parallel and normal to the particle axis, respectively. The rotational velocities can be obtained from Eq. (5.25) or (5.14) as

$$\boldsymbol{\omega}_i^{\parallel} = \boldsymbol{\Omega}^{\parallel} + \frac{1}{\eta X^C}\mathbf{T}_i^{\parallel}, \tag{7.11}$$

$$\boldsymbol{\omega}_i^{\perp} = \boldsymbol{\Omega}^{\perp} + \frac{1}{\eta Y^C}\mathbf{T}_i^{\perp} - \frac{Y^H}{Y^C}(\boldsymbol{\varepsilon} \cdot \mathbf{n}_i\mathbf{n}_i) : \mathbf{E}. \tag{7.12}$$

The relation of $\{(\boldsymbol{\varepsilon}\cdot\mathbf{n}_i\mathbf{n}_i):\mathbf{E}\}\cdot\mathbf{n}_i=0$ has been taken into account to obtain the above equations. The following relations, similar to Eqs. (7.9) and (7.10), also have to be taken into account for the torques and angular velocities:

$$\mathbf{T}_i^{\parallel} = (\mathbf{T}_i \cdot \mathbf{n}_i)\mathbf{n}_i, \quad \mathbf{T}_i^{\perp} = \mathbf{T}_i - (\mathbf{T}_i \cdot \mathbf{n}_i)\mathbf{n}_i, \tag{7.13}$$

$$\boldsymbol{\Omega}^{\parallel} = (\boldsymbol{\Omega} \cdot \mathbf{n}_i)\mathbf{n}_i, \quad \boldsymbol{\Omega}^{\perp} = \boldsymbol{\Omega} - (\boldsymbol{\Omega} \cdot \mathbf{n}_i)\mathbf{n}_i. \tag{7.14}$$

As in the case of forces, if the particles have magnetic properties, the torque exerted on the ambient fluid by particle *i*, $\mathbf{T}_i$, is equivalent to the magnetic torque exerted on particle *i* by the ambient particles.

Since the velocities $\mathbf{v}_i$ (i=1,2,...,N) and angular velocities $\boldsymbol{\omega}_i$ (i=1,2,...,N) at a certain time *t* can be evaluated from Eqs. (7.7), (7.8), (7.11), and (7.12), the position $\mathbf{r}_i(t+\Delta t)$ and direction $\mathbf{n}_i(t+\Delta t)$ of particle i at the next time step $(t+\Delta t)$ can be obtained from the following equations:

$$\mathbf{r}_i(t+\Delta t)=\mathbf{r}_i(t)+\Delta t\mathbf{v}_i(t) \qquad (i=1,2,\cdots,N), \tag{7.15}$$

$$\mathbf{n}_i(t+\Delta t)=\mathbf{n}_i(t)+\Delta t\boldsymbol{\omega}_i(t)\times\mathbf{n}_i(t) \qquad (i=1,2,\cdots,N), \tag{7.16}$$

which are the finite difference expressions of the equations:

$$\mathbf{v}_i=\frac{d\mathbf{r}_i}{dt}, \quad \frac{d\mathbf{n}_i}{dt}=\boldsymbol{\omega}_i\times\mathbf{n}_i \qquad (i=1,2,\cdots,N). \tag{7.17}$$

As pointed out, the influence of temperature only appears through the viscosity of the base liquid and the forces and torques acting between particles.

7.3 Molecular Dynamics with the Inertia Term for Spherical Particles

Although the molecular dynamics method neglecting the inertia term is applicable to almost all colloidal systems, we briefly explain the molecular dynamics method which takes into account the inertia terms.

We rewrite Eq. (7.1) as

$$m\frac{d\mathbf{v}}{dt}=\mathbf{F}-\xi\mathbf{v}, \tag{7.18}$$

in which the subscript i for the particle discrimination is omitted for simplicity. If the velocity at time t is denoted by $\mathbf{v}(t)$ and the force $\mathbf{F}$ can be regarded as constant during the short time interval $(t\text{-}t_0)$, the velocity at time t, $\mathbf{v}(t)$, can be derived from Eq. (7.18) as

$$\mathbf{v}(t)=\mathbf{v}(t_0)e^{-\frac{\xi}{m}(t-t_0)}+\frac{1}{\xi}\mathbf{F}(t_0)\left\{1-e^{-\frac{\xi}{m}(t-t_0)}\right\}. \tag{7.19}$$

The derivation of this equation should be referred to in Exercise 9.1. If the integration of Eq. (7.19) is carried out, the particle position $\mathbf{r}(t)$ can be obtained as

$$\begin{aligned}\mathbf{r}(t)=\mathbf{r}(t_0)+\int_{t_0}^{t}\mathbf{v}(t')dt'&=\mathbf{r}(t_0)+\frac{m}{\xi}\mathbf{v}(t_0)\left\{1-e^{-(\xi/m)(t-t_0)}\right\}\\&+\frac{1}{\xi}\mathbf{F}(t_0)\left\{(t-t_0)-\frac{m}{\xi}\left(1-e^{-(\xi/m)(t-t_0)}\right)\right\},\end{aligned} \tag{7.20}$$

If we rewrite Eqs. (7.19) and (7.20) for the relation between the positions at the present time t and at the next time step $(t+h)$, the particle position $\mathbf{r}(t+h)$ and velocity $\mathbf{v}(t+h)$ can be expressed, respectively, as

$$\mathbf{r}(t+h)=\mathbf{r}(t)+\frac{m}{\xi}\mathbf{v}(t)\left\{1-e^{-(\xi/m)h}\right\}+\frac{1}{\xi}\mathbf{F}(t)\left\{h-\frac{m}{\xi}\left(1-e^{-(\xi/m)h}\right)\right\}, \tag{7.21}$$

$$\mathbf{v}(t+h)=\mathbf{v}(t)e^{-(\xi/m)h}+\frac{1}{\xi}\mathbf{F}(t)\left\{1-e^{-(\xi/m)h}\right\}. \tag{7.22}$$

The simulation can advance one time step from these equations. The friction coefficient $\xi=6\pi\eta a$ (in which η is the viscosity of the base liquid and a is the particle radius) is frequently used in simulations of a spherical particle system; this friction expression is Stokes' drag formula.

In the limit of $\xi\to 0$, Eqs. (7.21) and (7.22) reduce to, respectively,

$$\mathbf{r}(t+h)=\mathbf{r}(t)+h\mathbf{v}(t)+\frac{h^2}{2m}\mathbf{F}(t), \tag{7.23}$$

$$\mathbf{v}(t+h)=\mathbf{v}(t)+\frac{h}{m}\mathbf{F}(t). \tag{7.24}$$

It can be seen from Exercise 7.2 that the expression for the velocity in Eq. (7.24) does not have sufficient accuracy. A more accurate expression for $\mathbf{v}(t+h)$ than Eq. (7.24) can be obtained by regarding the force as changing linearly with time, i.e. not constant during the time interval:

$$\mathbf{v}(t+h)=\mathbf{v}(t)e^{-(\xi/m)h}+\frac{1}{\xi}\mathbf{F}(t)\left\{1-e^{-(\xi/m)h}\right\}+\frac{1}{\xi h}(\mathbf{F}(t+h)-\mathbf{F}(t)) \times\left\{h-\frac{m}{\xi}\left(1-e^{-(\xi/m)h}\right)\right\}. \tag{7.25}$$

Hence Eqs. (7.23) and (7.25) can be used when the friction coefficient ξ is small. Lastly, it should be noted that Eq. (7.25) reduces to the velocity expression of the velocity Verlet algorithm in Eq. (7.31) for a molecular system in the limit of $\xi\to 0$.

Exercises

7.1 If $f(t)$ is a function of t, and also if h is sufficiently small, derive the following finite difference approximations by means of a Taylor series expansion.

$$\frac{df(t)}{dt}=\frac{f(t+h)-f(t)}{h}+O(h), \tag{7.26}$$

$$\frac{df(t)}{dt}=\frac{f(t)-f(t-h)}{h}+O(h), \tag{7.27}$$

$$\frac{df(t)}{dt}=\frac{f(t+h)-f(t-h)}{2h}+O(h^2). \tag{7.28}$$

Equations (7.26) and (7.27) are called the forward and backward finite difference approximations, respectively, and have first-order accuracy. In contrast, Eq. (7.28) is called the central finite difference approximation, and has second-order accuracy.

7.2 We use the notations m for the mass of particle i, $\mathbf{r}_i$ for the particle position, $\mathbf{v}_i$ for the particle velocity, and $\mathbf{f}_i$ for the force acting on particle i. With the above finite difference expressions, derive the following molecular dynamics algorithms which are widely used for a molecular system.

(1) Verlet algorithm:

$$\mathbf{r}_i(t+h) = 2\mathbf{r}_i(t) - \mathbf{r}_i(t-h) + h^2\mathbf{f}_i(t)/m. \tag{7.29}$$

(2) Velocity Verlet algorithm:

$$\mathbf{r}_i(t+h) = \mathbf{r}_i(t) + h\mathbf{v}_i(t) + \frac{h^2}{2m}\mathbf{f}_i(t), \tag{7.30}$$

$$\mathbf{v}_i(t+h) = \mathbf{v}_i(t) + \frac{h}{2m}(\mathbf{f}_i(t+h) + \mathbf{f}_i(t)). \tag{7.31}$$

(3) Leapfrog algorithm:

$$\mathbf{r}_i(t+h) = \mathbf{r}_i(t) + h\mathbf{v}_i(t+h/2), \tag{7.32}$$

$$\mathbf{v}_i(t+h/2) = \mathbf{v}_i(t-h/2) + \frac{h}{m}\mathbf{f}_i(t). \tag{7.33}$$

7.3 Prove that the condition shown in Eq. (7.4) is equivalent to that in Eq. (7.5).

7.4 With the relations in Eqs. (7.9) and (7.10), show that Eq. (5.24) can be decomposed into Eqs. (7.7) and (7.8)

7.5 Similarly, with the relations in Eqs. (7.13) and (7.14), show that Eq. (5.25) can be decomposed into Eqs. (7.11) and (7.12).

CHAPTER 8

STOKESIAN DYNAMICS METHODS

The Stokesian dynamics method takes into account multi-body hydrodynamic interactions among particles, but not the Brownian motion of particles. Hence this method is applicable to a non-dilute colloidal dispersion, for which standard molecular dynamics methods cannot be applied. The approximations of the additivity of forces and the additivity of velocities, which have been explained in Chapter 6, are the key concepts of the Stokesian dynamics method. In this chapter we consider a typical flow field, such as a simple shear flow, and explain the Stokesian dynamics method for this flow field. Since nondimensional quantities are generally used in simulations, we will explain the nondimensionalization method in the last section of this chapter. The methodology, such as setting an initial configuration, generating a simple shear flow, and dealing with boundary conditions, will be explained later in Chapter 11 and is therefore not discussed here. In the following, we consider a spherical particle system for simplicity.

8.1 The Approximation of Additivity of Forces

We here consider two cases for using the approximation of additivity of forces in the Stokesian dynamics method. The first case is relatively simple and is applicable to simulations that only need to take into account the translational motion of particles. The second case is more complicated and takes into account not only the translational but also rotational motion of spherical particles.

8.1.1 Stokesian dynamics for only the translational motion of spherical particles

We consider a simple shear flow directed in the x-direction with a shear rate $\dot{\gamma}$, as shown in Fig. 8.1. In this case the flow field $\mathbf{U}(\mathbf{r})$ is expressed as

$$\mathbf{U}(\mathbf{r}) = \dot{\gamma} y \boldsymbol{\delta}_x . \tag{8.1}$$

With this flow field, the angular velocity $\boldsymbol{\Omega}$ and the rate-of-strain tensor $\mathbf{E}$ are derived from the definition in Eq. (4.28) as

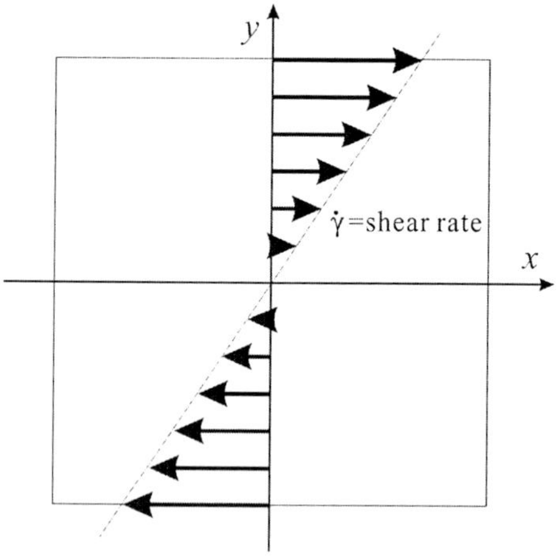

Figure 8.1 A simple shear flow in the x-direction with the shear rate $\dot{\gamma}$.

$$\mathbf{\Omega} = -\frac{\dot{\gamma}}{2}\boldsymbol{\delta}_z, \qquad \mathbf{E} = \frac{\dot{\gamma}}{2}\begin{bmatrix} 0 & 1 & 0 \\ 1 & 0 & 0 \\ 0 & 0 & 0 \end{bmatrix}, \tag{8.2}$$

in which ($\boldsymbol{\delta}_x$, $\boldsymbol{\delta}_y$, $\boldsymbol{\delta}_z$) are the unit vectors for each axis direction (or the fundamental vectors).

If we use the approximation of the additivity of forces, which is expressed in Eq. (6.6), and take into account the expressions for the resistance tensors, then the force exerted on the ambient fluid by particle α, $\mathbf{F}_\alpha$, can be written as

$$\mathbf{F}_\alpha = 6\pi a\eta(\mathbf{v}_\alpha - \mathbf{U}) + \eta\left\{ \sum_{\beta=1(\neq\alpha)}^{N} (\mathbf{A}_{\alpha\alpha} - 6\pi a\mathbf{I})\cdot(\mathbf{v}_\alpha - \mathbf{U}) + \sum_{\beta=1(\neq\alpha)}^{N} \mathbf{A}_{\alpha\beta}\cdot(\mathbf{v}_\beta - \mathbf{U}) \right\} \\ -\eta\widetilde{\mathbf{G}}'_\alpha : \mathbf{E}, \tag{8.3}$$

in which $\mathbf{A}_{\alpha\alpha}$ and $\mathbf{A}_{\alpha\beta}$ can be expressed in similar equations to Eq. (5.71), with the notation $\mathbf{e}=(\mathbf{r}_\beta - \mathbf{r}_\alpha)/|\mathbf{r}_\beta - \mathbf{r}_\alpha|$, as

$$\mathbf{A}_{\alpha\alpha} = X^A_{\alpha\alpha}\mathbf{ee} + Y^A_{\alpha\alpha}(\mathbf{I} - \mathbf{ee}), \qquad \mathbf{A}_{\alpha\beta} = X^A_{\alpha\beta}\mathbf{ee} + Y^A_{\alpha\beta}(\mathbf{I} - \mathbf{ee}). \tag{8.4}$$

As already pointed out, $\mathbf{A}_{\alpha\alpha}$ is dependent on the position of particle β, although the subscript β is not attached. This regularity concerning the subscripts is applied to the resistance and mobility tensors, and also the resistance and mobility functions. The expressions for the resistance functions $X_{\alpha\alpha}{}^a$, etc., should be referred to in Sec. 5.3.5 and Appendix A4. By taking into account Eqs. (6.11) and (6.15), $\widetilde{\mathbf{G}}_\alpha{}'$ can be written as

$$\widetilde{\mathbf{G}}'_\alpha = \sum_{\beta=1(\neq\alpha)}^{N} (\widetilde{\mathbf{G}}_{\alpha\alpha} + \widetilde{\mathbf{G}}_{\alpha\beta}) . \tag{8.5}$$

The last term on the right-hand side in Eq. (8.3) is a first-rank tensor, or a vector, since $\widetilde{\mathbf{G}}_\alpha'$ is a third-rank tensor and the rate-of-strain tensor **E** is a second-rank tensor. Now we derive a final expression for the last term in Eq. (8.3). As shown before in Sec. 5.3, the tensor $\widetilde{\mathbf{G}}_{\alpha\beta}$ is expressed in component form as

$$\widetilde{G}_{kij}^{\alpha\beta} = G_{ijk}^{\beta\alpha} , \quad \widetilde{G}_{kij}^{\alpha\alpha} = G_{ijk}^{\alpha\alpha} , \tag{8.6}$$

in which $G_{ijk}^{\alpha\beta}$ has already be shown in Eq. (5.72) and is rewritten as follows:

$$G_{ijk}^{\beta\alpha} = X_{\beta\alpha}^{G} (e_i e_j - \frac{1}{3}\delta_{ij}) e_k + Y_{\beta\alpha}^{G} (e_i \delta_{jk} + e_j \delta_{ik} - 2e_i e_j e_k) . \tag{8.7}$$

The expression for $\mathbf{G}_{\alpha\alpha}$ can be obtained by setting $\beta=\alpha$ in this equation. Since the *ij*-component E_{ij} of the rate-of-strain tensor **E** is zero except for E_{12} and E_{21}, which are equal to $\dot{\gamma}/2$, $(\widetilde{\mathbf{G}}_{\alpha\alpha} + \widetilde{\mathbf{G}}_{\alpha\beta}):\mathbf{E}$ can be evaluated straightforwardly and finally reduces to the following equation:

$$(\widetilde{\mathbf{G}}_{\alpha\alpha} + \widetilde{\mathbf{G}}_{\alpha\beta}) : \mathbf{E} = \frac{\dot{\gamma}}{2} \sum_{i=1}^{3} (G_{12i}^{\alpha\alpha} + G_{12i}^{\beta\alpha} + G_{21i}^{\alpha\alpha} + G_{21i}^{\beta\alpha}) \boldsymbol{\delta}_i , \tag{8.8}$$

in which

$$\left. \begin{aligned} G_{12i}^{\beta\alpha} &= X_{\beta\alpha}^{G} e_1 e_2 e_i + Y_{\beta\alpha}^{G} (e_1 \delta_{2i} + e_2 \delta_{1i} - 2e_1 e_2 e_i), \\ G_{21i}^{\beta\alpha} &= X_{\beta\alpha}^{G} e_2 e_1 e_i + Y_{\beta\alpha}^{G} (e_2 \delta_{1i} + e_1 \delta_{2i} - 2e_2 e_1 e_i). \end{aligned} \right\} \tag{8.9}$$

The expressions for $G_{12i}^{\alpha\alpha}$ and $G_{21i}^{\alpha\alpha}$ can be obtained by setting $\beta=\alpha$. Hence Eq. (8.8) reduces to the following equation:

$$\begin{aligned} (\widetilde{\mathbf{G}}_{\alpha\alpha} + \widetilde{\mathbf{G}}_{\alpha\beta}) : \mathbf{E} = \dot{\gamma} \Big[&\big\{ (X_{\alpha\alpha}^{G} - X_{\alpha\beta}^{G}) e_1^{\,2} e_2 + (Y_{\alpha\alpha}^{G} - Y_{\alpha\beta}^{G})(e_2 - 2e_1^{\,2} e_2) \big\} \boldsymbol{\delta}_1 \\ &+ \big\{ (X_{\alpha\alpha}^{G} - X_{\alpha\beta}^{G}) e_1 e_2^{\,2} + (Y_{\alpha\alpha}^{G} - Y_{\alpha\beta}^{G})(e_1 - 2e_1 e_2^{\,2}) \big\} \boldsymbol{\delta}_2 \\ &+ \big\{ (X_{\alpha\alpha}^{G} - X_{\alpha\beta}^{G}) e_1 e_2 e_3 + (Y_{\alpha\alpha}^{G} - Y_{\alpha\beta}^{G})(-2e_1 e_2 e_3) \big\} \boldsymbol{\delta}_3 \Big]. \end{aligned} \tag{8.10}$$

To obtain this equation, we have used the symmetry concerning particles α and β. That is, Eq. (8.7) is unchanged by changing the sign of each component of the vector **e**:

$$X_{\beta\beta}^{G} = -X_{\alpha\alpha}^{G}, \quad X_{\beta\alpha}^{G} = -X_{\alpha\beta}^{G}, \quad Y_{\beta\beta}^{G} = -Y_{\alpha\alpha}^{G}, \quad Y_{\beta\alpha}^{G} = -Y_{\alpha\beta}^{G} \quad (\alpha \neq \beta) . \tag{8.11}$$

Finally, the last term on the right-hand side in Eq. (8.3) can be obtained from the following relation:

$$\widetilde{\mathbf{G}}'_\alpha : \mathbf{E} = \sum_{\beta=1(\neq\alpha)}^{N} (\widetilde{\mathbf{G}}_{\alpha\alpha} + \widetilde{\mathbf{G}}_{\alpha\beta}) : \mathbf{E} . \tag{8.12}$$

The main part of the Stokesian dynamics algorithm for this method is shown below:

1. Specify an initial configuration of colloidal particles
2. Compute the forces acting on each particle
3. Compute the resistance functions
4. Write down Eq. (8.3) for all particles, compute the inverse of the resistance matrix, and then compute the velocities of all particles
5. Evaluate the positions of particles at the next step from Eq. (7.6)
6. Follow the Lees-Edwards periodic boundary condition and move the replicas of the main cell, located at the neighboring positions normal to the shear flow direction, in the shear flow direction by an specified time interval
7. Return to step 2

A face-centered lattice or simple cubic lattice is frequently used as an initial configuration. In a similar way to the molecular dynamics method, the nearest image convention is used to calculate the interaction energies or forces between particles. These techniques will be explained in Chapter 11.

8.1.2 Stokesian dynamics for both the translational and rotational motion of spherical particles

Magnetic and electro-rheological fluids have magnetic and electric properties, respectively, so that particles do not necessarily rotate with the local velocity of the flow field, but with a certain angular velocity which depends on the magnetic or electric field. Hence, even for a spherical particle system, rotational motion has to be taken into account together with the translational motion.

As in the preceding section, we consider a simple shear flow problem which is indicated in Fig. 8.1. The force $\mathbf{F}_\alpha$ and torque $\mathbf{T}_\alpha$ exerted on the ambient fluid by particle α can be written on the level of the additivity of forces, from Eq. (6.6), as

$$\begin{aligned}\mathbf{F}_\alpha = 6\pi a\eta(\mathbf{v}_\alpha - \mathbf{U}) + \eta\Bigg\{ &\sum_{\beta=1(\neq\alpha)}^{N}(\mathbf{A}_{\alpha\alpha} - 6\pi a\mathbf{I})\cdot(\mathbf{v}_\alpha - \mathbf{U}) + \sum_{\beta=1(\neq\alpha)}^{N}\mathbf{A}_{\alpha\beta}\cdot(\mathbf{v}_\beta - \mathbf{U}) \\ &+ \sum_{\beta=1(\neq\alpha)}^{N}\widetilde{\mathbf{B}}_{\alpha\alpha}\cdot(\boldsymbol{\omega}_\alpha - \boldsymbol{\Omega}) + \sum_{\beta=1(\neq\alpha)}^{N}\widetilde{\mathbf{B}}_{\alpha\beta}\cdot(\boldsymbol{\omega}_\beta - \boldsymbol{\Omega})\Bigg\} - \eta\widetilde{\mathbf{G}}'_\alpha : \mathbf{E},\end{aligned} \tag{8.13}$$

$$\mathbf{T}_\alpha = 8\pi a^3 \eta(\boldsymbol{\omega}_\alpha - \boldsymbol{\Omega}) + \eta \left\{ \sum_{\beta=1(\neq\alpha)}^{N} \mathbf{B}_{\alpha\alpha} \cdot (\mathbf{v}_\alpha - \mathbf{U}) + \sum_{\beta=1(\neq\alpha)}^{N} \mathbf{B}_{\alpha\beta} \cdot (\mathbf{v}_\beta - \mathbf{U}) \right.$$
$$\left. + \sum_{\beta=1(\neq\alpha)}^{N} (\mathbf{C}_{\alpha\alpha} - 8\pi a^3 \mathbf{I}) \cdot (\boldsymbol{\omega}_\alpha - \boldsymbol{\Omega}) + \sum_{\beta=1(\neq\alpha)}^{N} \mathbf{C}_{\alpha\beta} \cdot (\boldsymbol{\omega}_\beta - \boldsymbol{\Omega}) \right\} - \eta \widetilde{\mathbf{H}}'_\alpha : \mathbf{E}, \tag{8.14}$$

in which the expressions for $\mathbf{A}_{\alpha\alpha}$, $\mathbf{A}_{\alpha\beta}$, $\widetilde{\mathbf{G}}_\alpha'$, etc., have already been given in the preceding section. The resistance tensors $\mathbf{C}_{\alpha\alpha}$, $\mathbf{C}_{\alpha\beta}$, $\mathbf{B}_{\alpha\alpha}$, and $\mathbf{B}_{\alpha\beta}$ are expressed from Eq. (5.71) as

$$\mathbf{C}_{\alpha\alpha} = X^C_{\alpha\alpha}\mathbf{ee} + Y^C_{\alpha\alpha}(\mathbf{I} - \mathbf{ee}), \qquad \mathbf{C}_{\alpha\beta} = X^C_{\alpha\beta}\mathbf{ee} + Y^C_{\alpha\beta}(\mathbf{I} - \mathbf{ee}), \tag{8.15}$$

$$\mathbf{B}_{\alpha\alpha} = Y^B_{\alpha\alpha}\boldsymbol{\varepsilon} \cdot \mathbf{e}, \qquad \mathbf{B}_{\alpha\beta} = Y^B_{\alpha\beta}\boldsymbol{\varepsilon} \cdot \mathbf{e}. \tag{8.16}$$

The expressions for $X_{\alpha\alpha}^C$, etc., are shown in Appendix A4. $\widetilde{\mathbf{H}}_\alpha'$ can be expressed from Eqs. (6.12) and (6.15) as

$$\widetilde{\mathbf{H}}'_\alpha = \sum_{\beta=1(\neq\alpha)}^{N} (\widetilde{\mathbf{H}}_{\alpha\alpha} + \widetilde{\mathbf{H}}_{\alpha\beta}). \tag{8.17}$$

Now we show how the last term in Eq. (8.14) can finally be written for a spherical particle system. As shown before, the tensor $\widetilde{\mathbf{H}}_{\alpha\beta}$ can be written in component form as

$$\widetilde{H}^{\alpha\beta}_{kij} = H^{\beta\alpha}_{ijk}, \qquad \widetilde{H}^{\alpha\alpha}_{kij} = H^{\alpha\alpha}_{ijk}, \tag{8.18}$$

in which $H_{ijk}^{\beta\alpha}$ has already been shown and written as

$$H^{\beta\alpha}_{ijk} = Y^H_{\beta\alpha} \sum_{l=1}^{3} (\varepsilon_{ikl} e_l e_j + \varepsilon_{jkl} e_l e_i). \tag{8.19}$$

The expression for $H_{ijk}^{\alpha\alpha}$ can be obtained by setting $\beta=\alpha$ in this equation. Hence, from a similar procedure to the previous case for $(\widetilde{\mathbf{G}}_{\alpha\alpha} + \widetilde{\mathbf{G}}_{\alpha\beta})$:$\mathbf{E}$, $(\widetilde{\mathbf{H}}_{\alpha\alpha} + \widetilde{\mathbf{H}}_{\alpha\beta})$:$\mathbf{E}$ can be reformed as

$$(\widetilde{\mathbf{H}}_{\alpha\alpha} + \widetilde{\mathbf{H}}_{\alpha\beta}) : \mathbf{E} = \frac{\dot{\gamma}}{2} \sum_{i=1}^{3} (H^{\alpha\alpha}_{12i} + H^{\beta\alpha}_{12i} + H^{\alpha\alpha}_{21i} + H^{\beta\alpha}_{21i}) \boldsymbol{\delta}_i, \tag{8.20}$$

in which

$$\left. \begin{aligned} H^{\beta\alpha}_{12i} &= Y^H_{\beta\alpha} \sum_{l=1}^{3} (e_1 \varepsilon_{2il} e_l + e_2 \varepsilon_{1il} e_l), \\ H^{\beta\alpha}_{21i} &= Y^H_{\beta\alpha} \sum_{l=1}^{3} (e_2 \varepsilon_{1il} e_l + e_1 \varepsilon_{2il} e_l). \end{aligned} \right\} \tag{8.21}$$

If we use $\beta=\alpha$ in this equation, the expressions for $H_{12i}^{\alpha\alpha}$ and $H_{21i}^{\alpha\alpha}$ can be obtained. The substitution of these expressions into Eq. (8.20) leads to the following equation:

$$(\widetilde{\mathbf{H}}_{\alpha\alpha} + \widetilde{\mathbf{H}}_{\alpha\beta}) : \mathbf{E} = \dot{\gamma}(Y^H_{\alpha\alpha} + Y^H_{\alpha\beta}) \{ -e_1 e_3 \boldsymbol{\delta}_1 + e_2 e_3 \boldsymbol{\delta}_2 + (e_1 e_1 - e_2 e_2) \boldsymbol{\delta}_3 \}. \tag{8.22}$$

In obtaining this equation, we have used the symmetry concerning particles α and β. That is,

Eq. (8.19) is unchanged by changing the sign of each component of the vector **e**:

$$Y^H_{\alpha\beta} = Y^H_{\beta\alpha}, \qquad Y^H_{\beta\beta} = Y^H_{\alpha\alpha}. \tag{8.23}$$

Finally, the last term on the right-hand side in Eq. (8.14) reduces to the following relation:

$$\widetilde{\mathbf{H}}'_\alpha : \mathbf{E} = \sum_{\beta=1(\neq\alpha)}^{N} (\widetilde{\mathbf{H}}_{\alpha\alpha} + \widetilde{\mathbf{H}}_{\alpha\beta}) : \mathbf{E}. \tag{8.24}$$

If particles have magnetic properties, the force $\mathbf{F}_\alpha$ and torque $\mathbf{T}_\alpha$ exerted on the ambient fluid by particle α are equal to the total magnetic force and torque exerted on particle α by the other particles.

The main part of the Stokesian dynamics algorithm for this case is shown below.

1. Specify an initial configuration of particles and their directions
2. Compute the forces and torques of all particles
3. Compute the resistance functions
4. Write down Eqs. (8.13) and (8.14) for all particles, compute the inverse of the resistance matrix, and then compute the velocities and angular velocities of all particles
5. Evaluate the positions and directions of particles at the next time step from Eqs. (7.15) and (7.16)
6. Follow the Lees-Edwards periodic boundary condition and move the replicas of the main cell, located at the neighboring positions normal to the shear flow direction, in the shear flow direction by the specified time interval
7. Return to step 2

The method of setting an initial configuration, etc., will be explained in Chapter 11.

8.2 The Approximation of Additivity of Velocities

As in Sec. 8.1, we consider two cases for using the approximation of additivity of velocities in the Stokesian dynamics method. The first case is for taking into account the translational motion of particles alone, and the second case is for taking account of both the translational and rotational motion of spherical particles.

8.2.1 Stokesian dynamics for only the translational motion of spherical particles

As in the preceding section, we consider a simple shear flow directed in the *x*-direction with shear rate $\dot{\gamma}$, shown in Fig. 8.1. On the level of the additivity of velocities, the velocity of a particle in a shear flow has already been expressed in Eq. (6.21) and the velocity of a spherical particle α, $\mathbf{v}_\alpha$, can be written as

$$\mathbf{v}_\alpha = \dot{\gamma} y \boldsymbol{\delta}_x + \frac{1}{\eta}\left\{\frac{1}{6\pi a}\mathbf{F}_\alpha + \sum_{\beta=1(\neq\alpha)}^{N}(\mathbf{a}_{\alpha\alpha} - \frac{1}{6\pi a}\mathbf{I})\cdot\mathbf{F}_\alpha + \sum_{\beta=1(\neq\alpha)}^{N}\mathbf{a}_{\alpha\beta}\cdot\mathbf{F}_\beta\right\} + \tilde{\mathbf{g}}'_\alpha : \mathbf{E}, \tag{8.25}$$

in which $\mathbf{a}_{\alpha\alpha}$ and $\mathbf{a}_{\alpha\beta}$ are expressed, from Eq. (5.73), as

$$\mathbf{a}_{\alpha\alpha} = x^a_{\alpha\alpha}\mathbf{ee} + y^a_{\alpha\alpha}(\mathbf{I}-\mathbf{ee}), \qquad \mathbf{a}_{\alpha\beta} = x^a_{\alpha\beta}\mathbf{ee} + y^a_{\alpha\beta}(\mathbf{I}-\mathbf{ee}). \tag{8.26}$$

The explicit expressions and numerical results for the mobility functions such as $x_{\alpha\alpha}{}^a$, etc., may be found in Sec. 5.3.5 and Appendix A4. Also $\tilde{\mathbf{g}}_\alpha$ is written from Eqs. (6.26) and (6.29) as

$$\tilde{\mathbf{g}}'_\alpha = \sum_{\beta=1(\neq\alpha)}^{N}\tilde{\mathbf{g}}_\alpha . \tag{8.27}$$

Since $\tilde{\mathbf{g}}_\alpha$ is a third-rank tensor, and the rate-of-strain tensor $\mathbf{E}$ is a second-rank tensor, the last term on the right-hand side in Eq. (8.25) is a first-rank tensor or a vector. Now we derive the final form of the last term in Eq. (8.25). As shown in the preceding section, the tensor $\tilde{\mathbf{g}}_\alpha$ is expressed in component form as

$$\tilde{g}^\alpha_{kij} = g^{\alpha\alpha}_{ijk} + g^{\beta\alpha}_{ijk}, \tag{8.28}$$

in which the expression for $g_{ijk}{}^{\beta\alpha}$ has already been shown and is written as

$$g^{\beta\alpha}_{ijk} = x^g_{\beta\alpha}(e_ie_j - \frac{1}{3}\delta_{ij})e_k + y^g_{\beta\alpha}(e_i\delta_{jk} + e_j\delta_{ik} - 2e_ie_je_k). \tag{8.29}$$

The expression for $\mathbf{g}_{\alpha\alpha}$ can be obtained by setting β=α. The *ij*-component E_{ij} of the rate-of-strain tensor $\mathbf{E}$ is zero except for E_{12} and E_{21}, which are equal to $\dot{\gamma}/2$, so that $\tilde{\mathbf{g}}_\alpha$:$\mathbf{E}$ can straightforwardly be calculated and finally reduces to the following expression:

$$\tilde{\mathbf{g}}_\alpha : \mathbf{E} = \frac{\dot{\gamma}}{2}\sum_{i=1}^{3}(g^{\alpha\alpha}_{12i} + g^{\beta\alpha}_{12i} + g^{\alpha\alpha}_{21i} + g^{\beta\alpha}_{21i})\boldsymbol{\delta}_i , \tag{8.30}$$

in which

$$\left.\begin{aligned} g^{\beta\alpha}_{12i} &= x^g_{\beta\alpha}e_1e_2e_i + y^g_{\beta\alpha}(e_1\delta_{2i} + e_2\delta_{1i} - 2e_1e_2e_i), \\ g^{\beta\alpha}_{21i} &= x^g_{\beta\alpha}e_2e_1e_i + y^g_{\beta\alpha}(e_2\delta_{1i} + e_1\delta_{2i} - 2e_2e_1e_i). \end{aligned}\right\} \tag{8.31}$$

The expressions for $g_{12i}{}^{\alpha\alpha}$ and $g_{21i}{}^{\alpha\alpha}$ can be obtained by setting β=α. Hence Eq. (8.30) reduces to the following expression:

$$\begin{aligned} \tilde{\mathbf{g}}_\alpha : \mathbf{E} = \dot{\gamma}\Big[&\big\{(x^g_{\alpha\alpha} - x^g_{\alpha\beta})e_1{}^2e_2 + (y^g_{\alpha\alpha} - y^g_{\alpha\beta})(e_2 - 2e_1{}^2e_2)\big\}\boldsymbol{\delta}_1 \\ &+ \big\{(x^g_{\alpha\alpha} - x^g_{\alpha\beta})e_1e_2{}^2 + (y^g_{\alpha\alpha} - y^g_{\alpha\beta})(e_1 - 2e_1e_2{}^2)\big\}\boldsymbol{\delta}_2 \\ &+ \big\{(x^g_{\alpha\alpha} - x^g_{\alpha\beta})e_1e_2e_3 + (y^g_{\alpha\alpha} - y^g_{\alpha\beta})(-2e_1e_2e_3)\big\}\boldsymbol{\delta}_3\Big]. \end{aligned} \tag{8.32}$$

In obtaining this equation, we have used the symmetry concerning particles α and β. That is, Eq. (8.29) is unchanged by changing the sign of each component of the vector **e**:

$$x^{g}_{\alpha\beta} = -x^{g}_{\beta\alpha}, \quad y^{g}_{\alpha\beta} = -y^{g}_{\beta\alpha} \qquad (\alpha \neq \beta). \tag{8.33}$$

Finally, the third term on the right-hand side in Eq. (8.25) can be obtained from the following relation:

$$\tilde{\mathbf{g}}'_{\alpha} : \mathbf{E} = \sum_{\beta=1(\neq\alpha)}^{N} \tilde{\mathbf{g}}_{\alpha} : \mathbf{E}. \tag{8.34}$$

The main part of the Stokesian dynamics algorithm for this method is shown below.

1. Specify an initial configuration of colloidal particles
2. Compute the forces acting on each particle
3. Compute the mobility functions
4. Compute the velocities of all particles from Eq. (8.25)
5. Compute the particle positions at the next step from Eq. (7.6)
6. Follow the Lees-Edwards periodic boundary condition and move the replicas of the main cell, located at the neighboring positions normal to the shear flow direction, in the shear flow direction by the specified time interval
7. Return to step 2

8.2.2 Stokesian dynamics for both the translational and rotational motion of spherical particles

As before, we consider the simple shear flow shown in Fig. 8.1, but in this section we treat both the translational and rotational motion of spherical particles. On the level of the additivity of velocities, the velocity $\mathbf{v}_{\alpha}$ and angular velocity $\boldsymbol{\omega}_{\alpha}$ of particle α can be written from Eq. (6.21) as

$$\begin{aligned}\mathbf{v}_{\alpha} = \dot{\gamma} y \boldsymbol{\delta}_x + \frac{1}{\eta}\Bigg\{ &\frac{1}{6\pi a}\mathbf{F}_{\alpha} + \sum_{\beta=1(\neq\alpha)}^{N}\Big(\mathbf{a}_{\alpha\alpha} - \frac{1}{6\pi a}\mathbf{I}\Big)\cdot\mathbf{F}_{\alpha} + \sum_{\beta=1(\neq\alpha)}^{N}\mathbf{a}_{\alpha\beta}\cdot\mathbf{F}_{\beta} \\ &+ \sum_{\beta=1(\neq\alpha)}^{N}\tilde{\mathbf{b}}_{\alpha\alpha}\cdot\mathbf{T}_{\alpha} + \sum_{\beta=1(\neq\alpha)}^{N}\tilde{\mathbf{b}}_{\alpha\beta}\cdot\mathbf{T}_{\beta}\Bigg\} + \tilde{\mathbf{g}}'_{\alpha} : \mathbf{E},\end{aligned} \tag{8.35}$$

$$\begin{aligned}\boldsymbol{\omega}_{\alpha} = -\frac{1}{2}\dot{\gamma}\boldsymbol{\delta}_z + \frac{1}{\eta}\Bigg\{ &\sum_{\beta=1(\neq\alpha)}^{N}\mathbf{b}_{\alpha\alpha}\cdot\mathbf{F}_{\alpha} + \sum_{\beta=1(\neq\alpha)}^{N}\mathbf{b}_{\alpha\beta}\cdot\mathbf{F}_{\beta} + \frac{1}{8\pi a^{3}}\cdot\mathbf{T}_{\alpha} \\ &+ \sum_{\beta=1(\neq\alpha)}^{N}\Big(\mathbf{c}_{\alpha\alpha} - \frac{1}{8\pi a^{3}}\mathbf{I}\Big)\cdot\mathbf{T}_{\alpha} + \sum_{\beta=1(\neq\alpha)}^{N}\mathbf{c}_{\alpha\beta}\cdot\mathbf{T}_{\beta}\Bigg\} + \tilde{\mathbf{h}}'_{\alpha} : \mathbf{E},\end{aligned} \tag{8.36}$$

in which the expressions for $\mathbf{a}_{\alpha\alpha}$, $\mathbf{a}_{\alpha\beta}$, and $\widetilde{\mathbf{g}}_\alpha{}'$ have already been given in Sec. 8.2.1. The expressions for $\mathbf{c}_{\alpha\alpha}$, $\mathbf{c}_{\alpha\beta}$, $\mathbf{b}_{\alpha\alpha}$, and $\mathbf{b}_{\alpha\beta}$ are written from Eq. (5.73) as

$$\mathbf{c}_{\alpha\alpha} = x^c_{\alpha\alpha}\mathbf{ee} + y^c_{\alpha\alpha}(\mathbf{I}-\mathbf{ee}), \qquad \mathbf{c}_{\alpha\beta} = x^c_{\alpha\beta}\mathbf{ee} + y^c_{\alpha\beta}(\mathbf{I}-\mathbf{ee}), \tag{8.37}$$

$$\mathbf{b}_{\alpha\alpha} = y^b_{\alpha\alpha}\boldsymbol{\varepsilon}\cdot\mathbf{e}, \qquad \mathbf{b}_{\alpha\beta} = y^b_{\alpha\beta}\boldsymbol{\varepsilon}\cdot\mathbf{e}\,. \tag{8.38}$$

The expressions for the mobility functions such as $x_{\alpha\alpha}{}^c$ may be found in Appendix A4. Additionally the expressions for $\widetilde{\mathbf{b}}_{\alpha\alpha}$ and $\widetilde{\mathbf{b}}_{\alpha\beta}$ are written from Eq. (5.66) as

$$\widetilde{\mathbf{b}}_{\alpha\alpha} = \mathbf{b}^t{}_{\alpha\alpha}, \qquad \widetilde{\mathbf{b}}_{\alpha\beta} = \mathbf{b}^t{}_{\beta\alpha}\,. \tag{8.39}$$

Also $\widetilde{\mathbf{h}}_\alpha{}'$ can be expressed from Eqs. (6.27) and (6.29) as

$$\widetilde{\mathbf{h}}'_\alpha = \sum_{\beta=1(\neq\alpha)}^{N} \widetilde{\mathbf{h}}_\alpha\,. \tag{8.40}$$

Now we derive the final expression for the last term on the right-hand side in Eq. (8.36). As shown before, the tensor $\widetilde{\mathbf{h}}_\alpha$ can be written in component form as

$$\widetilde{h}^\alpha_{kij} = h^{\alpha\alpha}_{ijk} + h^{\beta\alpha}_{ijk}\,, \tag{8.41}$$

in which $h_{ijk}{}^{\beta\alpha}$ has already been shown and is written as

$$h^{\beta\alpha}_{ijk} = y^h_{\beta\alpha}\sum_{l=1}^{3}(\varepsilon_{ikl}e_l e_j + \varepsilon_{jkl}e_l e_i)\,. \tag{8.42}$$

The expression for $h_{ijk}{}^{\alpha\alpha}$ is obtained by setting β=α in this expression. Hence from a similar procedure to the previous case for $\widetilde{\mathbf{g}}_\alpha$:**E**, $\widetilde{\mathbf{h}}_\alpha$:**E** can be reformed as

$$\widetilde{\mathbf{h}}_\alpha : \mathbf{E} = \frac{\dot{\gamma}}{2}\sum_{i=1}^{3}(h^{\alpha\alpha}_{12i} + h^{\beta\alpha}_{12i} + h^{\alpha\alpha}_{21i} + h^{\beta\alpha}_{21i})\boldsymbol{\delta}_i\,, \tag{8.43}$$

in which

$$\left.\begin{aligned} h^{\beta\alpha}_{12i} &= y^h_{\beta\alpha}\sum_{l=1}^{3}(e_1\varepsilon_{2il}e_l + e_2\varepsilon_{1il}e_l), \\ h^{\beta\alpha}_{21i} &= y^h_{\beta\alpha}\sum_{l=1}^{3}(e_2\varepsilon_{1il}e_l + e_1\varepsilon_{2il}e_l). \end{aligned}\right\} \tag{8.44}$$

If we set β=α in this equation, the expressions for $h_{12i}{}^{\alpha\alpha}$ and $h_{21i}{}^{\alpha\alpha}$ are obtained. The substitution of these expressions into Eq. (8.43) leads to the following equation:

$$\widetilde{\mathbf{h}}_\alpha : \mathbf{E} = \dot{\gamma}(y^h_{\alpha\alpha} + y^h_{\alpha\beta})\{-e_1e_3\boldsymbol{\delta}_1 + e_2e_3\boldsymbol{\delta}_2 + (e_1e_1 - e_2e_2)\boldsymbol{\delta}_3\}. \tag{8.45}$$

In obtaining this equation, we have used the symmetry concerning particles α and β, that is, Eq. (8.42) is unchanged by changing the sign of each component of the vector **e**:

$$y^h_{\alpha\beta} = y^h_{\beta\alpha}\,. \tag{8.46}$$

Hence we obtain the final form of the last term in Eq. (8.36) from the following equation:

$$\widetilde{\mathbf{h}}'_\alpha : \mathbf{E} = \sum_{\beta=1(\neq\alpha)}^{N} \widetilde{\mathbf{h}}_\alpha : \mathbf{E}\,. \tag{8.47}$$

If the particles have magnetic properties, the force $\mathbf{F}_\alpha$ and torque $\mathbf{T}_\alpha$ exerted on the ambient fluid by particle α are equal to the total magnetic force and torque exerted on particle α by the other particles.

The main part of the Stokesian dynamics algorithm for this method is shown below.

1. Specify an initial configuration of particles and their directions
2. Compute the forces and torques of all particles
3. Compute the mobility functions
4. Evaluate the velocities and angular velocities of particles at the present time step from Eqs. (8.35) and (8.36)
5. Evaluate them at the next time step from Eqs. (7.15) and (7.16)
6. Follow the Lees-Edwards periodic boundary condition and move the replicas of the main cell, located at the neighboring positions normal to the shear flow direction, in the shear flow direction by the specified time interval
7. Return to step 2

8.3 The Nondimensionalization Method

Dimensional quantities are seldom treated in simulations for physical problems. Rather, physical qualities are usually nondimensionalized by the corresponding representative quantities which characterize a physical phenomenon. In this section we show the method of nondimensionalization which is generally used for the case of a spherical particle system subjected to a simple shear flow.

The particle radius a is here used as a representative value for distances, but the particle diameter $d(=2a)$ may also be used in some cases. There are two possible representative values for time. The first is the characteristic time representing a flow field, $1/\dot{\gamma}$, and the second is the relaxation time of the particle motion due to the frictional force. The former representative value is usually used. A combined value of the representative time and length, $\dot{\gamma}\,a$, is used as a representative value for the velocity. With this representative velocity, the representative force is obtained as $(6\pi\eta a\,\dot{\gamma}\,a)$ from Stokes' drag formula for a spherical particle. If we use $\dot{\gamma}$ for the representative angular velocity, the representative value for torque becomes ($8\pi\eta a^3\dot{\gamma}$) from Eq. (5.2). A representative value for any other quantity can be derived from a similar procedure.

Next, we show the standard nondimensional expressions for the resistance and mobility tensors:

$$\begin{aligned}&\mathbf{A}^*_{\alpha\beta} = \mathbf{A}_{\alpha\beta}/6\pi a, \ \ \mathbf{B}^*_{\alpha\beta} = \mathbf{B}_{\alpha\beta}/4\pi a^2, \ \ \mathbf{C}^*_{\alpha\beta} = \mathbf{C}_{\alpha\beta}/8\pi a^3, \\ &\mathbf{G}^*_{\alpha\beta} = \mathbf{G}_{\alpha\beta}/4\pi a^2, \ \ \mathbf{H}^*_{\alpha\beta} = \mathbf{H}_{\alpha\beta}/8\pi a^3, \ \ \mathbf{K}^*_{\alpha\beta} = \mathbf{K}_{\alpha\beta}/(20\pi a^3/3),\end{aligned} \tag{8.48}$$

$$\begin{aligned}&\mathbf{a}^*_{\alpha\beta} = 6\pi a\mathbf{a}_{\alpha\beta}, \ \ \mathbf{b}^*_{\alpha\beta} = 4\pi a^2\mathbf{b}_{\alpha\beta}, \ \ \mathbf{c}^*_{\alpha\beta} = 8\pi a^3\mathbf{c}_{\alpha\beta}, \\ &\mathbf{g}^*_{\alpha\beta} = \mathbf{g}_{\alpha\beta}/2a, \ \ \mathbf{h}^*_{\alpha\beta} = \mathbf{h}_{\alpha\beta}, \ \ \mathbf{k}^*_{\alpha\beta} = \mathbf{k}_{\alpha\beta}/(20\pi a^3/3).\end{aligned} \tag{8.49}$$

In these expressions, the qualities with the superscript * are dimensionless. If we replace β with α, the respective nondimensional quantities such as $\mathbf{A}_{\alpha\alpha}^*$ can be obtained. The rate-of-strain tensor **E** is nondimensionalized as $\mathbf{E}^*=\mathbf{E}/\dot{\gamma}$.

If we use the above-mentioned representative values and the nondimensional resistance and mobility tensors, then Eqs. (8.13), (8.14), (8.35), and (8.36) can be written in nondimensional form as

$$\begin{aligned}\mathbf{F}^*_\alpha = (\mathbf{v}^*_\alpha - \mathbf{U}^*) + &\left\{ \sum_{\beta=1(\neq\alpha)}^{N} (\mathbf{A}^*_{\alpha\alpha} - \mathbf{I})\cdot(\mathbf{v}^*_\alpha - \mathbf{U}^*) + \sum_{\beta=1(\neq\alpha)}^{N} \mathbf{A}^*_{\alpha\beta}\cdot(\mathbf{v}^*_\beta - \mathbf{U}^*) \right\} \\ &+ \frac{2}{3}\left\{ \sum_{\beta=1(\neq\alpha)}^{N} \widetilde{\mathbf{B}}^*_{\alpha\alpha}\cdot(\boldsymbol{\omega}^*_\alpha - \boldsymbol{\Omega}^*) + \sum_{\beta=1(\neq\alpha)}^{N} \widetilde{\mathbf{B}}^*_{\alpha\beta}\cdot(\boldsymbol{\omega}^*_\beta - \boldsymbol{\Omega}^*) \right\} - \frac{2}{3}\widetilde{\mathbf{G}}'^*_\alpha : \mathbf{E}^*,\end{aligned} \tag{8.50}$$

$$\begin{aligned}\mathbf{T}^*_\alpha = (\boldsymbol{\omega}^*_\alpha - \boldsymbol{\Omega}^*) + &\frac{1}{2}\left\{ \sum_{\beta=1(\neq\alpha)}^{N} \mathbf{B}^*_{\alpha\alpha}\cdot(\mathbf{v}^*_\alpha - \mathbf{U}^*) + \sum_{\beta=1(\neq\alpha)}^{N} \mathbf{B}^*_{\alpha\beta}\cdot(\mathbf{v}^*_\beta - \mathbf{U}^*) \right\} \\ &+ \left\{ \sum_{\beta=1(\neq\alpha)}^{N} (\mathbf{C}^*_{\alpha\alpha} - \mathbf{I})\cdot(\boldsymbol{\omega}^*_\alpha - \boldsymbol{\Omega}^*) + \sum_{\beta=1(\neq\alpha)}^{N} \mathbf{C}^*_{\alpha\beta}\cdot(\boldsymbol{\omega}^*_\beta - \boldsymbol{\Omega}^*) \right\} - \widetilde{\mathbf{H}}'^*_\alpha : \mathbf{E}^*,\end{aligned} \tag{8.51}$$

$$\begin{aligned}\mathbf{v}^*_\alpha = y^*\boldsymbol{\delta}_x + \mathbf{F}^*_\alpha + &\sum_{\beta=1(\neq\alpha)}^{N} (\mathbf{a}^*_{\alpha\alpha} - \mathbf{I})\cdot\mathbf{F}^*_\alpha + \sum_{\beta=1(\neq\alpha)}^{N} \mathbf{a}^*_{\alpha\beta}\cdot\mathbf{F}^*_\beta \\ &+ 2\left(\sum_{\beta=1(\neq\alpha)}^{N} \widetilde{\mathbf{b}}^*_{\alpha\alpha}\cdot\mathbf{T}^*_\alpha + \sum_{\beta=1(\neq\alpha)}^{N} \widetilde{\mathbf{b}}^*_{\alpha\beta}\cdot\mathbf{T}^*_\beta \right) + 2\widetilde{\mathbf{g}}'^*_\alpha : \mathbf{E}^*,\end{aligned} \tag{8.52}$$

$$\begin{aligned}\boldsymbol{\omega}^*_\alpha = -\frac{1}{2}\boldsymbol{\delta}_z + &\frac{3}{2}\left(\sum_{\beta=1(\neq\alpha)}^{N} \mathbf{b}^*_{\alpha\alpha}\cdot\mathbf{F}^*_\alpha + \sum_{\beta=1(\neq\alpha)}^{N} \mathbf{b}^*_{\alpha\beta}\cdot\mathbf{F}^*_\beta \right) + \mathbf{T}^*_\alpha \\ &+ \sum_{\beta=1(\neq\alpha)}^{N} \left(\mathbf{c}^*_{\alpha\alpha} - \mathbf{I}\right)\cdot\mathbf{T}^*_\alpha + \sum_{\beta=1(\neq\alpha)}^{N} \mathbf{c}^*_{\alpha\beta}\cdot\mathbf{T}^*_\beta + \widetilde{\mathbf{h}}'^*_\alpha : \mathbf{E}^*.\end{aligned} \tag{8.53}$$

It should be noted that, for ferromagnetic colloidal dispersions, since forces are nondimensionalized by the viscous shear force, the expression for the nondimensional force includes the nondimensional number representing the strength of the magnetic force relative to the viscous shear force. Another point to be noted is that the nondimensional shear rate is constant and unity, that is, $\dot{\gamma}^*=1$. Thus one may have a question, "How can a shear flow with an arbitrary strength of the shear rate be generated?" As will be shown in Chapter 12, the dimensional shear rate can be changed through the nondimensional number which appears in the expressions of the nondimensional forces and torques.

Exercises

8.1 Derive Eq. (8.8) using the following relations:

$$\begin{aligned}(\tilde{\mathbf{G}}_{\alpha\alpha}+\tilde{\mathbf{G}}_{\alpha\beta}):\mathbf{E} &= \sum_{i=1}^{3}\left(\sum_{j=1}^{3}\sum_{k=1}^{3}(\tilde{G}_{ijk}^{\alpha\alpha}+\tilde{G}_{ijk}^{\alpha\beta})E_{kj}\right)\boldsymbol{\delta}_i \\ &= \sum_{i=1}^{3}(\tilde{G}_{i12}^{\alpha\alpha}E_{21}+\tilde{G}_{i21}^{\alpha\alpha}E_{12}+\tilde{G}_{i21}^{\alpha\beta}E_{12}+\tilde{G}_{i12}^{\alpha\beta}E_{21})\boldsymbol{\delta}_i .\end{aligned} \tag{8.54}$$

8.2 Derive Eqs. (8.22) and (8.45) using the following relations:

$$\left.\begin{aligned}&\sum_{l=1}^{3}\varepsilon_{21l}e_l=-e_3,\quad \sum_{l=1}^{3}\varepsilon_{22l}e_l=0,\quad \sum_{l=1}^{3}\varepsilon_{23l}e_l=e_1,\\ &\sum_{l=1}^{3}\varepsilon_{11l}e_l=0,\quad \sum_{l=1}^{3}\varepsilon_{12l}e_l=e_3,\quad \sum_{l=1}^{3}\varepsilon_{13l}e_l=-e_2.\end{aligned}\right\} \tag{8.55}$$

.

8.3 Derive Eq. (8.30) using the following relations:

$$\tilde{\mathbf{g}}_{\alpha}:\mathbf{E}=\sum_{i=1}^{3}\left(\sum_{j=1}^{3}\sum_{k=1}^{3}\tilde{g}_{ijk}^{\alpha}E_{kj}\right)\boldsymbol{\delta}_i=\sum_{i=1}^{3}\left(\tilde{g}_{i12}^{\alpha}E_{21}+\tilde{g}_{i21}^{\alpha}E_{12}\right)\boldsymbol{\delta}_i . \tag{8.56}$$

8.4 Derive the nondimensional equations in Eqs. (8.50) to (8.53) from Eqs. (8.13), (8.14), (8.35), and (8.36) by applying the method of nondimensionalization in Sec. 8.3.

CHAPTER 9

BROWNIAN DYNAMICS METHODS

A light particle, such as a pollen grain, randomly moves with zigzag motion in a liquid medium. This irregular motion is induced by the motion of molecules composing the dispersion medium and is well known as Brownian motion. A particle moving with this motion is called a Brownian particle. In engineering fields, Brownian motion can be observed in a functional fluid such as a ferrofluid. Hence a simulation which takes into account the particle Brownian motion plays an important role in investigating the stability, internal structure, and rheological properties of a dispersion. Such studies of colloidal dispersions are very important in generating new functional fluids. If Brownian particles are not much larger than molecules of a dispersion medium, ordinary molecular dynamics methods are applicable. However, Brownian particles are generally much larger than such molecules. In this situation, the representative time characterizing the Brownian motion is much longer than that characterizing the motion of molecules. This implies that molecular dynamics methods are not a realistic technique for simulating the colloidal dispersion with these two representative times. To circumvent this difficulty, the exact motion of molecules of a dispersion medium is not concentrated on, but rather such molecules are regarded as a continuum. In this approach, the random force inducing the Brownian motion is treated as a stochastic force. Firstly we show the Brownian dynamics method based on the Langevin equation and then discuss more complicated Brownian dynamics methods in which hydrodynamic interactions are taken into account.

9.1 The Langevin Equation

If a colloidal dispersion is dilute enough to neglect hydrodynamic interactions between particles, the Brownian motion of particles is generally described by the following Langevin equation:

$$m\frac{d\mathbf{v}}{dt} = \mathbf{F} - m\zeta\mathbf{v} + \mathbf{F}^B , \tag{9.1}$$

in which m is the mass of Brownian particles, $\mathbf{v}$ the particle velocity, $\mathbf{F}$ the sum of the force exerted on the particle of interest by the others and the force by the outer field, ξ (=$m\zeta$) the friction coefficient, and $\mathbf{F}^B$ the random force inducing the Brownian motion of particles. We here consider a multi-particle system in which particles interact with each other through the interparticle interactions, although hydrodynamic interactions are neglected in this section. The subscript for discriminating particles has been dropped for simplicity in each quantity such as **v** and **F**.

The force $\mathbf{F}^B$ inducing the random motion of particles should be independent of the particle, i.e. the particle position and velocity. Since it is observed with an optical microscope that the amplitude and direction of the velocity of a particle abruptly change in the Brownian motion, the random force $\mathbf{F}^B$ can be described by the following properties:

$$\left.\begin{aligned} &\left\langle \mathbf{F}^B(t)\right\rangle = 0, \\ &\left\langle \mathbf{F}^B(t)\cdot\mathbf{F}^B(t')\right\rangle = A\delta(t-t'), \end{aligned}\right\} \tag{9.2}$$

in which T is the system temperature, and A is a constant and is expressed as A=6$m\zeta kT$, which will be derived later.

If the velocity at time t_0 is denoted by $\mathbf{v}(t_0)$ and the force $\mathbf{F}$ can be regarded as constant during the short time interval (t-t_0), then the velocity at time t, $\mathbf{v}(t)$, can be derived from Eq. (9.1) as

$$\mathbf{v}(t) = \mathbf{v}(t_0)e^{-\zeta(t-t_0)} + \frac{1}{m\zeta}\mathbf{F}(t_0)\left\{1-e^{-\zeta(t-t_0)}\right\} + \frac{1}{m}\int_{t_0}^{t}\mathbf{F}^B(\tau)e^{-\zeta(t-\tau)}d\tau\,. \tag{9.3}$$

The derivation of this equation should be referred to in Exercise 9.1. If the integration of Eq. (9.3) is carried out, the particle position $\mathbf{r}(t)$ can be obtained as

$$\begin{aligned} \mathbf{r}(t) &= \mathbf{r}(t_0) + \int_{t_0}^{t}\mathbf{v}(t')dt' = \mathbf{r}(t_0) + \frac{1}{\zeta}\mathbf{v}(t_0)\left\{1-e^{-\zeta(t-t_0)}\right\} \\ &+ \frac{1}{m\zeta}\mathbf{F}(t_0)\left\{(t-t_0) - \frac{1}{\zeta}\left(1-e^{-\zeta(t-t_0)}\right)\right\} + \frac{1}{m\zeta}\int_{t_0}^{t}\mathbf{F}^B(\tau)(1-e^{-\zeta(t-\tau)})d\tau, \end{aligned} \tag{9.4}$$

in which the last term on the right-hand side has been obtained by integration by parts.

The random force $\mathbf{F}^B$ should be independent of the force acting on the particle. Hence, if we neglect $\mathbf{F}$ in Eq. (9.3) and take into account the relations in Eq. (9.2), the following equation can be obtained:

$$\begin{aligned} &\left\langle \left(\mathbf{v}(t) - \mathbf{v}(t_0)e^{-\zeta(t-t_0)}\right)^2\right\rangle = \frac{1}{m^2}\int_{t_0}^{t}\int_{t_0}^{t}\left\langle \mathbf{F}^B(\tau')\cdot\mathbf{F}^B(\tau)\right\rangle e^{-\zeta\{(t-\tau')+(t-\tau)\}}d\tau d\tau' \\ &= \frac{1}{m^2}\int_{t_0}^{t}\int_{t_0}^{t}A\delta(\tau'-\tau)e^{-\zeta\{(t-\tau')+(t-\tau)\}}d\tau d\tau' = \frac{A}{m^2}\int_{t_0}^{t}e^{-2\zeta(t-\tau)}d\tau = \frac{A}{2m^2\zeta}\left(1-e^{-2\zeta(t-t_0)}\right). \end{aligned} \tag{9.5}$$

By taking the limit of $t \to \infty$ and considering the following definition of temperature:

$$3kT/2 = m\langle v^2 \rangle /2, \tag{9.6}$$

the constant A can be obtained as

$$A = 6m\zeta kT. \tag{9.7}$$

Since the random variable or the random force $\mathbf{F}^B$ is included in Eqs. (9.3) and (9.4), the final expressions for $\mathbf{v}(t)$ and $\mathbf{r}(t)$ cannot be obtained unless the integration concerning the random force is dealt with. This procedure was discussed in Chandrasekhar's paper [1] in detail, and we therefore just outline the method below.

We consider a general case concerning the random variable $\mathbf{F}^B$ and define the following quantities $\mathbf{B}(t)$ and $\mathbf{C}(t)$:

$$\mathbf{B}(t) = \int_0^t \alpha(\tau)\mathbf{F}^B(\tau)d\tau, \qquad \mathbf{C}(t) = \int_0^t \beta(\tau)\mathbf{F}^B(\tau)d\tau. \tag{9.8}$$

The reader may understand the following discussion more easily by regarding $\mathbf{B}(t)$ and $\mathbf{C}(t)$ as quantities such as $\mathbf{r}(t)$ and $\mathbf{v}(t)$. Then $\mathbf{B}(t)$ and $\mathbf{C}(t)$ are specified by the following probability density function $\rho(\mathbf{B}(t), \mathbf{C}(t))$:

$$\rho(\mathbf{B}(t), \mathbf{C}(t)) = \frac{1}{8\pi^3(EG-H^2)^{3/2}} \exp\left\{ -\frac{G\mathbf{B}^2 - 2H\mathbf{B}\cdot\mathbf{C} + E\mathbf{C}^2}{2(EG-H^2)} \right\}, \tag{9.9}$$

in which

$$E = 2m\zeta kT \int_0^t \alpha^2(\tau)d\tau, \quad G = 2m\zeta kT \int_0^t \beta^2(\tau)d\tau, \quad H = 2m\zeta kT \int_0^t \alpha(\tau)\beta(\tau)d\tau. \tag{9.10}$$

Now we apply this method to the present problem.

If we define $\delta\mathbf{r}^B$ and $\delta\mathbf{v}^B$ as

$$\delta\mathbf{r}^B(t) = \mathbf{r}(t) - \mathbf{r}(t_0) - \frac{1}{\zeta}\mathbf{v}(t_0)\left\{1 - e^{-\zeta(t-t_0)}\right\} - \frac{1}{m\zeta}\mathbf{F}(t_0)\left\{(t-t_0) - \frac{1}{\zeta}\left(1 - e^{-\zeta(t-t_0)}\right)\right\}, \tag{9.11}$$

$$\delta\mathbf{v}^B(t) = \mathbf{v}(t) - \mathbf{v}(t_0)e^{-\zeta(t-t_0)} - \frac{1}{m\zeta}\mathbf{F}(t_0)\left\{1 - e^{-\zeta(t-t_0)}\right\}, \tag{9.12}$$

then Eqs. (9.4) and (9.3) reduce to the following simple expressions, respectively:

$$\delta\mathbf{r}^B(t) = \int_{t_0}^t \frac{1}{m\zeta}\left(1 - e^{-\zeta(t-\tau)}\right)\mathbf{F}^B(\tau)d\tau, \tag{9.13}$$

$$\delta\mathbf{v}^B(t) = \int_{t_0}^{t} \frac{1}{m} e^{-\zeta(t-\tau)} \mathbf{F}^B(\tau) d\tau . \tag{9.14}$$

By changing the integral range from $t_0 \sim t$ to $0 \sim t$, Eqs. (9.13) and (9.14) are put in similar form to Eq. (9.8). Hence the stochastic properties of $\delta\mathbf{r}^B$ and $\delta\mathbf{v}^B$ can be specified by the following probability density:

$$\rho(\delta\mathbf{r}^B(t), \delta\mathbf{v}^B(t)) = \frac{1}{8\pi^3 (EG - H^2)^{3/2}} \times \exp\left\{ -\frac{G(\delta\mathbf{r}^B(t))^2 - 2H\delta\mathbf{r}^B(t) \cdot \delta\mathbf{v}^B(t) + E(\delta\mathbf{v}^B(t))^2}{2(EG - H^2)} \right\}, \tag{9.15}$$

in which

$$E = 2m\zeta kT \int_{t_0}^{t} \frac{1}{m^2 \zeta^2} \left(1 - e^{-\zeta(t-\tau)}\right)^2 d\tau = \frac{kT}{m\zeta^2} \left\{ 2\zeta(t - t_0) - 3 + 4e^{-\zeta(t-t_0)} - e^{-2\zeta(t-t_0)} \right\}, \tag{9.16}$$

$$G = 2m\zeta kT \int_{t_0}^{t} \frac{1}{m^2} e^{-2\zeta(t-\tau)} d\tau = \frac{kT}{m} \left(1 - e^{-2\zeta(t-t_0)}\right), \tag{9.17}$$

$$H = 2m\zeta kT \int_{t_0}^{t} \frac{1}{m^2 \zeta} e^{-\zeta(t-\tau)} \left(1 - e^{-\zeta(t-\tau)}\right) d\tau = \frac{kT}{m\zeta} \left(1 - e^{-\zeta(t-t_0)}\right)^2 . \tag{9.18}$$

If $\rho(\delta\mathbf{r}^B, \delta\mathbf{v}^B)$ is decomposed into $\hat{\rho}(\delta x^B, \delta v_x^B)\hat{\rho}(\delta y^B, \delta v_y^B)\hat{\rho}(\delta z^B, \delta v_z^B)$, then $\hat{\rho}(\delta x^B, \delta v_x^B)$, for example, can be expressed as

$$\hat{\rho}(\delta x^B, \delta v_x^B) = \frac{1}{\left\{4\pi^2 (EG - H^2)\right\}^{1/2}} \exp\left\{ -\frac{G(\delta x^B)^2 - 2H\delta x^B \delta v_x^B + E(\delta v_x^B)^2}{2(EG - H^2)} \right\}. \tag{9.19}$$

By using the following new notations

$$\sigma_r^2 = E, \quad \sigma_v^2 = G, \quad c_{rv} = H/(EG)^{1/2}, \tag{9.20}$$

Eq. (9.19) reduces to the bivariate normal distribution [2] written as

$$\hat{\rho}(\delta x^B, \delta v_x^B) = \frac{1}{2\pi\sigma_r\sigma_v(1 - c_{rv}^2)^{1/2}} \exp\left[-\frac{1}{2(1 - c_{rv}^2)} \left\{ \left(\frac{\delta x^B}{\sigma_r}\right)^2 - 2c_{rv}\left(\frac{\delta x^B}{\sigma_r}\right)\left(\frac{\delta v_x^B}{\sigma_v}\right) + \left(\frac{\delta v_x^B}{\sigma_v}\right)^2 \right\} \right]. \tag{9.21}$$

Similarly, the expressions for the other components can straightforwardly be obtained from a similar procedure.

Now we can show a Brownian dynamics algorithm for the particle motion from time t to time $(t+h)$. Firstly, according to the method of generating random numbers which is explained in Appendix A9, $\delta\mathbf{r}^B(t+h)$ and $\delta\mathbf{v}^B(t+h)$ are sampled from the bivariate normal distribution shown in Eq. (9.21). In the present case, the following expressions have to be used:

$$\sigma_r^2 = E = \frac{kT}{m\zeta^2}\left(2\zeta h - 3 + 4e^{-\zeta h} - e^{-2\zeta h}\right), \tag{9.22}$$

$$\sigma_v^2 = G = \frac{kT}{m}\left(1 - e^{-2\zeta h}\right), \tag{9.23}$$

$$c_{rv} = H/(EG)^{1/2} = \frac{1}{\sigma_r\sigma_v}\cdot\frac{kT}{m\zeta}\left(1 - e^{-\zeta h}\right)^2. \tag{9.24}$$

The terms $\delta\mathbf{r}^B(t+h)$ and $\delta\mathbf{v}^B(t+h)$ are quantities due to the random forces. With the sampled values of $\delta\mathbf{r}^B(t+h)$ and $\delta\mathbf{v}^B(t+h)$, the particle position $\mathbf{r}(t+h)$ and velocity $\mathbf{v}(t+h)$ at the next time step $(t+h)$ can be obtained from the following equations:

$$\mathbf{r}(t+h) = \mathbf{r}(t) + \frac{1}{\zeta}\mathbf{v}(t)\left(1 - e^{-\zeta h}\right) + \frac{1}{m\zeta}\mathbf{F}(t)\left\{h - \frac{1}{\zeta}\left(1 - e^{-\zeta h}\right)\right\} + \delta\mathbf{r}^B(t+h), \tag{9.25}$$

$$\mathbf{v}(t+h) = \mathbf{v}(t)e^{-\zeta h} + \frac{1}{m\zeta}\mathbf{F}(t)\left(1 - e^{-\zeta h}\right) + \delta\mathbf{v}^B(t+h). \tag{9.26}$$

If the above procedure is conducted for all the Brownian particles in the dispersion, the simulation advances just by one time step. The friction coefficient $m\zeta$ is usually used as $\zeta=6\pi\eta a/m$ (η the viscosity and a the particle radius) which is from Stokes' drag formula.

If we take the limit of $\zeta\to 0$, Eqs (9.25) and (9.26) reduce to the following equations, respectively:

$$\mathbf{r}(t+h) = \mathbf{r}(t) + h\mathbf{v}(t) + \frac{h^2}{2m}\mathbf{F}(t), \tag{9.27}$$

$$\mathbf{v}(t+h) = \mathbf{v}(t) + \frac{h}{m}\mathbf{F}(t). \tag{9.28}$$

These equations are completely equivalent to Eqs. (7.23) and (7.24). As already pointed out, this shows, therefore, that Eq. (9.26) does not have sufficient accuracy when the friction coefficient is very small. By assuming that the force $\mathbf{F}$ is not constant, but changes lineally during the time interval, a more accurate equation for $\mathbf{v}(t+h)$ can be obtained. That is,

$$\mathbf{v}(t+h)=\mathbf{v}(t)e^{-\zeta h}+\frac{1}{m\zeta}\mathbf{F}(t)\left(1-e^{-\zeta h}\right)+\frac{1}{m\zeta h}(\mathbf{F}(t+h)-\mathbf{F}(t))$$
$$\times\left\{h-\frac{1}{\zeta}\left(1-e^{-\zeta h}\right)\right\}+\mathbf{v}^{B}(t+h). \tag{9.29}$$

Hence, when ζ is small, Eq. (9.29) is used together with Eq. (9.25) instead of Eq. (9.26).

9.2 The Generalized Langevin Equation

In the Langevin equation, hydrodynamic interactions between Brownian particles are not taken into account and the correlation between random forces at different time is assumed to be negligible. However, if colloidal particles are not sufficiently large compared with molecules of the dispersion medium, this approximation of uncorrelated random forces is not acceptable and the generalization of the random forces is necessary. The basic equation governing Brownian motion with generalized random forces is called a generalized Langevin equation and written as

$$m\frac{d\mathbf{v}}{dt}=\mathbf{F}-m\int_{-\infty}^{t}M(t-\tau)\mathbf{v}(\tau)d\tau+\mathbf{F}^{B}, \tag{9.30}$$

in which $M(t)$ is the memory function. If we set $M(t)=2\zeta\delta(t)$, Eq. (9.30) reduces to Eq. (9.1), so that Eq. (9.1) is a special case of Eq. (9.30). The other symbols have already been defined in Eq. (9.1). It should be noted that hydrodynamic interaction between particles is neglected even in Eq. (9.30). Equation (9.30) can be written in component form as

$$m\frac{dv_{\alpha}}{dt}=F_{\alpha}-m\int_{-\infty}^{t}M(t-\tau)v_{\alpha}(\tau)d\tau+F_{\alpha}^{B} \qquad (\alpha=x,y,z). \tag{9.31}$$

The memory function $M(t)$ is related to the random force $\mathbf{F}^{B}(t)$ through the fluctuation-dissipation theorem [3]:

$$\left\langle F_{\alpha}^{B}(t)F_{\alpha}^{B}(0)\right\rangle=M(t)\left\langle(mv_{\alpha})^{2}\right\rangle=mkTM(t) \qquad (\alpha=x,y,z), \tag{9.32}$$

in which T is the system temperature.

Now we show an example of the Brownian dynamics algorithm [4]. If we use the notations $\mathbf{a}(t)$ for the acceleration, $\hat{\mathbf{F}}(t)$ for $\mathbf{F}(t)/m$, $\hat{\mathbf{F}}^{B}(t)$ for $\mathbf{F}^{B}(t)/m$, and $\hat{\mathbf{F}}^{F}(t)$ for $\mathbf{F}^{F}(t)/m$, then Eq. (9.30) is written as

$$\mathbf{a}(t+\tau)=\hat{\mathbf{F}}(t+\tau)-\hat{\mathbf{F}}^{F}(t+\tau)+\hat{\mathbf{F}}^{B}(t+\tau), \tag{9.33}$$

in which $\mathbf{F}^{F}(t)$ is the friction force and the second term, excluding the negative sign, in Eq.

(9.30). $\hat{\mathbf{F}}(t+\tau)$ is expanded in a Taylor series as

$$\hat{\mathbf{F}}(t+\tau)=\hat{\mathbf{F}}(t)+\tau\frac{d\hat{\mathbf{F}}(t)}{dt}+O(\tau^2). \tag{9.34}$$

By taking account of this equation and conducting the integration of Eq. (9.33) from $-\Delta t/2$ to $\Delta t/2$ with respect to τ, the following equation can be obtained:

$$\mathbf{v}(t+\frac{1}{2}\Delta t)=\mathbf{v}(t-\frac{1}{2}\Delta t)+\hat{\mathbf{F}}(t)\Delta t-\int_{-\Delta t/2}^{\Delta t/2}\hat{\mathbf{F}}^F(t+\tau)d\tau+\mathbf{S}(t)+O((\Delta t)^3). \tag{9.35}$$

On the other hand, by conducting firstly the integration of Eq. (9.33) from $-\tau'$ to τ' with respect to τ and secondly from 0 to Δt with respect to τ', the expression for the position $\mathbf{r}(t)$ can be written as

$$\mathbf{r}(t+\Delta t)=2\mathbf{r}(t)-\mathbf{r}(t-\Delta t)+\hat{\mathbf{F}}(t)(\Delta t)^2-\int_0^{\Delta t}\int_{-\tau'}^{\tau'}\hat{\mathbf{F}}^F(t+\tau)d\tau d\tau'+\mathbf{T}(t)+O((\Delta t)^4), \tag{9.36}$$

in which

$$\left.\begin{aligned}\mathbf{S}(t)&=\int_{-\Delta t/2}^{\Delta t/2}\hat{\mathbf{F}}^B(t+\tau)d\tau,\\ \mathbf{T}(t)&=\int_0^{\Delta t}\int_{-\tau'}^{\tau'}\hat{\mathbf{F}}^B(t+\tau)d\tau d\tau'.\end{aligned}\right\} \tag{9.37}$$

Furthermore, $\hat{\mathbf{F}}^B(t+\tau)$ is expanded in a Taylor series as

$$\hat{\mathbf{F}}^B(t+\tau)=\hat{\mathbf{F}}^B(t)+\tau\frac{d\hat{\mathbf{F}}^B(t)}{dt}+\frac{\tau^2}{2}\cdot\frac{d^2\hat{\mathbf{F}}^B(t)}{dt^2}+O(\tau^3). \tag{9.38}$$

The substitution of this equation into Eq. (9.37) leads to the following equation:

$$\left.\begin{aligned}\mathbf{S}(t)&=\hat{\mathbf{F}}^B(t)\Delta t+O((\Delta t)^3),\\ \mathbf{T}(t)&=\hat{\mathbf{F}}^B(t)(\Delta t)^2+O((\Delta t)^4).\end{aligned}\right\} \tag{9.39}$$

On the other hand, the friction force $\hat{\mathbf{F}}^F(t+\tau)$ can be reformed as

$$\begin{aligned}\hat{\mathbf{F}}^F(t+\tau)&=\int_{-\infty}^{t+\tau}M(t+\tau-t')\mathbf{v}(t')dt'=\int_{-\infty}^{\Delta t/2}M(\frac{1}{2}\Delta t-t'')\mathbf{v}(t+\tau-\frac{1}{2}\Delta t+t'')dt''\\ &=\sum_{k=0}^{\infty}\int_{-\Delta t/2}^{\Delta t/2}M(k\Delta t+\frac{1}{2}\Delta t-t')\mathbf{v}(t+\tau-k\Delta t-\frac{1}{2}\Delta t+t')dt'.\end{aligned} \tag{9.40}$$

If the velocity $\mathbf{v}$ and the memory function M are expanded in Taylor series about $(t+\tau-k\Delta t-\Delta t/2)$ and $(k\Delta t+\Delta t/2)$, respectively, we obtain the following equation:

$$\hat{\mathbf{F}}^F(t+\tau)=\Delta t\sum_{k=0}^{\infty}M(k\Delta t+\frac{1}{2}\Delta t)\mathbf{v}(t+\tau-k\Delta t-\frac{1}{2}\Delta t)+\mathrm{O}((\Delta t)^2)\,. \tag{9.41}$$

Additionally, after expanding **v** in a Taylor series about (t-$k\Delta t$-$\Delta t/2$) in this equation, the integral terms in Eqs. (9.35) and (9.36) are evaluated and expressed as

$$\left.\begin{aligned}&\int_{-\Delta t/2}^{\Delta t/2}\hat{\mathbf{F}}^F(t+\tau)d\tau=\hat{\mathbf{F}}^F(t)\Delta t+\mathrm{O}((\Delta t)^3),\\&\int_0^{\Delta t}\int_{-\tau'}^{\tau'}\hat{\mathbf{F}}^F(t+\tau)d\tau d\tau'=\hat{\mathbf{F}}^F(t)(\Delta t)^2+\mathrm{O}((\Delta t)^4).\end{aligned}\right\} \tag{9.42}$$

Thus, by substituting Eqs. (9.39) and (9.42) into Eqs. (9.36) and (9.35), we obtain the final results as

$$\mathbf{r}(t+\Delta t)=2\mathbf{r}(t)-\mathbf{r}(t-\Delta t)+\mathbf{a}(t)(\Delta t)^2+\mathrm{O}((\Delta t)^4)\,, \tag{9.43}$$

$$\mathbf{v}(t+\Delta t/2)=\mathbf{v}(t-\Delta t/2)+\mathbf{a}(t)\Delta t+\mathrm{O}((\Delta t)^3)\,, \tag{9.44}$$

$$\mathbf{a}(t)=\hat{\mathbf{F}}(t)-\hat{\mathbf{F}}^F(t)+\hat{\mathbf{F}}^B(t)\,, \tag{9.45}$$

in which

$$\hat{\mathbf{F}}^F(t)=\Delta t\sum_{k=0}^{\infty}M(k\Delta t+\Delta t/2)\mathbf{v}(t-k\Delta t-\Delta t/2)+\mathrm{O}((\Delta t)^2)\,, \tag{9.46}$$

$$\left\langle \hat{F}_\alpha^B(t)\hat{F}_\alpha^B(0)\right\rangle=\frac{kT}{m}M(t)\qquad(\alpha=x,y,z)\,. \tag{9.47}$$

The Brownian motion of particles can now be simulated according to Eqs. (9.43)~(9.45). It is seen that Eq. (9.43) is equivalent to the Verlet algorithm and Eq. (9.44) equivalent to the leapfrog algorithm in molecular dynamics methods. If $M(t)$ is approximated as an exponential function, the treatment of the memory function becomes straightforward in simulations [5].

The random force $\hat{F}_\alpha^B(t)$ (α=x,y,z) can be regarded as a stochastic variable obeying the normal distribution. If the memory function M can be regarded as zero after time ($n\Delta t$), that is, $\langle \hat{F}_\alpha^B(n\Delta t)\hat{F}_\alpha^B(0)\rangle=kTM(n\Delta t)/m=0$, then the random forces $\hat{F}_{\alpha 1}^B$, $\hat{F}_{\alpha 2}^B,\ldots,$ $\hat{F}_{\alpha n}^B$ have to be obey the following multivariate normal distribution $\rho(\hat{F}_{\alpha 1}^B,\ \hat{F}_{\alpha 2}^B,\ldots,\ \hat{F}_{\alpha n}^B)$:

$$\rho(\hat{F}_{\alpha 1}^B,\hat{F}_{\alpha 2}^B,\cdots,\hat{F}_{\alpha n}^B)=\frac{1}{\left\{(2\pi)^n|\mathbf{D}|\right\}^{1/2}}\exp\left(-\frac{1}{2}\mathbf{x}\cdot\mathbf{D}^{-1}\cdot\mathbf{x}\right), \tag{9.48}$$

in which $\hat{F}_{\alpha i}^B$ means the random force arising at time (t_0+$i\Delta t$), **x** is the vector as **x**=[$\hat{F}_{\alpha 1}^B$, $\hat{F}_{\alpha 2}^B,\ldots,$ $\hat{F}_{\alpha n}^B$], **D** is the $n\times n$ matrix as **D**=[D_{ij}], in which D_{ij}=$\left\langle\hat{F}_{\alpha i}^B\hat{F}_{\alpha j}^B\right\rangle$, $\mathbf{D}^{-1}$ is the inverse matrix of **D**, and $|\mathbf{D}|$ is the determinant. If $\hat{F}_{\alpha 1}^B$, $\hat{F}_{\alpha 2}^B,\ldots,$ $\hat{F}_{\alpha n-1}^B$ are known, $\hat{F}_{\alpha n}^B$ satisfying Eq.

(9.48) can be generated using uniform random numbers by the method shown in Appendix A9.

9.3 The Diffusion Tensor

Before we start discussing a Brownian dynamics method which takes into account the hydrodynamic interactions between particles, we explain the relationship between the particle Brownian motion and the diffusion coefficients. Through this explanation, it is clarified how the diffusion, resistance, and mobility tensors can be related to each other.

9.3.1 The diffusion coefficient for translational motion

We now consider the Brownian motion of a single spherical particle in a quiescent flow field. If a particle is at rest and located at the origin of a coordinate at time t=0, the expression for $\mathbf{r}(t)\mathbf{r}(t)$ can be written from Eq. (9.4) as

$$\mathbf{r}(t)\mathbf{r}(t)=\frac{1}{m^2\zeta^2}\int_0^t\int_0^t \mathbf{F}^B(\tau)\mathbf{F}^B(\tau')(1-e^{-\zeta(t-\tau)})(1-e^{-\zeta(t-\tau')})d\tau d\tau' , \tag{9.49}$$

in which we have assumed that no external force acts on the particle, that is, $\mathbf{F}$=0. It is seen from Eq. (9.2) that the following equation is satisfied:

$$\left\langle \mathbf{F}^B(\tau)\mathbf{F}^B(\tau')\right\rangle = 2m\zeta kT\delta(\tau-\tau')\mathbf{I} . \tag{9.50}$$

The ensemble average of Eq. (9.49) leads to the following equation:

$$\begin{aligned}\left\langle \mathbf{r}(t)\mathbf{r}(t)\right\rangle &= \frac{1}{m^2\zeta^2}\int_0^t\int_0^t 2m\zeta kT(1-e^{-\zeta(t-\tau)})(1-e^{-\zeta(t-\tau')})\delta(\tau-\tau')\mathbf{I}d\tau d\tau' \\ &= \frac{2kT}{m\zeta}\int_0^t (1-e^{-\zeta(t-\tau)})^2\,\mathbf{I}d\tau.\end{aligned} \tag{9.51}$$

If we take t as ζt>>1 and take account of $\xi=m\zeta$, Eq. (9.51) reduces to

$$\left\langle \mathbf{r}(t)\mathbf{r}(t)\right\rangle = 2\frac{kT}{\xi}t\mathbf{I} . \tag{9.52}$$

By applying this equation to the general case that the particle located at $\mathbf{r}(t)$ at time t moves to a new position ($\mathbf{r}(t)+\Delta\mathbf{r}$) at time ($t+\Delta t$), Eq. (9.52) can be reformed as

$$\left\langle \Delta\mathbf{r}\Delta\mathbf{r}\right\rangle = 2\frac{kT}{\xi}\Delta t\mathbf{I} . \tag{9.53}$$

Now we introduce the transition probability $p(\Delta\mathbf{r},\Delta t)$ for the particle movement from $\mathbf{r}(t)$ at time t to ($\mathbf{r}(t)+\Delta\mathbf{r}$) at time ($t+\Delta t$). Firstly, $p(\Delta\mathbf{r},\Delta t)$ has to satisfy the following normalization condition:

$$\int p(\Delta\mathbf{r},\Delta t)d(\Delta\mathbf{r})=1. \tag{9.54}$$

Hence it is clear that the following equation is satisfied:

$$\langle\Delta\mathbf{r}\Delta\mathbf{r}\rangle=\int p(\Delta\mathbf{r},\Delta t)\Delta\mathbf{r}\Delta\mathbf{r}d(\Delta\mathbf{r}). \tag{9.55}$$

Next we introduce the probability density function $n(\mathbf{r},t)$ for the particle position. Since the probability of the particle being found in the infinitesimal range of $\mathbf{r}\sim(\mathbf{r}+\Delta\mathbf{r})$ is denoted by $n(\mathbf{r},t)d\mathbf{r}$, we can write the following expression:

$$n(\mathbf{r},t+\Delta t)=\int n(\mathbf{r}-\Delta\mathbf{r},t)p(\Delta\mathbf{r},\Delta t)d(\Delta\mathbf{r}). \tag{9.56}$$

With the following Taylor series expansions:

$$\left.\begin{aligned} n(\mathbf{r},t+\Delta t)&=n(\mathbf{r},t)+\Delta t\frac{\partial n(\mathbf{r},t)}{\partial t}+\cdots,\\ n(\mathbf{r}-\Delta\mathbf{r},t)&=n(\mathbf{r},t)-\Delta\mathbf{r}\cdot\nabla n+\frac{1}{2}\Delta\mathbf{r}\Delta\mathbf{r}:\nabla\nabla n+\cdots, \end{aligned}\right\} \tag{9.57}$$

Eq. (9.56) can be reformed as

$$\Delta t\frac{\partial n}{\partial t}=-\int(\Delta\mathbf{r}\cdot\Delta n)p(\Delta\mathbf{r},\Delta t)d(\Delta\mathbf{r})+\frac{1}{2}\int(\Delta\mathbf{r}\Delta\mathbf{r}:\nabla\nabla n)p(\Delta\mathbf{r},\Delta t)d(\Delta\mathbf{r})+\cdots, \tag{9.58}$$

in which the first term on the right-hand side is found to be zero because of $p(-\Delta\mathbf{r},t)=p(\Delta\mathbf{r},t)$. The second term can straightforwardly be reformed by considering Eqs. (9.55) and (9.53), and finally Eq. (9.58) reduces to the following diffusion equation:

$$\frac{\partial n}{\partial t}=\frac{kT}{\xi}\nabla^2 n. \tag{9.59}$$

Hence, if the diffusion coefficient is denoted by D, D is related to the friction coefficient $\xi(=m\,\zeta=6\pi\eta a)$ by the following equation:

$$D=kT/\xi. \tag{9.60}$$

With this relation, Eq. (9.53) can be rewritten as

$$\langle\Delta\mathbf{r}\Delta\mathbf{r}\rangle=2D\Delta t\mathbf{I}. \tag{9.61}$$

9.3.2 The diffusion coefficient for rotational motion

Similar expressions to Eqs. (9.60) and (9.61) exist for the rotational Brownian motion. If we define the change in the particle orientation $d\boldsymbol{\varphi}$ by $\boldsymbol{\omega}=d\boldsymbol{\varphi}/dt$, the expression $d\boldsymbol{\varphi}$ can be written as

$$d\boldsymbol{\varphi}=d\varphi_x\boldsymbol{\delta}_x+d\varphi_y\boldsymbol{\delta}_y+d\varphi_z\boldsymbol{\delta}_z. \tag{9.62}$$

It is noted that the relationship between **r** and **v** is very similar to that between **φ** and **ω**. Hence the following relation, which corresponds to Eq. (9.61), is satisfied:

$$\langle \Delta\boldsymbol{\varphi}\Delta\boldsymbol{\varphi} \rangle = 2D^R \Delta t \mathbf{I} . \tag{9.63}$$

Thus the diffusion coefficient D^R for the rotational Brownian motion can be expressed as

$$D^R = kT / \xi^R , \tag{9.64}$$

in which ξ^R is the rotational friction coefficient and expressed as $\xi^R = 8\pi\eta a^3$ for spherical particles.

9.3.3 The diffusion tensor for spherical particles

For spherical particles, the translational and rotational motion are not coupled and there is, therefore, no correlation between $\Delta\mathbf{r}$ and $\Delta\boldsymbol{\varphi}$. That is,

$$\langle \Delta\mathbf{r}\Delta\boldsymbol{\varphi} \rangle = 0 . \tag{9.65}$$

Hence, if the diffusion coefficient for the translational motion is denoted by D^T, then the diffusion tensor $\mathbf{D}^T$ for the translational motion can be expressed, using Eq. (9.61), as

$$\mathbf{D}^T = \begin{bmatrix} D^T_{xx} & D^T_{xy} & D^T_{xz} \\ D^T_{yx} & D^T_{yy} & D^T_{yz} \\ D^T_{zx} & D^T_{zy} & D^T_{zz} \end{bmatrix} = \begin{bmatrix} D^T & 0 & 0 \\ 0 & D^T & 0 \\ 0 & 0 & D^T \end{bmatrix} . \tag{9.66}$$

By replacing the superscripts T with R in this equation, the diffusion tensor $\mathbf{D}^R$ for the rotational motion can be obtained, in which the rotational diffusion coefficient is denoted by D^R. We now use the general notations q_i (i=1,2, …, 6) for the particle position and orientation to write down simple form instead of Eqs. (9.61) and (9.63). That is,

$$\langle \Delta q_i \Delta q_j \rangle = 2D_{ij}\Delta t . \tag{9.67}$$

It is noted that $(q_1,q_2,q_3,q_4,q_5,q_6)$ means $(x,y,z,\varphi_x,\varphi_y,\varphi_z)$. Hence the diffusion matrix **D** for the particle translational and rotational motion can be written as

$$\mathbf{D} = \begin{bmatrix} D_{11} & D_{12} & D_{13} & D_{14} & D_{15} & D_{16} \\ D_{21} & D_{22} & D_{23} & D_{24} & D_{25} & D_{26} \\ D_{31} & D_{32} & D_{33} & D_{34} & D_{35} & D_{36} \\ D_{41} & D_{42} & D_{43} & D_{44} & D_{45} & D_{46} \\ D_{51} & D_{52} & D_{53} & D_{54} & D_{55} & D_{56} \\ D_{61} & D_{62} & D_{63} & D_{64} & D_{65} & D_{66} \end{bmatrix} = \begin{bmatrix} D^T & 0 & 0 & 0 & 0 & 0 \\ 0 & D^T & 0 & 0 & 0 & 0 \\ 0 & 0 & D^T & 0 & 0 & 0 \\ 0 & 0 & 0 & D^R & 0 & 0 \\ 0 & 0 & 0 & 0 & D^R & 0 \\ 0 & 0 & 0 & 0 & 0 & D^R \end{bmatrix} . \tag{9.68}$$

It is seen from the relation in Eq. (9.67) that the diffusion matrix is symmetric.

The mobility matrix **M**, for the motion of a single particle in a quiescent flow field, can be written from Eqs. (5.45), (5.1), and (5.2) as

$$\mathbf{M}=\begin{bmatrix} M_{11} & M_{12} & M_{13} & M_{14} & M_{15} & M_{16} \\ M_{21} & M_{22} & M_{23} & M_{24} & M_{25} & M_{26} \\ M_{31} & M_{32} & M_{33} & M_{34} & M_{35} & M_{36} \\ M_{41} & M_{42} & M_{43} & M_{44} & M_{45} & M_{46} \\ M_{51} & M_{52} & M_{53} & M_{54} & M_{55} & M_{56} \\ M_{61} & M_{62} & M_{63} & M_{64} & M_{65} & M_{66} \end{bmatrix}=\begin{bmatrix} M^T & 0 & 0 & 0 & 0 & 0 \\ 0 & M^T & 0 & 0 & 0 & 0 \\ 0 & 0 & M^T & 0 & 0 & 0 \\ 0 & 0 & 0 & M^R & 0 & 0 \\ 0 & 0 & 0 & 0 & M^R & 0 \\ 0 & 0 & 0 & 0 & 0 & M^R \end{bmatrix}, \tag{9.69}$$

in which M^T and M^R are written as

$$M^T=1/6\pi a, \quad M^R=1/8\pi a^3. \tag{9.70}$$

Similarly, the resistance matrix **R** can be expressed as

$$\mathbf{R}=\begin{bmatrix} R_{11} & R_{12} & R_{13} & R_{14} & R_{15} & R_{16} \\ R_{21} & R_{22} & R_{23} & R_{24} & R_{25} & R_{26} \\ R_{31} & R_{32} & R_{33} & R_{34} & R_{35} & R_{36} \\ R_{41} & R_{42} & R_{43} & R_{44} & R_{45} & R_{46} \\ R_{51} & R_{52} & R_{53} & R_{54} & R_{55} & R_{56} \\ R_{61} & R_{62} & R_{63} & R_{64} & R_{65} & R_{66} \end{bmatrix}=\begin{bmatrix} R^T & 0 & 0 & 0 & 0 & 0 \\ 0 & R^T & 0 & 0 & 0 & 0 \\ 0 & 0 & R^T & 0 & 0 & 0 \\ 0 & 0 & 0 & R^R & 0 & 0 \\ 0 & 0 & 0 & 0 & R^R & 0 \\ 0 & 0 & 0 & 0 & 0 & R^R \end{bmatrix}, \tag{9.71}$$

in which R^T and R^R are written as

$$R^T=6\pi a, \quad R^R=8\pi a^3. \tag{9.72}$$

It is clear that the diffusion matrix **D**, mobility matrix **M**, and resistance matrix **R** are related to each other as follows:

$$\mathbf{D}=\frac{kT}{\eta}\mathbf{M}=\frac{kT}{\eta}\mathbf{R}^{-1}. \tag{9.73}$$

9.3.4 The general relationship between the diffusion, mobility, and resistance matrices

In the preceding section, we have discussed the motion of a single particle. We here consider a system composed of N particles with an arbitrary geometric shape. As before, we use the general notation q_i (i=1,2,...,6N) for the particle positions and orientations, in which (q_1,q_2,q_3) means the position (x,y,z) of particle 1, and $(q_{3N+1},q_{3N+2},q_{3N+3})$ means the direction $(\varphi_x,\varphi_y,\varphi_z)$ of particle 1, etc. With these notations, the diffusion tensor **D** composed of diffusion coefficients can be written as a $6N{\times}6N$ matrix. That is,

$$\mathbf{D} = \begin{bmatrix} D_{11} & D_{12} & \cdots & D_{1,3N} & D_{1,3N+1} & D_{1,3N+2} & \cdots & D_{1,6N} \\ D_{21} & D_{22} & \cdots & D_{2,3N} & D_{2,3N+1} & D_{2,3N+2} & \cdots & D_{2,6N} \\ \vdots & \vdots & & \vdots & \vdots & \vdots & & \vdots \\ D_{3N,1} & D_{3N,2} & \cdots & D_{3N,3N} & D_{3N,3N+1} & D_{3N,3N+2} & \cdots & D_{3N,6N} \\ D_{3N+1,1} & D_{3N+1,2} & \cdots & D_{3N+1,3N} & D_{3N+1,3N+1} & D_{3N+1,3N+2} & \cdots & D_{3N+1,6N} \\ D_{3N+2,1} & D_{3N+2,2} & \cdots & D_{3N+2,3N} & D_{3N+2,3N+1} & D_{3N+2,3N+2} & \cdots & D_{3N+2,6N} \\ \vdots & \vdots & & \vdots & \vdots & \vdots & & \vdots \\ D_{6N,1} & D_{6N,2} & \cdots & D_{6N,3N} & D_{6N,3N+1} & D_{6N,3N+2} & \cdots & D_{6N,6N} \end{bmatrix}. \tag{9.74}$$

Similarly, the mobility matrix **M** and resistance matrix **R** are expressed as

$$\mathbf{M} = \begin{bmatrix} M_{11} & M_{12} & \cdots & M_{1,3N} & M_{1,3N+1} & M_{1,3N+2} & \cdots & M_{1,6N} \\ M_{21} & M_{22} & \cdots & M_{2,3N} & M_{2,3N+1} & M_{2,3N+2} & \cdots & M_{2,6N} \\ \vdots & \vdots & & \vdots & \vdots & \vdots & & \vdots \\ M_{3N,1} & M_{3N,2} & \cdots & M_{3N,3N} & M_{3N,3N+1} & M_{3N,3N+2} & \cdots & M_{3N,6N} \\ M_{3N+1,1} & M_{3N+1,2} & \cdots & M_{3N+1,3N} & M_{3N+1,3N+1} & M_{3N+1,3N+2} & \cdots & M_{3N+1,6N} \\ M_{3N+2,1} & M_{3N+2,2} & \cdots & M_{3N+2,3N} & M_{3N+2,3N+1} & M_{3N+2,3N+2} & \cdots & M_{3N+2,6N} \\ \vdots & \vdots & & \vdots & \vdots & \vdots & & \vdots \\ M_{6N,1} & M_{6N,2} & \cdots & M_{6N,3N} & M_{6N,3N+1} & M_{6N,3N+2} & \cdots & M_{6N,6N} \end{bmatrix}, \tag{9.75}$$

$$\mathbf{R} = \begin{bmatrix} R_{11} & R_{12} & \cdots & R_{1,3N} & R_{1,3N+1} & R_{1,3N+2} & \cdots & R_{1,6N} \\ R_{21} & R_{22} & \cdots & R_{2,3N} & R_{2,3N+1} & R_{2,3N+2} & \cdots & R_{2,6N} \\ \vdots & \vdots & & \vdots & \vdots & \vdots & & \vdots \\ R_{3N,1} & R_{3N,2} & \cdots & R_{3N,3N} & R_{3N,3N+1} & R_{3N,3N+2} & \cdots & R_{3N,6N} \\ R_{3N+1,1} & R_{3N+1,2} & \cdots & R_{3N+1,3N} & R_{3N+1,3N+1} & R_{3N+1,3N+2} & \cdots & R_{3N+1,6N} \\ R_{3N+2,1} & R_{3N+2,2} & \cdots & R_{3N+2,3N} & R_{3N+2,3N+1} & R_{3N+2,3N+2} & \cdots & R_{3N+2,6N} \\ \vdots & \vdots & & \vdots & \vdots & \vdots & & \vdots \\ R_{6N,1} & R_{6N,2} & \cdots & R_{6N,3N} & R_{6N,3N+1} & R_{6N,3N+2} & \cdots & R_{6N,6N} \end{bmatrix}. \tag{9.76}$$

As for the single particle motion, these diffusion, mobility, and resistance matrices can be related to each other by

$$\mathbf{D} = \frac{kT}{\eta}\mathbf{M} = \frac{kT}{\eta}\mathbf{R}^{-1}. \tag{9.77}$$

The displacements Δq_i and Δq_j due to the Brownian motion can be related to the diffusion coefficient D_{ij} by

$$\langle \Delta q_i \Delta q_j \rangle = 2D_{ij}\Delta t . \tag{9.78}$$

It is clear from this relation that the diffusion, mobility, and resistance matrices are all symmetric. The diffusion matrix **D** shown in Eq. (9.74) can be rewritten, using the translational and rotational diffusion tensors $\mathbf{D}_{ij}{}^{T}$ and $\mathbf{D}_{ij}{}^{R}$ for particles i and j, as

$$\mathbf{D} = \begin{bmatrix} \mathbf{D}_{11}^{T} & \mathbf{D}_{12}^{T} & \cdots & \mathbf{D}_{1N}^{T} & \widetilde{\mathbf{D}}_{11}^{C} & \widetilde{\mathbf{D}}_{12}^{C} & \cdots & \widetilde{\mathbf{D}}_{1N}^{C} \\ \mathbf{D}_{21}^{T} & \mathbf{D}_{22}^{T} & \cdots & \mathbf{D}_{2N}^{T} & \widetilde{\mathbf{D}}_{21}^{C} & \widetilde{\mathbf{D}}_{22}^{C} & \cdots & \widetilde{\mathbf{D}}_{2N}^{C} \\ \vdots & \vdots & & \vdots & \vdots & \vdots & & \vdots \\ \mathbf{D}_{N1}^{T} & \mathbf{D}_{N2}^{T} & \cdots & \mathbf{D}_{NN}^{T} & \widetilde{\mathbf{D}}_{N1}^{C} & \widetilde{\mathbf{D}}_{N2}^{C} & \cdots & \widetilde{\mathbf{D}}_{NN}^{C} \\ \mathbf{D}_{11}^{C} & \mathbf{D}_{12}^{C} & \cdots & \mathbf{D}_{1N}^{C} & \mathbf{D}_{11}^{R} & \mathbf{D}_{12}^{R} & \cdots & \mathbf{D}_{1N}^{R} \\ \mathbf{D}_{21}^{C} & \mathbf{D}_{22}^{C} & \cdots & \mathbf{D}_{2N}^{C} & \mathbf{D}_{21}^{R} & \mathbf{D}_{22}^{R} & \cdots & \mathbf{D}_{2N}^{R} \\ \vdots & \vdots & & \vdots & \vdots & \vdots & & \vdots \\ \mathbf{D}_{N1}^{C} & \mathbf{D}_{N2}^{C} & \cdots & \mathbf{D}_{NN}^{C} & \mathbf{D}_{N1}^{R} & \mathbf{D}_{N2}^{R} & \cdots & \mathbf{D}_{NN}^{R} \end{bmatrix}, \tag{9.79}$$

in which the tensors with the superscript C are quantities related to the coupling of the translational and rotational motion, and $\widetilde{\mathbf{D}}_{ij}^{C}$ is related to $\mathbf{D}_{ij}^{C}$ by the following relation:

$$\widetilde{\mathbf{D}}_{ij}^{C} = (\mathbf{D}_{ji}^{C})^{t} . \tag{9.80}$$

In this equation, the superscript t means a transposed tensor. The mobility and resistance matrices can be rewritten in similar form to Eq. (9.79). With the expression for the diffusion matrix shown in Eq. (9.79), the displacements of particles, $\Delta\mathbf{r}_i$ and $\Delta\mathbf{r}_j$, can be related to the diffusion tensor $\mathbf{D}_{ij}^{T}$ by the following expression:

$$\langle \Delta\mathbf{r}_i \Delta\mathbf{r}_j \rangle = 2\mathbf{D}_{ij}^{T}\Delta t . \tag{9.81}$$

Similarly, the changes in the orientation of particles i and j, $\Delta\boldsymbol{\varphi}_i$ and $\Delta\boldsymbol{\varphi}_j$, can be related to the diffusion tensor $\mathbf{D}_{ij}^{R}$ by the following equation:

$$\langle \Delta\boldsymbol{\varphi}_i \Delta\boldsymbol{\varphi}_j \rangle = 2\mathbf{D}_{ij}^{R}\Delta t . \tag{9.82}$$

Furthermore, $\Delta\mathbf{r}_i$ and $\Delta\boldsymbol{\varphi}_j$ are related by

$$\langle \Delta\mathbf{r}_i \Delta\boldsymbol{\varphi}_j \rangle = 2\widetilde{\mathbf{D}}_{ij}^{C}\Delta t . \tag{9.83}$$

Finally, we show the relationship between the diffusion coefficients in Eq. (9.74). For arbitrary constants a and b, the following equation of variance is identically satisfied:

$$\sigma^2(a\Delta q_i + b\Delta q_j) = \langle a^2(\Delta q_i)^2 + b^2(\Delta q_j)^2 + 2ab\Delta q_i \Delta q_j \rangle \geq 0 . \tag{9.84}$$

Hence, by setting $a=(D_{jj})^{1/2}$ and $b=\pm(D_{ii})^{1/2}$ in this equation, the following inequality can be derived:

$$(D_{ii})^{1/2}(D_{jj})^{1/2} \geq D_{ij} \geq -(D_{ii})^{1/2}(D_{jj})^{1/2}. \tag{9.85}$$

9.4 Brownian Dynamics Algorithms with Hydrodynamic Interactions between Particles

The methods explained in Secs. 9.1 and 9.2 are for a dilute colloidal dispersion in which hydrodynamic interactions between Brownian particles are negligible. In the present section, we discuss a Brownian dynamics method which takes into account the hydrodynamic interactions between particles. Hence this Brownian dynamics method is very useful for simulating a non-dilute dispersion. For simplicity, we consider the linear flow field which has been shown in Eq. (4.29).

9.4.1 The translational motion

If rotational Brownian motion is negligible, then the equation of motion for N Brownian particles, which hydrodynamically interact with each other, can be written as a generalized expression of Eq. (9.1):

$$m_i \frac{d\mathbf{v}_i}{dt} = \mathbf{F}_i^H + \mathbf{F}_i^P + \mathbf{F}_i^B, \tag{9.86}$$

in which $\mathbf{F}_i^H$ is the force exerted on particle i by the ambient fluid, $\mathbf{F}_i^P$ is a non-hydrodynamic force due to the potential energy between particles, and $\mathbf{F}_i^B$ is the random force inducing the Brownian motion. If Eq.(6.6) is taken into account for $\mathbf{F}_i^H$, the above three forces can be written as

$$\mathbf{F}_i^H = -\eta \sum_{j=1}^{N} \mathbf{R}_{ij} \cdot (\mathbf{v}_j - \mathbf{U}(\mathbf{r}_j)) + \eta \widetilde{\mathbf{G}}_i' : \mathbf{E}, \tag{9.87}$$

$$\mathbf{F}_i^P = -\sum_{j=1(\neq i)}^{N} \frac{\partial}{\partial \mathbf{r}_i} u_{ij} + \mathbf{F}_i^{(ext)}, \tag{9.88}$$

$$\left\langle \mathbf{F}_i^B(t) \right\rangle = 0, \quad \left\langle \mathbf{F}_i^B(t)\mathbf{F}_j^B(t') \right\rangle = 2kT\eta \mathbf{R}_{ij}\delta(t-t'), \tag{9.89}$$

in which $\mathbf{R}_{ij}$ is the resistance tensor, u_{ij} is the potential energy between particles i and j, $\mathbf{F}_i^{(ext)}$ is the external force acting on particle i, k is Boltzmann's constant, and T is the temperature of the system. The random force $\mathbf{F}_i^B$ can also be written as

$$\mathbf{F}_i^B = \sum_{j=1}^{N} \boldsymbol{\alpha}_{ij} \cdot \hat{\mathbf{F}}_j^B, \tag{9.90}$$

in which the tensor $\boldsymbol{\alpha}_{ij}$ and the vector $\hat{\mathbf{F}}_i^B$ have to satisfy the following relations:

$$\mathbf{R}_{ij} = \frac{1}{\eta kT}\sum_{l=1}^{N}\boldsymbol{\alpha}_{il}\cdot\boldsymbol{\alpha}_{jl}\,, \tag{9.91}$$

$$\left\langle \hat{\mathbf{F}}_i^B(t)\right\rangle = 0, \quad \left\langle \hat{\mathbf{F}}_i^B(t)\hat{\mathbf{F}}_j^B(t')\right\rangle = 2\delta_{ij}\delta(t-t')\mathbf{I}\,, \tag{9.92}$$

in which δ_{ij} is Kronecker's delta, and **I** is the unit tensor. It is straightforward to verify that if the tensor $\boldsymbol{\alpha}_{ij}$ is symmetric, $\mathbf{F}_i^B$ expressed in Eq. (9.90) satisfies Eq. (9.89).

By replacing the superscripts *i* with *j* in Eq. (9.86), multiplying this equation by $\mathbf{D}_{ij}/kT$, and summing both sides with respect to *j*, the following equation is obtained:

$$\sum_{j=1}^{N}\boldsymbol{\tau}_{ij}\cdot\frac{d\mathbf{v}_j}{dt} = -(\mathbf{v}_i - \mathbf{U}(\mathbf{r}_i)) + \frac{\eta}{kT}\sum_{j=1}^{N}\mathbf{D}_{ij}\cdot(\widetilde{\mathbf{G}}'_j:\mathbf{E}) + \frac{1}{kT}\sum_{j=1}^{N}\mathbf{D}_{ij}\cdot\mathbf{F}_j^P + \sum_{j=1}^{N}\boldsymbol{\sigma}_{ij}\cdot\hat{\mathbf{F}}_j^B\,, \tag{9.93}$$

in which

$$\sum_{l=1}^{N}\mathbf{R}_{il}\cdot\mathbf{D}_{lj} = \sum_{l=1}^{N}\mathbf{D}_{il}\cdot\mathbf{R}_{lj} = \frac{kT}{\eta}\delta_{ij}\mathbf{I}\,, \tag{9.94}$$

$$\boldsymbol{\sigma}_{ij} = \frac{1}{kT}\sum_{l=1}^{N}\mathbf{D}_{il}\cdot\boldsymbol{\alpha}_{lj}\,, \quad \boldsymbol{\tau}_{ij} = \frac{m_j}{kT}\mathbf{D}_{ij}\,, \tag{9.95}$$

in which $\mathbf{D}_{ij}$ is the diffusion tensor. Equation (9.94) is due to the fact that the diffusion matrix **D** (composed of small matrices $\mathbf{D}_{ij}$) and the resistance matrix **R** (composed of small matrices $\mathbf{R}_{ij}$) are related by $\mathbf{D}=kT\mathbf{R}^{-1}/\eta$. **R** and **D**, for example, for a three particle system, are expressed as

$$\mathbf{R} = \begin{bmatrix}\mathbf{R}_{11} & \mathbf{R}_{12} & \mathbf{R}_{13}\\ \mathbf{R}_{21} & \mathbf{R}_{22} & \mathbf{R}_{23}\\ \mathbf{R}_{31} & \mathbf{R}_{32} & \mathbf{R}_{33}\end{bmatrix}, \quad \mathbf{D} = \begin{bmatrix}\mathbf{D}_{11} & \mathbf{D}_{12} & \mathbf{D}_{13}\\ \mathbf{D}_{21} & \mathbf{D}_{22} & \mathbf{D}_{23}\\ \mathbf{D}_{31} & \mathbf{D}_{32} & \mathbf{D}_{33}\end{bmatrix}. \tag{9.96}$$

According to Ermak and McCammon's mathematical manipulation [7] concerning Eq. (9.93), the following expression can be derived:

$$\begin{aligned}\mathbf{r}_i(t+\Delta t) = \mathbf{r}_i(t) + \mathbf{U}(\mathbf{r}_i)\Delta t + \frac{\eta}{kT}\sum_{j=1}^{N}\mathbf{D}_{ij}(t)\cdot(\widetilde{\mathbf{G}}'_j(t):\mathbf{E})\Delta t \\ + \frac{1}{kT}\sum_{j=1}^{N}\mathbf{D}_{ij}(t)\cdot\mathbf{F}_j^P(t)\Delta t + \sum_{j=1}^{N}\frac{\partial}{\partial\mathbf{r}_j}\cdot(\mathbf{D}_{ij}(t))\Delta t + \Delta\mathbf{r}_i^B(t),\end{aligned} \tag{9.97}$$

$$\left\langle \Delta\mathbf{r}_i^B(t)\right\rangle = 0, \quad \left\langle (\Delta\mathbf{r}_i^B(t))(\Delta\mathbf{r}_j^B(t))\right\rangle = 2\mathbf{D}_{ij}(t)\Delta t\,. \tag{9.98}$$

Equation (9.97) is valid under the condition that the time interval Δt is small so that the forces acting among particles and the gradient of the diffusion tensor can be regarded as constant, although Δt is sufficiently long compared with the relaxation time ($\approx m/6\pi\eta a$, a is the particle radius) for the Brownian motion. In other words, this equation has been derived under the assumption that the relaxation time of the particle momenta is much shorter than that of the change in the position. This means that the rapid change in the particle velocity due to the random force is sufficiently flatted during the time interval Δt.

By considering the forward finite difference approximation for the velocity $\mathbf{v}_i(t)$, it is easily seen that Eq. (9.97) can be written as

$$\begin{aligned}\mathbf{v}_i(t) = \mathbf{U}(\mathbf{r}_i) + \frac{1}{kT}\sum_{j=1}^{N}\mathbf{D}_{ij}(t)\cdot\mathbf{F}_j^P(t) + \frac{\eta}{kT}\sum_{j=1}^{N}\mathbf{D}_{ij}(t)\cdot(\widetilde{\mathbf{G}}'_j(t):\mathbf{E}) \\ + \sum_{j=1}^{N}\frac{\partial}{\partial\mathbf{r}_j}\cdot(\mathbf{D}_{ij}(t)) + \Delta\mathbf{v}_i^B(t),\end{aligned} \tag{9.99}$$

$$\left\langle \Delta\mathbf{v}_i^B(t)\right\rangle = 0, \qquad \left\langle (\Delta\mathbf{v}_i^B(t))(\Delta\mathbf{v}_j^B(t))\right\rangle = 2\mathbf{D}_{ij}(t)/\Delta t\,. \tag{9.100}$$

This is just the equation which Tough et al. [8] derived. Equation (9.99) is more useful than Eq. (9.97) for understanding the physical meaning of each term. The second term on the right-hand side in Eq. (9.99) is the sum of the friction term for i=j and the hydrodynamic forces exerted on particle i by the ambient fluid for i≠j, which originally result from the motion of the other particles. The third term is due to an applied linear flow field. It is noted that $\mathbf{v}_i(t)$ expressed in Eq. (9.99) doesn't seem to be equivalent to the physical velocity since it depends on the time interval through $\Delta\mathbf{v}_i^B$. If the Brownian motion is negligible, Eq. (9.99) reduces to Eq. (6.21). It is, therefore, clear that the expression in Eq. (9.97) is based on the approximation of the additivity of velocities.

If we use the Oseen tensor in Eq. (5.57) or the Rotne-Prager tensor in (5.61), the following relation is satisfied:

$$\sum_{j=1}^{N}\frac{\partial}{\partial\mathbf{r}_j}\cdot(\mathbf{D}_{ij}) = 0\,, \tag{9.101}$$

so that the treatment of Eq. (9.99) becomes easier. However, more exact diffusion tensors do not always satisfy this characteristic. Since stochastic variables satisfying the condition in Eq. (9.100) obey the multivariate normal distribution, we have to use the method of generating such stochastic variables in an actual simulation, which is explained in Appendix A9.

9.4.2 The translational and rotational motion

If both translational and rotational motion of spherical colloidal particles are taken into account, the equation of motion can be written in simple matrix form as

$$
\begin{bmatrix} m_1 d\mathbf{v}_1/dt \\ m_2 d\mathbf{v}_2/dt \\ \vdots \\ m_N d\mathbf{v}_N/dt \\ I_1 d\boldsymbol{\omega}_1/dt \\ I_2 d\boldsymbol{\omega}_2/dt \\ \vdots \\ I_N d\boldsymbol{\omega}_N/dt \end{bmatrix} = -\eta \begin{bmatrix} \mathbf{R}_{11}^T & \mathbf{R}_{12}^T & \cdots & \mathbf{R}_{1N}^T & \widetilde{\mathbf{R}}_{11}^C & \widetilde{\mathbf{R}}_{12}^C & \cdots & \widetilde{\mathbf{R}}_{1N}^C \\ \mathbf{R}_{21}^T & \mathbf{R}_{22}^T & \cdots & \mathbf{R}_{2N}^T & \widetilde{\mathbf{R}}_{21}^C & \widetilde{\mathbf{R}}_{22}^C & \cdots & \widetilde{\mathbf{R}}_{2N}^C \\ \vdots & \vdots & & \vdots & \vdots & \vdots & & \vdots \\ \mathbf{R}_{N1}^T & \mathbf{R}_{N2}^T & \cdots & \mathbf{R}_{NN}^T & \widetilde{\mathbf{R}}_{N1}^C & \widetilde{\mathbf{R}}_{N2}^C & \cdots & \widetilde{\mathbf{R}}_{NN}^C \\ \mathbf{R}_{11}^C & \mathbf{R}_{12}^C & \cdots & \mathbf{R}_{1N}^C & \mathbf{R}_{11}^R & \mathbf{R}_{12}^R & \cdots & \mathbf{R}_{1N}^R \\ \mathbf{R}_{21}^C & \mathbf{R}_{22}^C & \cdots & \mathbf{R}_{2N}^C & \mathbf{R}_{21}^R & \mathbf{R}_{22}^R & \cdots & \mathbf{R}_{2N}^R \\ \vdots & \vdots & & \vdots & \vdots & \vdots & & \vdots \\ \mathbf{R}_{N1}^C & \mathbf{R}_{N2}^C & \cdots & \mathbf{R}_{NN}^C & \mathbf{R}_{N1}^R & \mathbf{R}_{N2}^R & \cdots & \mathbf{R}_{NN}^R \end{bmatrix} \begin{bmatrix} \mathbf{v}_1 - \mathbf{U}(\mathbf{r}_1) \\ \mathbf{v}_2 - \mathbf{U}(\mathbf{r}_2) \\ \vdots \\ \mathbf{v}_N - \mathbf{U}(\mathbf{r}_N) \\ \boldsymbol{\omega}_1 - \boldsymbol{\Omega} \\ \boldsymbol{\omega}_2 - \boldsymbol{\Omega} \\ \vdots \\ \boldsymbol{\omega}_N - \boldsymbol{\Omega} \end{bmatrix}
$$

$$
+\eta \begin{bmatrix} \widetilde{\mathbf{G}}_1' : \mathbf{E} \\ \widetilde{\mathbf{G}}_2' : \mathbf{E} \\ \vdots \\ \widetilde{\mathbf{G}}_N' : \mathbf{E} \\ \widetilde{\mathbf{H}}_1' : \mathbf{E} \\ \widetilde{\mathbf{H}}_2' : \mathbf{E} \\ \vdots \\ \widetilde{\mathbf{H}}_N' : \mathbf{E} \end{bmatrix} + \begin{bmatrix} \mathbf{F}_1^P \\ \mathbf{F}_2^P \\ \vdots \\ \mathbf{F}_N^P \\ \mathbf{T}_1^P \\ \mathbf{T}_2^P \\ \vdots \\ \mathbf{T}_N^P \end{bmatrix} + \begin{bmatrix} \boldsymbol{\alpha}_{11}^T & \boldsymbol{\alpha}_{12}^T & \cdots & \boldsymbol{\alpha}_{1N}^T & \widetilde{\boldsymbol{\alpha}}_{11}^C & \widetilde{\boldsymbol{\alpha}}_{12}^C & \cdots & \widetilde{\boldsymbol{\alpha}}_{1N}^C \\ \boldsymbol{\alpha}_{21}^T & \boldsymbol{\alpha}_{22}^T & \cdots & \boldsymbol{\alpha}_{2N}^T & \widetilde{\boldsymbol{\alpha}}_{21}^C & \widetilde{\boldsymbol{\alpha}}_{22}^C & \cdots & \widetilde{\boldsymbol{\alpha}}_{2N}^C \\ \vdots & \vdots & & \vdots & \vdots & \vdots & & \vdots \\ \boldsymbol{\alpha}_{N1}^T & \boldsymbol{\alpha}_{N2}^T & \cdots & \boldsymbol{\alpha}_{NN}^T & \widetilde{\boldsymbol{\alpha}}_{N1}^C & \widetilde{\boldsymbol{\alpha}}_{N2}^C & \cdots & \widetilde{\boldsymbol{\alpha}}_{NN}^C \\ \boldsymbol{\alpha}_{11}^C & \boldsymbol{\alpha}_{12}^C & \cdots & \boldsymbol{\alpha}_{1N}^C & \boldsymbol{\alpha}_{11}^R & \boldsymbol{\alpha}_{12}^R & \cdots & \boldsymbol{\alpha}_{1N}^R \\ \boldsymbol{\alpha}_{21}^C & \boldsymbol{\alpha}_{22}^C & \cdots & \boldsymbol{\alpha}_{2N}^C & \boldsymbol{\alpha}_{21}^R & \boldsymbol{\alpha}_{22}^R & \cdots & \boldsymbol{\alpha}_{2N}^R \\ \vdots & \vdots & & \vdots & \vdots & \vdots & & \vdots \\ \boldsymbol{\alpha}_{N1}^C & \boldsymbol{\alpha}_{N2}^C & \cdots & \boldsymbol{\alpha}_{NN}^C & \boldsymbol{\alpha}_{N1}^R & \boldsymbol{\alpha}_{N2}^R & \cdots & \boldsymbol{\alpha}_{NN}^R \end{bmatrix} \begin{bmatrix} \hat{\mathbf{F}}_1^B \\ \hat{\mathbf{F}}_2^B \\ \vdots \\ \hat{\mathbf{F}}_N^B \\ \hat{\mathbf{T}}_1^B \\ \hat{\mathbf{T}}_2^B \\ \vdots \\ \hat{\mathbf{T}}_N^B \end{bmatrix}, \qquad (9.102)
$$

in which I_i (i=1,2,…,N) is the inertia moment of particle i which is a scalar quantity because we here consider a spherical particle system. For particles with a more general geometric shape, the inertia moment becomes a second-rank tensor and in this case Eq. (9.102) has to be written in more general form. As explained before, the first matrix on the right-hand side in Eq. (9.102) is the resistance matrix **R** composed of resistance tensors. If the last matrix in Eq. (9.102) which is composed of $\boldsymbol{\alpha}_{ij}^T$, $\boldsymbol{\alpha}_{ij}^C$, $\widetilde{\boldsymbol{\alpha}}_{ij}^C$, and $\boldsymbol{\alpha}_{ij}^R$ (i,j=1,2,…,N) is denoted by $\boldsymbol{\alpha}$, then $\boldsymbol{\alpha}$ can be related to the resistance matrix **R**, similarly to Eq. (9.91), as

$$
\mathbf{R} = \frac{1}{\eta kT}\boldsymbol{\alpha}\cdot\boldsymbol{\alpha}^t, \qquad (9.103)
$$

in which the superscript t means a transposed matrix, and $\boldsymbol{\alpha}_{ij}^T$, $\boldsymbol{\alpha}_{ij}^R$, $\boldsymbol{\alpha}_{ij}^C$ and $\widetilde{\boldsymbol{\alpha}}_{ij}^C$ have the following characteristics:

$$
\left.\begin{aligned} \boldsymbol{\alpha}_{ij}^T &= (\boldsymbol{\alpha}_{ji}^T)^t \quad (i,j=1,2,\cdots,N), \\ \boldsymbol{\alpha}_{ij}^R &= (\boldsymbol{\alpha}_{ji}^R)^t \quad (i,j=1,2,\cdots,N), \\ \widetilde{\boldsymbol{\alpha}}_{ij}^C &= (\boldsymbol{\alpha}_{ji}^C)^t \quad (i,j=1,2,\cdots,N). \end{aligned}\right\} \qquad (9.104)
$$

This means that the matrix $\boldsymbol{\alpha}$ is symmetric. $\hat{\mathbf{F}}_i^B$ and $\hat{\mathbf{T}}_i^B$ which induce the Brownian motion have the following properties:

$$\left.\begin{aligned}&\left\langle \hat{\mathbf{F}}_i^B(t)\hat{\mathbf{F}}_j^B(t')\right\rangle = 2\delta_{ij}\delta(t-t')\mathbf{I},\\&\left\langle \hat{\mathbf{T}}_i^B(t)\hat{\mathbf{T}}_j^B(t')\right\rangle = 2\delta_{ij}\delta(t-t')\mathbf{I},\\&\left\langle \hat{\mathbf{F}}_i^B(t)\hat{\mathbf{T}}_j^B(t')\right\rangle = 0.\end{aligned}\right\} \tag{9.105}$$

Hence it is straightforwardly shown that the generalized form of the Ermak-McCammon algorithm, taking into account the translational and rotational Brownian motion, can be expressed as [9,10]

$$\begin{aligned}\mathbf{r}_i(t+\Delta t) = {}& \mathbf{r}_i(t) + \mathbf{U}(\mathbf{r}_i)\Delta t + \frac{\eta}{kT}\sum_{j=1}^{N}\mathbf{D}_{ij}^T(t)\cdot(\widetilde{\mathbf{G}}'_j(t):\mathbf{E})\Delta t + \frac{\eta}{kT}\sum_{j=1}^{N}\widetilde{\mathbf{D}}_{ij}^C(t)\cdot(\widetilde{\mathbf{H}}'_j(t):\mathbf{E})\Delta t\\&+\frac{1}{kT}\sum_{j=1}^{N}\mathbf{D}_{ij}^T(t)\cdot\mathbf{F}_j^P(t)\Delta t + \frac{1}{kT}\sum_{j=1}^{N}\widetilde{\mathbf{D}}_{ij}^C(t)\cdot\mathbf{T}_j^P(t)\Delta t\\&+\sum_{j=1}^{N}\frac{\partial}{\partial\mathbf{r}_j}\cdot(\mathbf{D}_{ij}^T(t))\Delta t + \Delta\mathbf{r}_i^B(t) \qquad (i=1,2,\cdots,N),\end{aligned} \tag{9.106}$$

$$\begin{aligned}\boldsymbol{\varphi}_i(t+\Delta t) = {}& \boldsymbol{\varphi}_i(t) + \boldsymbol{\Omega}\Delta t + \frac{\eta}{kT}\sum_{j=1}^{N}\mathbf{D}_{ij}^C(t)\cdot(\widetilde{\mathbf{G}}'_j(t):\mathbf{E})\Delta t + \frac{\eta}{kT}\sum_{j=1}^{N}\mathbf{D}_{ij}^R(t)\cdot(\widetilde{\mathbf{H}}'_j(t):\mathbf{E})\Delta t\\&+\frac{1}{kT}\sum_{j=1}^{N}\mathbf{D}_{ij}^C(t)\cdot\mathbf{F}_j^P(t)\Delta t + \frac{1}{kT}\sum_{j=1}^{N}\mathbf{D}_{ij}^R(t)\cdot\mathbf{T}_j^P(t)\Delta t\\&+\sum_{j=1}^{N}\frac{\partial}{\partial\mathbf{r}_j}\cdot(\mathbf{D}_{ij}^C(t))\Delta t + \Delta\boldsymbol{\varphi}_i^B(t) \qquad (i=1,2,\cdots,N).\end{aligned} \tag{9.107}$$

The fact that the partial differential with respect to the orientation angle $\boldsymbol{\varphi}_i$ becomes zero has been taken into account in this equation since the diffusion tensors are not dependent on the orientation of particles for a spherical particle system. The displacements $\Delta\mathbf{r}_i^B$ and $\Delta\boldsymbol{\varphi}_i^B$ due to Brownian motion satisfy the following relations:

$$\left.\begin{aligned}&\left\langle \Delta\mathbf{r}_i^B(t)\right\rangle = 0, \quad \left\langle (\Delta\mathbf{r}_i^B(t))(\Delta\mathbf{r}_j^B(t))\right\rangle = 2\mathbf{D}_{ij}^T(t)\Delta t,\\&\left\langle \Delta\boldsymbol{\varphi}_i^B(t)\right\rangle = 0, \quad \left\langle (\Delta\boldsymbol{\varphi}_i^B(t))(\Delta\boldsymbol{\varphi}_j^B(t))\right\rangle = 2\mathbf{D}_{ij}^R(t)\Delta t,\\&\left\langle (\Delta\mathbf{r}_i^B(t))(\Delta\boldsymbol{\varphi}_j^B(t))\right\rangle = 2\widetilde{\mathbf{D}}_{ij}^C(t)\Delta t.\end{aligned}\right\} \tag{9.108}$$

The method of using Eqs. (9.106), (9.107), and (9.108) is the Ermak-MacCammon-type algorithm for the translational and rotational motion of colloidal particles. Hence, by determining the displacements $\Delta\mathbf{r}_i^B$ and $\Delta\boldsymbol{\varphi}_i^B$ according to Eq. (9.108) in terms of uniform random numbers, the particle positions and directions at the next time can be evaluated. Since the random variables are related to each other by Eq. (9.108), these variables have to obey the multivariate normal distribution. The method of generating such random variables may be found in Appendix A9.

It is seen from Eqs. (9.106), (9.107), and (A9.18) that generating the random numbers $\Delta\mathbf{r}_i^B$ and $\Delta\boldsymbol{\varphi}_i^B$ (i=1,2, ...,N) is a considerable task for a computer. Also, a computer will require a considerable memory capacity to save the components of the diffusion tensors and those of L_{ij} in Eq. (A9.19), even if the symmetry of the diffusion matrix and tensors is taken into account. Hence Brownian dynamics simulations are generally very expensive, from a computation point of view, and therefore usually restricted to a much smaller particle system than Stokesian dynamics simulations.

9.5 The nondimensionalization method

As in Sec. 8.3, we consider a spherical particle system subjected to a simple shear flow. The representative values which have already been shown in Sec. 8.3 and the method of nondimensionalizing the resistance and mobility tensors are also valid for Brownian dynamics simulations. However, since the temperature T which characterizes the Brownian motion is included in Brownian dynamics algorithms, the ratio of the representative viscous shear force to the representative Brownian random force appears as a nondimensional parameter. This nondimensional number is called the Péclet number and usually denoted by Pe, and is defined as

$$Pe = 6\pi a^3\eta\dot{\gamma} / kT . \tag{9.109}$$

This expression can straightforwardly be derived by nondimensionalizing Eq. (9.97) or (9.106). It is seen from this procedure (or Eqs. (9.137) and (9.138)) that Brownian motion is negligible for Pe>>1, but on the other hand, the particle behavior in a simple shear flow is governed by the Brownian motion for Pe<<1.

The diffusion tensors $\mathbf{D}_{ij}^T$, $\mathbf{D}_{ij}^R$, and $\mathbf{D}_{ij}^C$ are nondimensionalized as

$$\mathbf{D}_{ij}^{T*} = \frac{6\pi a\eta}{kT}\mathbf{D}_{ij}^T, \quad \mathbf{D}_{ij}^{R*} = \frac{8\pi a^3\eta}{kT}\mathbf{D}_{ij}^R, \quad \mathbf{D}_{ij}^{C*} = \frac{4\pi a^2\eta}{kT}\mathbf{D}_{ij}^C, \tag{9.110}$$

in which the superscript * means nondimensional quantities. These expressions are also valid for j=i.

9.6 The Brownian Dynamics Algorithm for Axisymmetric Particles

Finally, we explain a Brownian dynamics method for a system composed of axisymmetric particles such as spheroids and spherocylinders. Since the treatment of hydrodynamic interactions between non-spherical particles is still very difficult and under development, we here consider only the friction term between particles and the ambient fluid, not the hydrodynamic interactions between particles. These methods can be used for simulating a sufficiently dilute dispersion and the results, obtained by these methods, may be very useful as a first order approximation for a non-dilute dispersion.

We consider an axisymmetric particle shown in Fig.5.1(a) moving and rotating in a simple shear flow shown in Fig. 8.1. For this axisymmetric particle, the translational motion does not couple with the rotational one, so that the translational Brownian motion of the particle can be written from Eq. (9.106) as

$$\mathbf{r}(t+\Delta t)=\mathbf{r}(t)+\mathbf{U}(\mathbf{r})\Delta t+\frac{1}{kT}\mathbf{D}^{T}\cdot\mathbf{F}^{P}(t)\Delta t+\Delta\mathbf{r}^{B}(t), \tag{9.111}$$

in which $\Delta\mathbf{r}^B$ is the displacement due to the translational Brownian motion and has the following properties:

$$\left\langle\Delta\mathbf{r}^{B}(t)\right\rangle=0, \quad \left\langle(\Delta\mathbf{r}^{B}(t))(\Delta\mathbf{r}^{B}(t))\right\rangle=2\mathbf{D}^{T}(t)\Delta t. \tag{9.112}$$

In Eq. (9.111), $\mathbf{U}(\mathbf{r})$ is the simple shear flow field before dispersing colloidal particles, $\mathbf{F}^P$ is the non-hydrodynamic force acting on the particle such as magnetic forces, and $\mathbf{D}^T$ is the translational diffusion tensor of an axisymmetric particle, which is related to the resistance tensor $\mathbf{R}^T$ and the resistance functions by the following relations:

$$\mathbf{D}^{T}=\frac{kT}{\eta}(\mathbf{R}^{T})^{-1}=\frac{kT}{\eta}\left\{(X^{A})^{-1}\mathbf{nn}+(Y^{A})^{-1}(\mathbf{I}-\mathbf{nn})\right\}, \tag{9.113}$$

in which $\mathbf{n}$ is a unit vector denoting the orientation of the principal axis of the particle. It is noted that the subscript i has been dropped since a dilute system is here considered and hydrodynamic interactions between particles are not taken into account.

Additionally the expression for the rotational Brownian motion can be written from Eq. (9.107) as

$$\boldsymbol{\varphi}(t+\Delta t)=\boldsymbol{\varphi}(t)+\boldsymbol{\Omega}(t)\Delta t+\frac{1}{kT}\mathbf{D}^{R}\cdot\mathbf{T}^{P}(t)\Delta t-\frac{Y^{H}}{Y^{C}}(\boldsymbol{\varepsilon}\cdot\mathbf{nn}):\mathbf{E}\Delta t+\Delta\boldsymbol{\varphi}^{B}(t), \tag{9.114}$$

in which the vector $\boldsymbol{\varphi}$ has already been defined in Sec. 9.3.2 as $d\boldsymbol{\varphi}/dt=\boldsymbol{\omega}$. $\Delta\boldsymbol{\varphi}^B$ is the angular displacement due to the rotational Brownian motion and has the following properties:

$$\left\langle\Delta\boldsymbol{\varphi}^{B}(t)\right\rangle=0, \quad \left\langle(\Delta\boldsymbol{\varphi}^{B}(t))(\Delta\boldsymbol{\varphi}^{B}(t))\right\rangle=2\mathbf{D}^{R}(t)\Delta t. \tag{9.115}$$

In Eq. (9.114), $\mathbf{D}^R$ is the rotational diffusion tensor for an axisymmetric particle and related to the rotational resistance tensor $\mathbf{R}^R$:

$$\mathbf{D}^R = \frac{kT}{\eta}(\mathbf{R}^R)^{-1} = \frac{kT}{\eta}\left\{(X^C)^{-1}\mathbf{nn} + (Y^C)^{-1}(\mathbf{I}-\mathbf{nn})\right\}. \tag{9.116}$$

For an axisymmetric particle, the translational motion can be decomposed into the motion parallel and normal to the particle axis. If axisymmetric particles such as spherocylinders and spheroids have a large aspect ratio, then it is sufficient to take into account only the rotational motion about an axis through the center of mass that is normal to the particle axis of symmetry. With these characteristics of an axisymmetric particle, we show below how Eqs. (9.111) and (9.114) can be rewritten in different form.

First we consider the translational motion. If the mass center of the particle, $\mathbf{r}(t)$, is expressed as the sum of the vectors parallel and normal to the axis, $\mathbf{r}_{\parallel}(t)$ and $\mathbf{r}_{\perp}(t)$, that is, $\mathbf{r}(t)= \mathbf{r}_{\parallel}(t) + \mathbf{r}_{\perp}(t)$, then these position vectors can be written as

$$\left.\begin{aligned}
\mathbf{r}_{\parallel}(t+\Delta t) &= \mathbf{r}_{\parallel}(t) + \mathbf{U}_{\parallel}(t)\Delta t + \frac{1}{kT}D_{\parallel}^{T}\mathbf{F}_{\parallel}^{P}(t)\Delta t + \Delta r_{\parallel}^{B}\mathbf{n}(t), \\
\mathbf{r}_{\perp}(t+\Delta t) &= \mathbf{r}_{\perp}(t) + \mathbf{U}_{\perp}(t)\Delta t + \frac{1}{kT}D_{\perp}^{T}\mathbf{F}_{\perp}^{P}(t)\Delta t + \Delta r_{\perp 1}^{B}\mathbf{n}_{\perp 1}(t) + \Delta r_{\perp 2}^{B}\mathbf{n}_{\perp 2}(t),
\end{aligned}\right\} \tag{9.117}$$

in which $\mathbf{n}_{\perp 1}$ and $\mathbf{n}_{\perp 2}$ are one set of the unit vectors normal to the particle axis, then $\mathbf{F}_{\parallel}^{P}$ and $\mathbf{F}_{\perp}^{P}$ are the forces acting on the particle parallel and normal to the particle axis, respectively, and $\mathbf{U}_{\parallel}$ and $\mathbf{U}_{\perp}$ are the velocities of a given flow; the subscripts have the same meaning as $\mathbf{F}_{\parallel}^{P}$ and $\mathbf{F}_{\perp}^{P}$. These quantities are expressed as

$$\left.\begin{aligned}
\mathbf{F}_{\parallel}^{P} &= (\mathbf{F}^{P}\cdot\mathbf{n})\mathbf{n}, \qquad \mathbf{F}_{\perp}^{P} = \mathbf{F}^{P} - (\mathbf{F}^{P}\cdot\mathbf{n})\mathbf{n}, \\
\mathbf{U}_{\parallel} &= (\mathbf{U}\cdot\mathbf{n})\mathbf{n}, \qquad \mathbf{U}_{\perp} = \mathbf{U} - (\mathbf{U}\cdot\mathbf{n})\mathbf{n}.
\end{aligned}\right\} \tag{9.118}$$

Since the expression, after summing each side of the two equations in Eq. (9.117), has to be equal to Eq. (9.111), the following relations can be obtained:

$$\mathbf{D}^{T} = D_{\parallel}^{T}\mathbf{nn} + D_{\perp}^{T}(\mathbf{I}-\mathbf{nn}), \tag{9.119}$$

$$\Delta\mathbf{r}^{B} = \Delta r_{\parallel}^{B}\mathbf{n} + (\Delta r_{\perp 1}^{B}\mathbf{n}_{\perp 1} + \Delta r_{\perp 2}^{B}\mathbf{n}_{\perp 2}). \tag{9.120}$$

By taking into account the fact that there is no correlation among $\Delta r_{\parallel}^{B}$, $\Delta r_{\perp 1}^{B}$, and $\Delta r_{\perp 2}^{B}$, and substituting Eq. (9.120) into Eq. (9.112), the following expressions can be obtained:

$$\left.\begin{aligned}&\left\langle \Delta r_{\parallel}^{B}\right\rangle = \left\langle \Delta r_{\perp 1}^{B}\right\rangle = \left\langle \Delta r_{\perp 2}^{B}\right\rangle = 0, \\ &\left\langle (\Delta r_{\parallel}^{B})^2\right\rangle = 2D_{\parallel}^{T}\Delta t, \\ &\left\langle (\Delta r_{\perp 1}^{B})^2\right\rangle = \left\langle (\Delta r_{\perp 2}^{B})^2\right\rangle = 2D_{\perp}^{T}\Delta t.\end{aligned}\right\} \tag{9.121}$$

From comparing Eq. (9.119) with Eq. (9.113), the components of the diffusion tensor, $D_{\parallel}^{T}$ and $D_{\perp}^{T}$, can be related to the resistance functions:

$$D_{\parallel}^{T} = \frac{kT}{\eta}(X^{A})^{-1}, \quad D_{\perp}^{T} = \frac{kT}{\eta}(Y^{A})^{-1}. \tag{9.122}$$

It is seen from these discussions that the particle position at the next time step can be calculated after the particle displacement due to the translational Brownian motion is generated according to Eq. (9.121).

Next we consider the rotational Brownian motion. As pointed out before, the rotational motion about the axis through the mass center of the particle and normal to the principal axis is sufficient for determining the rotational motion of an axisymmetric particle with a large aspect ratio. From the definition of giving the particle axis direction by $d\boldsymbol{\varphi}/dt=\boldsymbol{\omega}$, it is seen that a small change in direction $\Delta\boldsymbol{\varphi}$ during a time interval Δt can be related by $\Delta\boldsymbol{\varphi}=\boldsymbol{\varphi}(t+\Delta t)-\boldsymbol{\varphi}(t)=\Delta t\boldsymbol{\omega}$. On the other hand, the change $\Delta\mathbf{n}$ in the vector $\mathbf{n}(t)$ denoting the particle direction during the time interval Δt can be related to $\boldsymbol{\omega}$ by the following relation:

$$\Delta\mathbf{n} = \mathbf{n}(t+\Delta t) - \mathbf{n}(t) = \Delta t\boldsymbol{\omega}\times\mathbf{n}. \tag{9.123}$$

With these relations, the following equation is obtained:

$$\mathbf{n}(t+\Delta t) - \mathbf{n}(t) = \big(\boldsymbol{\varphi}(t+\Delta t) - \boldsymbol{\varphi}(t)\big)\times\mathbf{n}. \tag{9.124}$$

Hence, if the both sides of Eq. (9.114) are multiplied by ($\times\mathbf{n}$), the equation giving the particle direction can be obtained as

$$\begin{aligned}\mathbf{n}(t+\Delta t) = \mathbf{n}(t) + \boldsymbol{\Omega}(t)\times\mathbf{n}\Delta t + \frac{1}{kT}D_{\perp}^{R}\mathbf{T}_{\perp}^{P}(t)\times\mathbf{n}(t)\Delta t \\ -\frac{Y^{H}}{Y^{C}}\big((\boldsymbol{\varepsilon}\cdot\mathbf{nn}):\mathbf{E}\big)\times\mathbf{n}\Delta t + \Delta\mathbf{n}^{B}(t),\end{aligned} \tag{9.125}$$

in which $D_{\perp}^{R}$ and $\mathbf{T}_{\perp}^{P}$ are the rotational diffusion coefficient and torque acting on the particle about the axis through the mass center of the particle and normal to the particle axis. Also $\Delta\mathbf{n}^{B}(t)$ is denoted by

$$\Delta\mathbf{n}^{B} = \Delta\boldsymbol{\varphi}^{B}\times\mathbf{n}. \tag{9.126}$$

If $\Delta\boldsymbol{\varphi}^{B}$ is expressed in a similar way to Eq. (9.120):

$$\Delta\boldsymbol{\varphi}^B = \Delta\varphi_{\|}^B \mathbf{n} + (\Delta\varphi_{\perp 1}^B \mathbf{n}_{\perp 1} + \Delta\varphi_{\perp 2}^B \mathbf{n}_{\perp 2}), \tag{9.127}$$

and if this equation is substituted into Eq. (9.115), we can obtain the following relations:

$$\left.\begin{aligned} &\left\langle \Delta\varphi_{\perp 1}^B \right\rangle = \left\langle \Delta\varphi_{\perp 2}^B \right\rangle = 0, \\ &\left\langle (\Delta\varphi_{\perp 1}^B)^2 \right\rangle = \left\langle (\Delta\varphi_{\perp 2}^B)^2 \right\rangle = 2D_{\perp}^R \Delta t = 2\frac{kT}{\eta}(Y^C)^{-1}\Delta t, \end{aligned}\right\} \tag{9.128}$$

in which the last expression on the right-hand side of the second equation has been obtained from the relation in Eq. (5.14). The rotational displacement $\Delta\mathbf{n}^B$ due to the rotational Brownian motion reduces to the following equation by substituting Eq. (9.127) into Eq. (9.126):

$$\Delta\mathbf{n}^B = \Delta\varphi_{\perp 1}^B (\mathbf{n}_{\perp 1} \times \mathbf{n}) + \Delta\varphi_{\perp 2}^B (\mathbf{n}_{\perp 2} \times \mathbf{n}), \tag{9.129}$$

in which $(\mathbf{n}_{\perp 1} \times \mathbf{n})$ and $(\mathbf{n}_{\perp 2} \times \mathbf{n})$ are one set of vectors normal to the unit vectors $\mathbf{n}$. If we use the notations $\mathbf{n}_{\perp 1}$ and $\mathbf{n}_{\perp 2}$ again for such a set of normal vectors, Eq. (9.129) can be rewritten as

$$\Delta\mathbf{n}^B = \Delta\varphi_{\perp 1}^B \mathbf{n}_{\perp 1} + \Delta\varphi_{\perp 2}^B \mathbf{n}_{\perp 2}. \tag{9.130}$$

It is seen from these discussions that the particle direction at the next time step can be evaluated by generating the displacements due to the rotational Brownian motion such that Eqs. (9.128) and (9.130) are satisfied.

References

1. S. Chandrasekhar, Stochastic Problems in Physics and Astronomy, Rev. Mod. Phys., 15 (1943) 1
2. A.B. Clarke and R.L. Disney, "Probability and Random Processes," 2nd Ed., John Wiley & Sons, New York, 1985.
3. G. Bossis, B. Quentrec, and J.P. Boon, Brownian Dynamics and the Fluctuation-Dissipation Theorem, Molec. Phys., 45 (1982) 191
4. L.G. Nilsson and J.A. Padro, A Time-Saving Algorithm for Generalized Langevin-Dynamics Simulations with Arbitrary Memory Kernels, Molec. Phys., 71 (1990) 355
5. D.L. Ermak and H. Buckholtz, Numerical Integration of the Langevin Equation: Monte Carlo Simulation, J. Comput. Phys., 35 (1980) 169
6. J.M Deutch and I. Oppenheim, Molecular Theory of Brownian Motion for Several Particles, 54 (1971) 3547
7. D.L. Ermak and J.A. McCammon, Brownian Dynamics with Hydrodynamic Interactions, J. Chem. Phys., 69 (1978) 1352

8. R.J.A. Tough, P.N. Pusey, H.N.W. Lekkerkerker, and C. Van den Broeck, Stochastic Descriptions of the Dynamics of Interacting Brownian Particles, Molec. Phys., 59 (1986) 595
9. E. Dickinson, Brownian Dynamics with Hydrodynamic Interactions: The Application to Protein Diffusional Problems, Chem. Soc. Rev., 14 (1985) 421
10. G. Bossis and J.F. Brady, Self-diffusion of Brownian Particles in Concentrated Suspensions under Shear, J. Chem. Phys., 87 (1987) 5437

Exercises

9.1 For the following first-rank differential equation:

$$\frac{dy}{dx} + P(x)y = Q(x), \tag{9.131}$$

the general solution can be obtained as

$$y = \exp(-\int P dx)\left\{\int Q \exp(\int P dx) dx + c\right\}. \tag{9.132}$$

Derive Eq. (9.3) using this formula.

9.2 With the following relation concerning integration by parts, derive Eq. (9.4):

$$\begin{aligned}\frac{1}{m}\int_{t_0}^{t}\left\{\int_{t_0}^{t'}\mathbf{F}^B(\tau)e^{\zeta\tau}d\tau\right\}e^{-\zeta t'}dt' &= -\frac{1}{m\zeta}\left[\int_{t_0}^{t'}\mathbf{F}^B(\tau)e^{\zeta\tau}d\tau\cdot e^{-\zeta t'}\right]_{t_0}^{t} \\ &+\frac{1}{m\zeta}\int_{t_0}^{t}\mathbf{F}^B(t')e^{\zeta t'}e^{-\zeta t'}dt' = \frac{1}{m\zeta}\int_{t_0}^{t}\mathbf{F}^B(\tau)\left\{1-e^{-\zeta(t-\tau)}\right\}d\tau.\end{aligned} \tag{9.133}$$

9.3 Using the following Maclaurin expansion of an exponential function:

$$e^{-\zeta h} = 1 - \zeta h + \frac{(\zeta h)^2}{2!} - \frac{(\zeta h)^3}{3!} + \cdots, \tag{9.134}$$

and taking the limit of $\zeta \to 0$, derive Eqs. (9.27) and (9.28).

9.4 With the following relations concerning tensors:

$$\nabla\nabla n = \sum_{i=1}^{3}\boldsymbol{\delta}_i \frac{\partial}{\partial x_i}\left(\sum_{j=1}^{3}\boldsymbol{\delta}_j \frac{\partial n}{\partial x_j}\right), \tag{9.135}$$

$$\mathbf{I}:\nabla\nabla n = \sum_{i=1}^{3}\frac{\partial^2 n}{\partial x_i^2} = \nabla^2 n, \tag{9.136}$$

derive the diffusion equation in Eq. (9.59).

9.5 Consider the physical meanings of $\Delta\varphi_x$, $\Delta\varphi_y$, and $\Delta\varphi_z$ defined in Eq. (9.62).

9.6 Derive the following equations by nondimensionalizing Eq. (9.106), (9.107), and (9.108).

$$\mathbf{r}_i^*(t+\Delta t) = \mathbf{r}_i^*(t) + \mathbf{U}^*(\mathbf{r}_i)\Delta t^* + \frac{2}{3}\sum_{j=1}^{N}\mathbf{D}_{ij}^{T*}(t)\cdot(\widetilde{\mathbf{G}}_j'^*:\mathbf{E}^*)\Delta t^* + 2\sum_{j=1}^{N}\widetilde{\mathbf{D}}_{ij}^{C*}(t)\cdot(\widetilde{\mathbf{H}}_j'^*:\mathbf{E}^*)\Delta t^*$$
$$+\sum_{j=1}^{N}\mathbf{D}_{ij}^{T*}(t)\cdot\mathbf{F}_j^{P*}(t)\Delta t^* + 2\sum_{j=1}^{N}\widetilde{\mathbf{D}}_{ij}^{C*}(t)\cdot\mathbf{T}_j^{P*}(t)\Delta t^*$$
$$+\frac{1}{Pe}\sum_{j=1}^{N}\frac{\partial}{\partial\mathbf{r}_j^*}\cdot(\mathbf{D}_{ij}^{T*}(t))\Delta t^* + \Delta\mathbf{r}_i^{B*}(t) \qquad (i=1,2,\cdots,N),$$

(9.137)

$$\boldsymbol{\varphi}_i(t+\Delta t) = \boldsymbol{\varphi}_i(t) + \boldsymbol{\Omega}^*\Delta t^* + \sum_{j=1}^{N}\mathbf{D}_{ij}^{C*}(t)\cdot(\widetilde{\mathbf{G}}_j'^*:\mathbf{E}^*)\Delta t^* + \sum_{j=1}^{N}\mathbf{D}_{ij}^{R*}(t)\cdot(\widetilde{\mathbf{H}}_j'^*:\mathbf{E}^*)\Delta t^*$$
$$+\frac{3}{2}\sum_{j=1}^{N}\mathbf{D}_{ij}^{C*}(t)\cdot\mathbf{F}_j^{P*}(t)\Delta t^* + \sum_{j=1}^{N}\mathbf{D}_{ij}^{R*}(t)\cdot\mathbf{T}_j^{P*}(t)\Delta t^* \qquad (9.138)$$
$$+\frac{3}{2Pe}\sum_{j=1}^{N}\frac{\partial}{\partial\mathbf{r}_j^*}\cdot(\mathbf{D}_{ij}^{C*}(t))\Delta t^* + \Delta\boldsymbol{\varphi}_i^{B}(t) \qquad (i=1,2,\cdots,N),$$

$$\left\langle\Delta\mathbf{r}_i^{B*}(t)\Delta\mathbf{r}_j^{B*}(t)\right\rangle = \frac{2}{Pe}\mathbf{D}_{ij}^{T*}(t)\Delta t^*, \qquad (9.139)$$

$$\left\langle\Delta\boldsymbol{\varphi}_i^{B*}(t)\Delta\boldsymbol{\varphi}_j^{B*}(t)\right\rangle = \frac{3}{2Pe}\mathbf{D}_{ij}^{R*}(t)\Delta t^*, \qquad (9.140)$$

$$\left\langle\Delta\mathbf{r}_i^{B*}(t)\Delta\boldsymbol{\varphi}_j^{B*}(t)\right\rangle = \frac{3}{Pe}\widetilde{\mathbf{D}}_{ij}^{C*}\Delta t^*. \qquad (9.141)$$

CHAPTER 10

TYPICAL PROPERTIES OF COLLOIDAL DISPERSIONS CALCULABLE BY MOLECULAR-MICROSIMULATIONS

10.1 The Pair Correlation Function

The pair correlation function or radial distribution function is usually used to describe quantitatively the internal structure of fluids [1-5]. This quantity can directly be compared with the structure factor which is obtained from experiments by means of neutron or X-ray scattering. The visualization of a snapshot with the aid of a commercial software and the quantitative evaluation of the pair correlation function are two very useful methods to investigate the internal structure of a colloidal dispersion.

The pair correlation function quantifies how the particle of interest is surrounded by other particles. For example, the pair correlation function or radial distribution function has a constant value for all radial distances for a rarefied gas in which there is no internal structure. In contrast, for a solid system in which molecules are almost regularly located, the pair correlation function has sharply peaked values at the positions of particles, and is nearly zero at the positions where particles are seldom located. That is, the pair correlation function has almost the characteristic of the Dirac delta function for a solid system.

In addition to the thermodynamic equilibrium, the pair correlation function is also very useful to investigate the time change in the internal structure of a colloidal dispersion. For example, the sub-average of the pair correlation function, during a certain time interval, gives useful information concerning the change in the internal structure of a colloidal dispersion in a flow field.

With the following definition of the pair density $\nu^{(2)}(\mathbf{r},\mathbf{r}')$:

$$\nu^{(2)}(\mathbf{r},\mathbf{r}') = \sum_{i}\sum_{\substack{j \\ (i\neq j)}} \delta(\mathbf{r}_i - \mathbf{r})\delta(\mathbf{r}_j - \mathbf{r}'), \qquad (10.1)$$

the pair correlation function $g^{(2)}(\mathbf{r},\mathbf{r}')$ can be defined as

$$g^{(2)}(\mathbf{r},\mathbf{r}') = \left\langle \nu^{(2)}(\mathbf{r},\mathbf{r}') \right\rangle \Big/ n^2, \qquad (10.2)$$

in which n is the particle number density and expressed as $n=N/V$, and $\mathbf{r}_i$ is the position vector of particle i. If a system is uniform, $g^{(2)}(\mathbf{r}',\mathbf{r}'+\mathbf{r})$ is independent of the position $\mathbf{r}'$. In this case, the pair correlation function $g^{(2)}(\mathbf{r})$ can be expressed as

$$g^{(2)}(\mathbf{r}) = \frac{1}{V}\int g^{(2)}(\mathbf{r}',\mathbf{r}'+\mathbf{r})d\mathbf{r}' = \frac{1}{Vn^2}\left\langle \int \sum_i \sum_{\substack{j \\ (i\neq j)}} \delta(\mathbf{r}_i - \mathbf{r}')\delta(\mathbf{r}_j - \mathbf{r}' - \mathbf{r})d\mathbf{r}' \right\rangle$$
$$= \frac{1}{Vn^2}\left\langle \sum_i \sum_{\substack{j \\ (i\neq j)}} \delta(\mathbf{r}_j - \mathbf{r}_i - \mathbf{r}) \right\rangle = \frac{1}{Vn^2}\left\langle \sum_i \sum_{\substack{j \\ (i\neq j)}} \delta(\mathbf{r}_{ji} - \mathbf{r}) \right\rangle, \tag{10.3}$$

in which $\mathbf{r}_{ji}=\mathbf{r}_j-\mathbf{r}_i$. In addition, if the force between particles is isotropic, $g^{(2)}(\mathbf{r})$ is dependent only on the amplitude of the distance r ($=|\mathbf{r}|$), that is, it reduces to the radial distribution function $g(r)$.

The physical meaning of the pair correlation function $g^{(2)}(\mathbf{r})$ can be interpreted as follows. If we concentrate our attention on a certain particle in a system, the number of particles existing within the range of $\mathbf{r}$ to ($\mathbf{r}+d\mathbf{r}$) from the particle, can be expressed as $ng^{(2)}(\mathbf{r})d\mathbf{r}$ on average. If we apply this interpretation of the pair correlation function to the following virial equation of state:

$$P = \frac{N}{V}kT + \frac{1}{3V}\left\langle \sum_i \sum_{\substack{j \\ (i<j)}} \mathbf{r}_{ij}\cdot\mathbf{f}_{ij} \right\rangle, \tag{10.4}$$

this equation can be expressed, using the radial distribution function, as

$$P = \frac{N}{V}kT + \frac{1}{3V}\left(\frac{N}{2}\cdot\int_0^\infty rf(r)\cdot ng(r)4\pi r^2 dr\right) = \frac{N}{V}kT + \frac{2\pi N^2}{3V^2}\int_0^\infty r^3 f(r)g(r)dr, \tag{10.5}$$

in which $\mathbf{f}_{ij}$ is the force exerted on particle i by particle j.

In Sec. 11.5, we will explain the method of calculating the pair correlation function in simulations and also discuss the results of the radial distribution function for a Lennard-Jones molecular system which were obtained by Monte Carlo methods.

10.2 Rheology

The stress of a colloidal dispersion can be expressed as the sum of the contributions from the dispersion medium itself and colloidal particles [6]. The effective stress $\boldsymbol{\tau}_{eff}$ can be obtained from the ensemble average or the volume average of an instant stress for a uniform dispersion. That is,

$$\boldsymbol{\tau}^{eff} = \frac{1}{V}\int_V \boldsymbol{\tau} dV = \frac{1}{V}\int_{V-\sum_i V_i} \boldsymbol{\tau} dV + \frac{1}{V}\int_{\sum_i V_i} \boldsymbol{\tau} dV , \tag{10.6}$$

in which the two terms on the right-hand side have appeared by decomposing the system volume into the fluid volume (the first term) and the volumes of particles (the second term). By taking into account the expression for the stress term in Eq. (4.17), the first term on the right-hand side in Eq. (10.6) can be reformed as

$$\begin{aligned}\frac{1}{V}\int_{V-\sum_i V_i} \boldsymbol{\tau} dV &= \frac{1}{V}\int_V \boldsymbol{\tau} dV - \frac{1}{V}\sum_{i=1}^{N}\int_{V_i} \boldsymbol{\tau} dV = -p\mathbf{I} + 2\eta\mathbf{E} - \frac{\eta}{V}\sum_i \int_{V_i}\{\nabla\mathbf{u} + (\nabla\mathbf{u})^t\} dV \\ &= -p\mathbf{I} + 2\eta\mathbf{E} - \frac{\eta}{V}\sum_i \int_{A_i}(\mathbf{nu} + \mathbf{un}) dA ,\end{aligned} \tag{10.7}$$

in which A_i is the surface area of particle i, $\mathbf{n}_i$ is the unit vector directed normally outward from the particle surface, and p and $\mathbf{E}$ are the volume-averaged quantities. In this transformation the divergence theorem in Appendix A1 has been used. For rigid particles, the velocity at the particle surface corresponds to rigid-body motion, so that the third term on the right-hand side of the last equation vanishes identically.

Next we consider the second term on the right-hand side in Eq. (10.6). Since the stress inside the particle is unknown, we have to reform the expression to evaluate it. The change from the surface integral to the volume integral leads to the possibility of conducting such an integration. If $\mathbf{r}\boldsymbol{\tau}^t$ is denoted by $\mathbf{A}$, then $\mathbf{A}$ is a third-rank tensor with the ijk-component A_{ijk}. It is seen from Eq. (5.43) that $\widetilde{A}_{ijk}=A_{jki}$. The divergence of $\widetilde{\mathbf{A}}$ can be expressed as

$$\nabla\cdot\widetilde{\mathbf{A}} = \boldsymbol{\tau} + \mathbf{r}(\nabla\cdot\boldsymbol{\tau}) . \tag{10.8}$$

From this relation, the second term in Eq. (10.6) can be reformed as

$$\begin{aligned}\frac{1}{V}\int_{\sum_i V_i} \boldsymbol{\tau} dV &= \frac{1}{V}\sum_{i=1}^{N}\int_{V_i}\left\{\nabla\cdot\widetilde{\mathbf{A}} - \mathbf{r}(\nabla\cdot\boldsymbol{\tau})\right\} dV \\ &= \frac{1}{V}\sum_{i=1}^{N}\int_{A_i}\mathbf{r}(\mathbf{n}\cdot\boldsymbol{\tau}) dA - \frac{1}{V}\sum_{i=1}^{N}\int_{V_i}\mathbf{r}(\nabla\cdot\boldsymbol{\tau}) dV = \frac{1}{2V}\sum_{i=1}^{N}\int_{A_i}\{\mathbf{r}(\mathbf{n}\cdot\boldsymbol{\tau}) + (\mathbf{n}\cdot\boldsymbol{\tau})\mathbf{r}\} dA \\ &\quad + \frac{1}{2V}\sum_{i=1}^{N}\int_{A_i}\{\mathbf{r}(\mathbf{n}\cdot\boldsymbol{\tau}) - (\mathbf{n}\cdot\boldsymbol{\tau})\mathbf{r}\} dA - \frac{1}{V}\sum_{i=1}^{N}\int_{V_i}\mathbf{r}(\nabla\cdot\boldsymbol{\tau}) dV .\end{aligned} \tag{10.9}$$

The last term on the right-hand side of this equation is due to the contribution from the force acting on the particle. By substituting Eqs. (10.7) and (10.9) into Eq. (10.6) with consideration of Eqs. (4.40), (4.41), and (4.44), the apparent stress $\boldsymbol{\tau}^{eff}$ can finally be expressed as [7,8]

$$\boldsymbol{\tau}^{eff} = -p^{eff}\mathbf{I} + 2\eta\mathbf{E} - \frac{1}{V}\sum_{i=1}^{N}\mathbf{S}_i - \frac{1}{V}\sum_{i=1}^{N}\mathbf{r}_i\mathbf{F}_i^P - \frac{1}{2V}\sum_{i=1}^{N}\boldsymbol{\varepsilon}\cdot\mathbf{T}_i^P , \tag{10.10}$$

in which the first term on the right-hand side is due to the apparent pressure in which the contribution from the last term in Eq. (4.40) is included, the second term is due to the fluid itself, the third term is due to the stresslets, and the fifth term is due to the non-hydrodynamic torques. In addition, the fourth term is due to the non-hydrodynamic interactions between particles or between particles and an external field, and has a similar expression to the viscosity for a molecular system [3].

If a dispersion is dilute enough to neglect hydrodynamic interactions between particles and a torque-free situation is considered, then the stresslet $\mathbf{S}$ exerted on the ambient fluid by a spherical particle with a radius a can be obtained from Eqs. (5.6), (5.7), and (5.18) as

$$\mathbf{S} = -\frac{20}{3}\pi\eta a^3\mathbf{E} . \tag{10.11}$$

In deriving this equation, we have assumed that the base liquid is incompressible and have taken into account the fact that the rate-of-strain tensor is symmetric for Newtonian liquids. With Eq. (10.11), the third term in Eq. (10.10) can be obtained as

$$-\frac{1}{V}\sum_{i=1}^{N}\mathbf{S}_i = \frac{20}{3}\cdot\frac{N}{V}\pi\eta a^3\mathbf{E} = 5\eta\varphi_V\mathbf{E} , \tag{10.12}$$

in which φ_V is the volumetric fraction and expressed as $\varphi_V=(4/3)\pi a^3 N/V$. It is the stresslets in Eq. (10.12) that give the second term in Eq. (4.36).

Next we explain the proper way to describe the rheological properties of colloidal dispersions. Since a simple shear flow is very useful for microsimulations of colloidal dispersions, we consider such a flow directed in the x-direction with a shear rate $\dot{\gamma}$ as shown in Fig.8.1. In this case, the effective or apparent viscosity η^{eff} of a colloidal dispersion is defined from the effective stress shown in Eq. (10.10) as

$$\eta_{yx}^{eff} = \tau_{yx}^{eff} / \dot{\gamma} . \tag{10.13}$$

The subscript yx, for example, τ_{yx}^{eff}, means the shear stress in the x-direction acting on the plane normal to the y-axis. The apparent viscosity η_{yx}^{eff} is generally a function of the shear rate $\dot{\gamma}$. If a dispersion does not respond to an external field, the following equation is generally satisfied:

$$\eta_{yx}^{eff} = \eta_{xy}^{eff} . \tag{10.14}$$

However, this relationship is seldom valid for ferrofluids, electro-rheological fluids, or magneto-rheological suspensions.

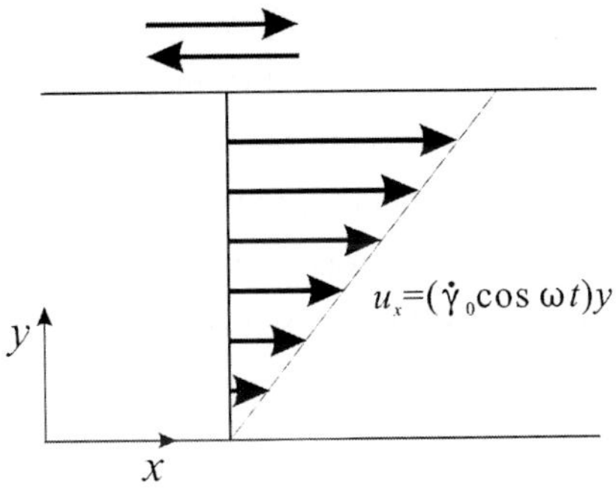

Figure 10.1 An oscillatory shear flow.

The elastic properties of a colloidal dispersion can be described by the following normal stress differences:

$$\left.\begin{aligned} \tau_{xx}^{eff} - \tau_{yy}^{eff} &= \Psi_1 \dot{\gamma}_{yx}^2, \\ \tau_{yy}^{eff} - \tau_{zz}^{eff} &= \Psi_2 \dot{\gamma}_{yx}^2, \end{aligned}\right\} \tag{10.15}$$

in which Ψ_1 and Ψ_2 are called the primary and secondary normal stress coefficients, respectively. After this analysis, the rheological properties of colloidal dispersions can now be described by the apparent viscosity and normal stress coefficients.

Besides a simple shear flow, a small-amplitude oscillatory shear flow is also very useful to investigate. We consider an oscillatory shear flow with a small amplitude $\dot{\gamma}_0$ as shown in Fig. 10.1 and in this case the flow field can be expressed as

$$u_x = (\dot{\gamma}_0 \cos \omega t) y . \tag{10.16}$$

Since the stress τ_{yx}^{eff} is not necessarily in phase with the shear rate, it is expressed as

$$\tau_{yx}^{eff} = \eta'_{yx} \dot{\gamma}_0 \cos \omega t + \eta''_{yx} \dot{\gamma}_0 \sin \omega t , \tag{10.17}$$

in which the η' term is an in-phase component, the η'' term is an out-of-phase component, and η' and η'' are functions of frequency . With the following notations:

$$\left.\begin{aligned} \tau_0^{eff} &= \sqrt{\eta'^{\,2}_{yx} + \eta''^{\,2}_{yx}}\, \dot{\gamma}_0, \\ \varphi &= \tan^{-1}(\eta''_{yx} / \eta'_{yx}) \qquad (0 \le \varphi \le \pi/2), \end{aligned}\right\} \tag{10.18}$$

the expression for τ_{yx}^{eff} can be written as

$$\tau_{yx}^{eff} = \tau_0^{eff} \cos(\omega t - \varphi) . \tag{10.19}$$

Hence η_{yx}' and η_{yx}'' have information about the amplitude and phase difference of the stress. If we take the limit of $\omega \to 0$, η_{yx}'' approaches zero, and η_{yx}' becomes equal to the usual viscosity for a simple shear flow. It is seen from Eq. (10.17) that the value of η_{yx}'' stands for the strength of out-of-phase by 90° to a flow field (in phase with the strain). The normal stress coefficients for a small-amplitude oscillatory shear flow can be defined in a similar way, but we here omit showing such expressions; they may be found in Ref. [9].

References

1. D.A. McQuarrie, "Statistical Mechanics," Happer & Row, New York, 1976
2. B.J. McClelland, "Statistical Thermodynamics," Chapman & Hall, London, 1973
3. M.P. Allen and D.J. Tildesley, "Computer Simulation of Liquids," Clarendon Press, Oxford, 1987
4. J.P. Hansen and I.R. McDonald, "Theory of Simple Liquids," 2nd Ed., Academic Press, London, 1986
5. P.A. Egelstaff, "An Introduction to the Liquid State," Academic Press, London, 1967
6. S. Kim and S.J. Karrila, "Microhydrodynamics: Principles and Selected Applications," Butterworth-Heinemann, Stoneham , 1991
7. J.F. Brady, R.J. Phillips, J.C. Lester, and G. Bossis, Dynamic Simulation of Hydrodynamically Interacting Suspensions, J. Fluid Mech., 195 (1988) 257
8. J.F. Brady, The Rheological Behavior of Concentrated Colloidal Dispersions, J. Chem. Phys., 99 (1993) 567
9. R.B. Bird, R.C. Armstrong, and O. Hassager, "Dynamics of Polymeric Liquids, Vol.1, Fluid Mechanics, " John Wiley & Sons, New York , 1977

Exercises

10.1 If the potential energy of a system composed of N particles in thermodynamic equilibrium is denoted by $U(\mathbf{r}_1, \mathbf{r}_2, \ldots, \mathbf{r}_N)$, and the probability density function $\rho(\mathbf{r}_1, \mathbf{r}_2, \ldots, \mathbf{r}_N)$ is defined by the following equation:

$$\rho(\mathbf{r}_1,\mathbf{r}_2,\cdots,\mathbf{r}_N) = \frac{\exp\left\{-\dfrac{1}{kT}U(\mathbf{r}_1,\mathbf{r}_2,\cdots,\mathbf{r}_N)\right\}}{\int\cdots\int\exp\left\{-\dfrac{1}{kT}U(\mathbf{r}_1,\mathbf{r}_2,\cdots,\mathbf{r}_N)\right\}d\mathbf{r}_1\cdots d\mathbf{r}_N}, \tag{10.20}$$

show that the pair correlation function defined by Eq. (10.2) is expressed as

$$g^{(2)}(\mathbf{r},\mathbf{r}') = \frac{1}{n^2} N(N-1) \frac{\int \cdots \int \exp\left\{-\frac{1}{kT} U(\mathbf{r}, \mathbf{r}', \mathbf{r}_3 \cdots, \mathbf{r}_N)\right\} d\mathbf{r}_3 \cdots d\mathbf{r}_N}{\int \cdots \int \exp\left\{-\frac{1}{kT} U(\mathbf{r}_1, \mathbf{r}_2, \mathbf{r}_3 \cdots, \mathbf{r}_N)\right\} d\mathbf{r}_1 \cdots d\mathbf{r}_N}. \tag{10.21}$$

10.2 We now set $n(\mathbf{r}) = \langle \nu^{(1)}(\mathbf{r}) \rangle$, in which $n(\mathbf{r})$ is the local number density at a position $\mathbf{r}$. In this case, show the expression for $\nu^{(1)}(\mathbf{r})$ which corresponds to Eq. (10.1) for $g^{(2)}(\mathbf{r},\mathbf{r}')$.

10.3 If the force exerted on particle i by particle j is denoted by $\mathbf{F}_{ij}^{P}$, show that the fourth term on the right-hand side in Eq. (10.10) can be expressed as

$$-\frac{1}{V}\sum_{i=1}^{N} \mathbf{r}_i \mathbf{F}_i^P = -\frac{1}{V}\sum_{i=1}^{N} \sum_{\substack{j=1 \\ (j>i)}}^{N} \mathbf{r}_{ij} \mathbf{F}_{ij}^P . \tag{10.22}$$

10.4 By treating the components in the orthogonal coordinate system, show that the fifth term on the right-hand side in Eq. (10.10) can be expressed as

$$-\frac{1}{2V}\sum_{i=1}^{N} \boldsymbol{\varepsilon} \cdot \mathbf{T}_i^P = -\frac{1}{2V}\sum_{i=1}^{N} \begin{bmatrix} 0 & T_{iz} & -T_{iy} \\ -T_{iz} & 0 & T_{ix} \\ T_{iy} & -T_{ix} & 0 \end{bmatrix}. \tag{10.23}$$

CHAPTER 11

THE METHODOLOGY OF SIMULATIONS

In this chapter we explain the methodology which is very useful for actually simulating colloidal dispersions [1-4]. We will also discuss the correction of results obtained by simulations and the methods of evaluating the accuracy of results.

11.1 The Initial Configuration of Particles

A finite simulation region, such as a cube, is usually used in actual simulations, although a cuboid may be preferred in some cases such as a non-spherical particle system. For cubic simulation regions, the initial positions of particles are specified randomly, using uniform random numbers, or are placed at regular points, such as the face-centered cubic lattices shown in Fig.11.1(a). This lattice is one of the close-packed lattices and useful for simulating a liquid or dense colloidal system. In the case of a cubic region, possible values of the particle number N are restricted as $N=4M^3$ (=32, 108, 256, 500, ...), in which the simulation box is made by replicating a unit cell M times in each direction. Hence, first the particle number N and the number density n are specified, and then the side length L of the cubic simulation box is determined from the relation $N=nL^3$. In a gas or dilute colloidal system, a simple cubic lattice shown in Fig. 11.1(b) may be used for an initial configuration and in this case the particle number N can be adopted from a relatively smooth choice, such as $N=M^3$ (=27, 64, 125, 216...).

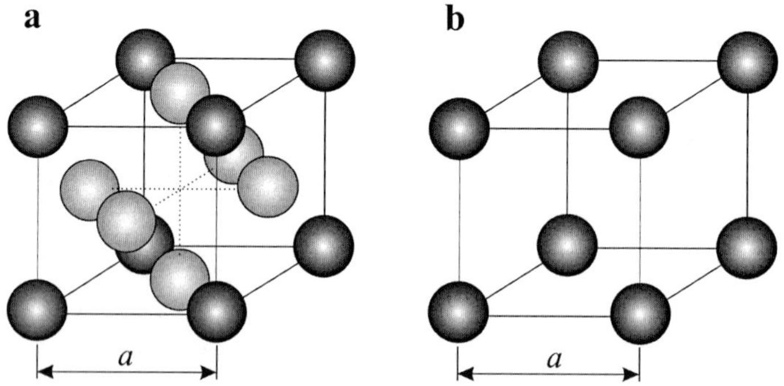

Figure 11.1 Initial configuration: (a) face-centered cubic lattice and (b)simple cubic lattice.

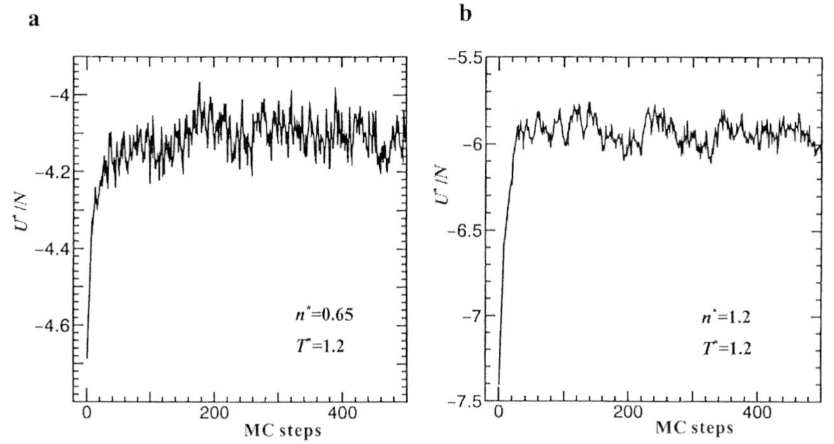

Figure 11.2 Transition from an initial configuration to equilibrium: (a) for n^*=0.65 and T^*=1.2, and (b) for n^*=1.2 and T^*=1.2.

In a dense colloidal system, especially, in a solid system, a physically reasonable configuration has to be given as an initial state; otherwise, a false crystal structure in equilibrium may be obtained from simulations. Hence, in some cases a cuboid simulation box may be more appropriate for a dense colloidal dispersion of non-spherical particles, or for a solid system of molecules.

An initial configuration of particles which is regularly given, promptly approaches a random configuration or thermodynamic equilibrium with time steps in a simulation. We now show an example of how an initial configuration approaches equilibrium in actual simulations. In this case, we simulated a molecular system composed of Lennard-Jones molecules, and a face-centered cubic lattice was used as an initial configuration. Figure 11.2 shows the transient process of the potential energy with MC steps (a MC step means N attempts of moving particles), which was obtained from Monte Carlo simulations. Figure 11.2(a) is for a liquid state with number density n^*=0.65 and the number of particles N=256, and Fig. 11.2 (b) is for a solid state of n^*=1.2 and N=500. Both results are for the common temperature T^*=1.2. It is clear from the figures that the regularity of the initial configuration rapidly decreases with MC steps and both curves may be regarded as having almost attained equilibrium after 500 MC steps. Since tens of thousands of MC steps are generally carried out as a minimum in a Monte Carlo simulation, the influence of the initial configuration can be regarded as appearing only in the early stage, and is soon lost.

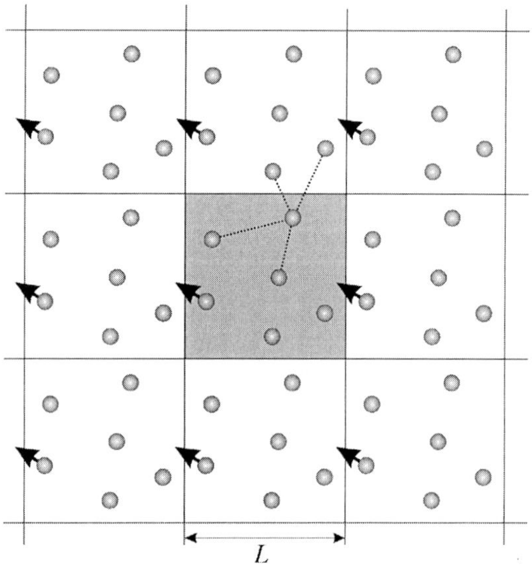

Figure 11.3 Periodic boundary condition.

Besides the system energy, the translational order parameter and the orientational order parameter for a non-spherical particle system are generally used as a criterion to judge whether the regularity of an initial configuration disappears and the system has attained equilibrium [2]

11.2 Boundary Conditions

11.2.1 The periodic boundary condition

Simulations are conducted for a finite simulation region, so an outer boundary condition has to be introduced. The periodic boundary condition, which is in general use at this time, is applicable to all states of a gas, liquid, solid, and colloidal system.

Since a cube or cuboid is generally used for the simulation box, we explain the periodic boundary condition for this geometry. Figure 11.3 shows the concept of the periodic boundary condition for a two-dimensional system. The central cell is the original simulation box and the ambient cells are virtual cells which are made by replicating the original cell. The application of the periodic boundary condition to this system means that, when a particle goes out of the central cell through the boundary, its image particle enters the central cell through the opposite boundary wall from a replicated virtual cell. Also, a particle in the

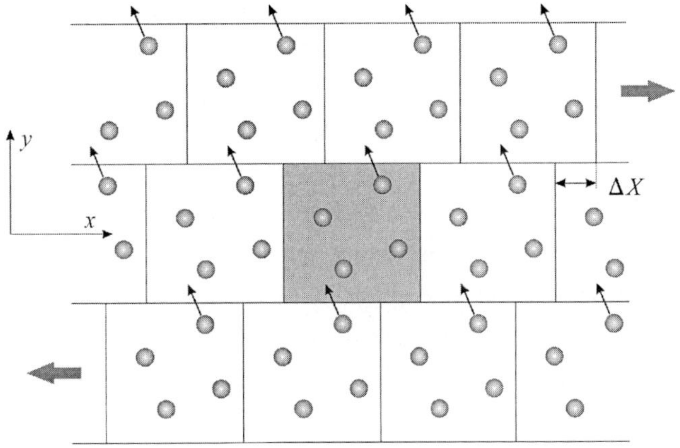

Figure 11.4 Lees-Edwards boundary condition for a simple shear flow.

central cell near the boundary interacts with both the real particles in the central cell and virtual particles in neighbor replicated cells. Hence, it may arise that a certain particle and its replicated virtual particles in the ambient cells have to be considered as interacting with the particle of interest. However, if the following minimum image convention [5] is adopted, then consideration of the closest particle, which may be a real particle or its virtual particle, is sufficient. If the length of the simulation box L is taken as $L>2r_{coff}$, in which r_{coff} is the cutoff radius for particle-particle interactions and will be explained in Sec. 11.3.1, then the interaction with either a real particle or its virtual particle is sufficient if calculated according to the minimum image convention. Hence the number of particles with which an arbitrary particle interacts becomes (N-1) at maximum, including real and virtual particles.

The periodic boundary condition has been used almost exclusively to date and it has been well known from those days that results obtained by simulations with this boundary condition can explain experimental results very well, even for a small simulation box.

11.2.2 The Lees-Edwards boundary condition

To generate a simple shear flow in simulations, the periodic boundary condition has to be modified in such a way that the replicated cells, located normal to the shear flow direction, are made to move in the shear flow direction [6,7]. This is known as the Lees-Edwards boundary condition. We now explain this boundary condition in more detail.

We use a cubic simulation box of length L centered at the origin, and consider the method of generating a simple shear flow $\mathbf{u}(\mathbf{r})=(\dot{\gamma}y,0,0)$ by making the replicated cells move in the shear flow direction, as shown in Fig. 11.4. In the Lees-Edwards boundary condition, the central cell and its replicated cells centered at y=0 are taken to be stationary, but the replicated cells centered at $y=L$ are made to move in the positive x-direction with speed $\dot{\gamma}L$ and the replicated cells centered at $y=-L$ are made to move in the negative x-direction with the same speed $\dot{\gamma}L$. The replicated cells in more remote layers are moved proportionally faster, relative to the central cell, in a similar way. The velocity $\mathbf{v}_i$ of an arbitrary particle i does not change even if it moves across a boundary surface normal to the x- or z-axis, but changes if it moves across a boundary surface normal to the y-axis (or shear plane). We explain this property more concretely using Fig. 11.4. If particle i located at $\mathbf{r}_i$ crosses the upper surface and moves to $\mathbf{r}_i'$, its replicated particle enters the central cell through the lower boundary surface and moves to $(x_i'-\Delta X,y'-L,z_i')$ with a velocity $(v_{ix}-\dot{\gamma}L,v_{iy},v_{iz})$, in which ΔX is the displacement of its replicated boxes relative to the central one and is taken as $0\leq X<L$. This procedure can be expressed using FORTRAN language as [2]

```
CORY=DNINT(RY(I)/L)
RX(I)=RX(I)-CORY*DX
RX(I)=RX(I)-DNINT(RX(I)/L)*L
RY(I)=RY(I)-CORY*L
RZ(I)=RZ(I)-DNINT(RZ(I)/L)*L
VX(I)=VX(I)-CORY*GAMMA*L
```

In addition to the boundary condition, the minimum image convention has to be modified. Although the procedures for the y- and z-directions are the same as usual, the procedure for the x-direction has to be corrected in the following manner. The replicated particles of an arbitrary particle j in the central box locate at ..., $x_j-2L+\Delta X$, $x_j-L+\Delta X$, $x_j+\Delta X$, $x_j+L+\Delta X$, $x_j+2L+\Delta X$, ..., for the x-coordinate in the virtual boxes centered at $y=L$ (in the upper layer), and locate at ..., $x_j-2L-\Delta X$, $x_j-L-\Delta X$, $x_j-\Delta X$, $x_j+L-\Delta X$, $x_j+2L-\Delta X$, ..., in the virtual boxes centered at $y=-L$ (in the lower layer). The replicated particles of particle j in the virtual boxes located at y=0 are treated in the same way as the ordinary minimum image convention. Hence, in the modified image convention, particle i interacts with only the nearest particle which is selected from particle j itself and its replicated particles, the position of which has been corrected according to the above treatment. This procedure can be expressed in FORTRAN language as [2]

```
RXJI=RX(J)-RX(I)
RYJI=RY(J)-RY(I)
RZJI=RZ(J)-RZ(I)
CORY=DNINT(RYJI/L)
RYJI=RYJI-CORY*L
RZJI=RZJI-DNINT(RZJI/L)*L
RXJI=RXJI-CORY*DX
RXJI=RXJI-DNINT(RXJI/L)*L
```

It is seen that the displacement ΔX ($0 \leq \Delta X < L$) of the upper and lower moving boxes has to be monitored to use the Lees-Edwards boundary condition.

11.2.3 The periodic-shell boundary condition

The periodic boundary condition which has been explained above is for a dispersion in thermodynamic equilibrium, and not applicable to certain types of non-equilibrium phenomena; for example, the number density is dependent on the position and time in sedimentation and diffusion phenomena in a colloidal suspension. It is the periodic-shell boundary condition that has been developed for simulating these non-equilibrium phenomena [8,9]. Originally, we developed this boundary condition for simulating a flow around an obstacle using molecular dynamics simulations, and simulations with the present boundary condition enable us to obtain very accurate numerical solutions of the flow field. It was then successfully applied to the generation of a normal shock wave by means of molecular dynamics simulations [10].

In the periodic-shell boundary condition, molecules inside a simulation region are made to interact with imaginary molecules outside the region. These molecules are images of the molecules existing near the boundary surfaces. Extrapolation conditions are usually used as an outer boundary condition to solve numerically the Navier-Stokes equation by finite difference methods. The periodic-shell boundary condition does apply the idea of this extrapolation procedure to MD simulations to diminish the influence of finite simulation regions; but it is noted that the present boundary condition does not mean conducting the exact extrapolation on a microscopic level.

The particles near the boundary surfaces of a simulation region must interact with imaginary particles outside the region; otherwise, the limited simulation region will significantly distort the flow field. The thin layer (shell) near the boundary is replicated next to the region itself so that the particles in the shell also act as imaginary particles. As shown in Fig. 11.5, particle

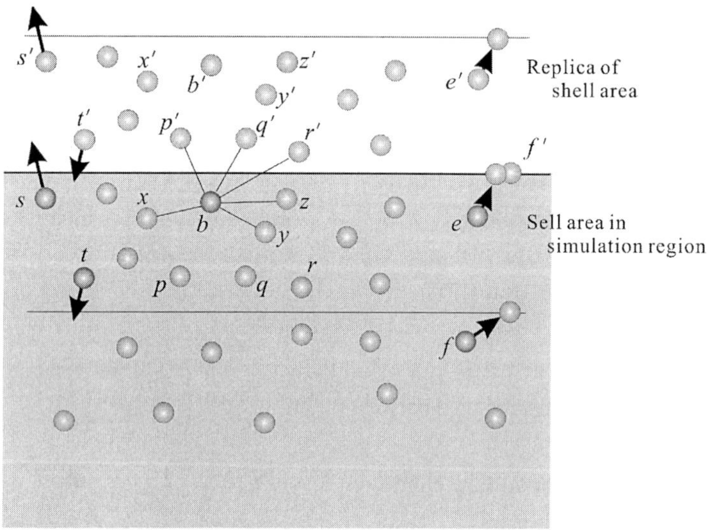

Figure 11.5 Periodic-shell boundary condition.

b interacts not only with the real particles *x*, *y*, *z*, etc., but also with the imaginary particles *p*', *q*', *r*', etc. Particle *s* in Fig. 11.5 is completely removed if it moves out of the simulation region. Incoming particles, on the other hand, are generated in the following manner. If particle *t* in Fig. 11.5 moves from the shell area to the inner region, the imaginary particle *t*' naturally enters into the shell area. This particle *t*' is regarded as a new incoming particle independent of particle *t*. Some problems may result from particle *f*, which crosses the boundary (the thin line in the figure) to enter the shell area. Suppose that at the next time step, particles *e* and *f* move to the head of the arrows. In this case, particle *f* moves to the upper region across the boundary (the thin line) and, at the same time, the imaginary particle *f*' suddenly appears in front of particle *e*. Such sudden appearances of imaginary particles may cause unreasonable overlap with real particles, so special attention has to be paid to this situation. In actual simulations, such unreasonable overlap may be relieved by shifting real particles an infinitesimal distance. For hard molecular systems, particle *e* may be displaced at random in its neighboring area. This simple procedure is not sufficient for soft molecular systems. We recommend the removal of these overlaps by Monte Carlo methods for soft molecular systems, which has more physical reasonability. However, we have found that in our simulations, this type of overlap rarely occurs after a stationary state has been attained. The reason is that particle *e* can notice the existence of particle *f*' to some degree from the interactions of the neighboring particles even if particle *f* does not enter the shell region. The shell area is usually taken to be much thinner than the side length of a simulation box, so that the use of this boundary condition does not lead to a significant increase in the computation time. Rather, even if a small simulation region is used, the flow field which is obtained by simulations using the periodic-shell boundary condition is not significantly distorted; in other words, a much shorter time is required to obtain an almost exact velocity field since a smaller simulation region can be used.

11.3 The Methodology to Reduce Computation Time

The most time consuming procedure in simulations is the calculation of interaction energies or forces between particles. In this section we explain some techniques for reducing the computation time in simulations.

11.3.1 The cutoff distance

With the above minimum image convention, the calculation of interaction energies, etc., is carried out (N-1) times per particle and in total $N(N$-1) times. Hence, if the number of particles with which a particle interacts can be reduced significantly, then the computation time drastically decreases. For a short-range potential as shown in Fig. A3.1, the potential energy can be regarded as zero at a certain distance r_{coff} over several times the particle diameter. This means that interactions between two particles with a separation over r_{coff} can be neglected and so do not need to be calculated. The particle separation for truncating the calculation of interactions is called the cutoff radius or cutoff distance. For example, in the Lennard-Jones potential, the cutoff radius is generally taken as $r_{coff}=2\sim3.5\sigma$.

The introduction of the cutoff radius alone does not lead to a great reduction of computation time, since the distance between two particles must be calculated $N(N$-1) times to judge whether or not a pair of particles exist within the range of r_{coff}. Hence the reduction of computation time can be further accomplished by using the cutoff radius together with the following neighbor list methods.

We note here that the averaged values, which are obtained using the cutoff radius, should be corrected as explained in Sec. 11.4.

11.3.2 Neighbor lists

If an arbitrary particle knows the identification of the ambient particles existing within the range of the cutoff radius, then it is unnecessary to check whether or not other particles are within this range by computing the distance between a pair of particles. Hence, if the number of particles existing within the range of the cutoff radius is approximately M ($<<N$), the calculation of the interactions between particles needs to be carried out just $N \times M$ times. This clearly shows that the method using a neighbor list and the cutoff radius drastically reduces the computation time compared with that without this technique, which would require $N(N-1)$ calculations. We explain two very useful neighbor list methods below.

A. The cell index method

We explain the concept of the cell index method [11,12] using a two-dimensional square simulation region shown in Fig. 11.6. In the cell index method, the square region is divided into $M \times M$ cells in such a way that the side length l ($=L/M$) of each cell is greater than the cutoff distance r_{coff}, that is, $l \geq r_{coff}$. In this situation, an arbitrary particle has the possibility of interacting with particles belonging to the same cell of the particle and its nearest neighbor cells; particles in the other cells are, therefore, unnecessary in calculating interaction energies or forces. For example, in Fig. 11.6, only particles existing in the cells 14, 15, 16, 20, 21, 22, 26, 27, and 28 are taken into account in calculating the interactions of an arbitrary particle in cell 21. Particles existing in the other cells do not need to be taken into account since such particles are not within the range of the cutoff distance.

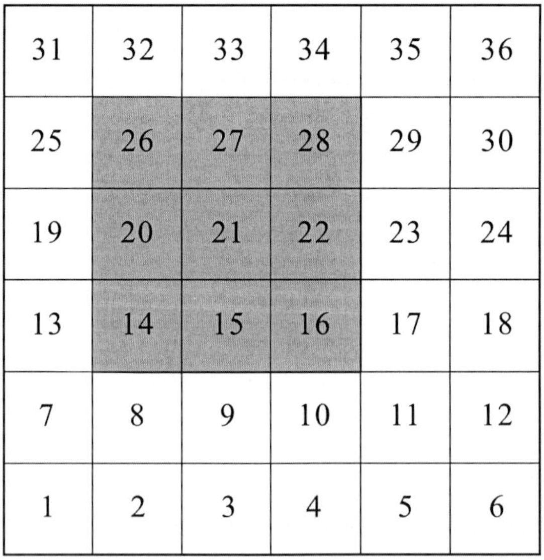

Figure 11.6 Cell index method.

Each cell must store the number and identification of particles existing in itself in order to make the cell index method function effectively. Although an effective method for saving this information in simulations is given in Ref. [2], we here explain a more understandable technique with slightly more memory area required. An arbitrary cell saves the names of the particles in the cell using the variable *NAME*(*J*,*K*) and the number of particles using *NMX*(*K*). In addition, an arbitrary particle *i* saves the name of its belonging cell using the variable *CELL*(*I*). In the above variables, *K* and *I* mean the names of a cell and particle, respectively. This method is quite understandable, but the dimension of the first array in *NAM*E has to be taken larger than necessary. This is because the number of particles in each cell in equilibrium is not known before the start of a simulation. The above method, therefore, requires a little more memory area on a computer than the minimum necessary.

In Monte Carlo simulations, after the movement of a particle, the old information about this particle is updated. In contrast, in molecular dynamics or Stokesian dynamics simulations, the information about all particles is updated.

B. The Verlet neighbor list method

The above-mentioned cell index method stores the particle position with the resolution of the side length l of each divided cell and each cell has the information concerning the names of particles belonging to itself. In contrast, in the Verlet neighbor list method [13], each particle has the information of the names of particles with which the particle interacts. We explain this method using Fig. 11.7. An arbitrary particle stores the names of particles interacting with itself within the range of r_l, larger than the cutoff radius r_{coff} (r_c in Fig. 11.7), in a list. In molecular-microsimulations, the displacement of a particle during one step is very small compared with the particle radius, so that the list of the names of interacting particles can be used over several time steps even if some particles move into and out of the cutoff area. This is the Verlet neighbor list method. If the neighbor lists only need to be renewed, for example, every ten MC steps, it is quite clear that a significant reduction in computation time can be accomplished

How often the neighbor lists have to be renewed and what an appropriate value for r_l is, strongly depend on the experiential aspects, but the following roughly criterion may be applied. If the maximum displacement of particles during one step is denoted by δr_{max}, an appropriate value of r_l has to be chosen as $r_l > r_{coff} + P\delta r_{max}$, in which the neighbor lists are renewed every P MC steps. This condition means that particles are never permitted to travel from the outer area of r_l to the cutoff region. However, a better way to determine appropriate values of r_l and

P is the method of trail and error, i.e. from trial simulations using such a roughly appropriate value as a trial guess.

11.4 Long-Range Correction

As pointed out, the cutoff radius r_{coff} is generally used in molecular-microsimulations. If the cutoff radius cannot be taken sufficiently large, results obtained by simulations need to be corrected.

We now consider a certain physical quantity X which is related to the respective quantity $x(r_{ij})$ through the ensemble average:

$$X = a\left\langle \sum_{i} \sum_{\substack{j \\ (i<j)}} x(r_{ij}) \right\rangle, \tag{11.1}$$

in which a is a constant. Similarly to Eq. (10.5), Eq. (11.1) can be written, using the radial distribution function, as

$$X = a\frac{N}{2}\int_{0}^{\infty} x(r)\cdot ng(r)4\pi r^2 dr = a\cdot 2\pi Nn\int_{0}^{\infty} x(r)g(r)r^2 dr , \tag{11.2}$$

in which N is the total number of particles, and n is the number density of particles. As will be discussed in Sec. 11.5, the radial distribution function $g(r)$ approaches unity as r increases. Hence, the quantity X can be expressed as the sum of X_{coff} which is evaluated with the cutoff radius and X_{LRC} which is the long-range correction term. That is,

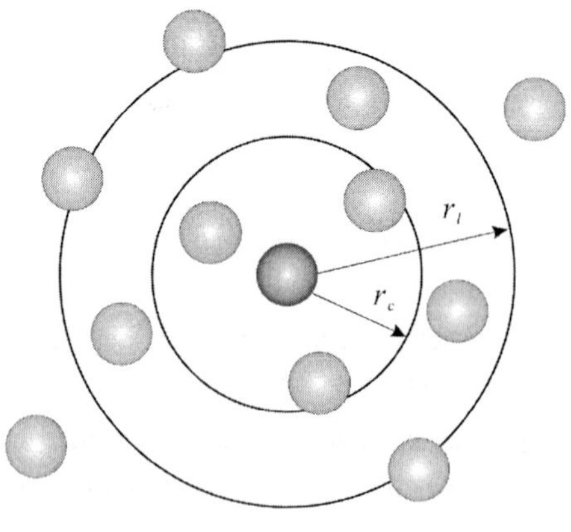

Figure 11.7 Verlet neighbor list method.

$$X \approx X_{coff} + X_{LRC} \,. \tag{11.3}$$

X_{LRC} can be expressed, by considering $g(r)\approx 1$ for $r>r_{coff}$ in Eq. (11.2), as

$$X_{LRC} = a \cdot 2\pi Nn \int_{r_{coff}}^{\infty} x(r) r^2 dr \,. \tag{11.4}$$

For example, the pressure P can be written from Eqs. (10.4) and (10.5) as

$$P = P_{coff} + \frac{2\pi N^2}{3V^2} \int_{r_{coff}}^{\infty} r^3 f(r) dr = P_{coff} - \frac{2\pi N^2}{3V^2} \int_{r_{coff}}^{\infty} r^3 \frac{du(r)}{dr} dr \,. \tag{11.5}$$

If the interaction energy between particles, $u(r)$, is expressed by the Lennard-Jones potential, then the second term on the right-hand side in Eq. (11.5) can be straightforwardly calculated.

11.5 The Evaluation of the Pair Correlation Function

Equation (10.3) clearly shows that the Dirac delta function has to be evaluated in order to calculate the pair correlation function $g^{(2)}(\mathbf{r})$. Since the discontinuous characteristic of the delta function cannot exactly be reproduced in the real world, a finite small volume is used in evaluating this discontinuous function. Hence, if a finite small volume at a position $\mathbf{r}$ is denoted by $\Delta V(\mathbf{r})$, and the number of particles existing in the volume is denoted by $\Delta N(\mathbf{r})$, the Dirac delta function $\delta(\mathbf{r})$ can be approximated as

$$\sum_{j=1}^{N} \delta(\mathbf{r} - \mathbf{r}_j) \approx \frac{\Delta N}{\Delta V} \,. \tag{11.6}$$

Hence, if the number of particles in the small volume $\Delta V(\mathbf{r})$ at the relative position $\mathbf{r}$ from the particle of interest is denoted by ΔN_i for particle i, then the pair correlation function $g^{(2)}(\mathbf{r})$ in Eq. (10.3) can be expressed as

$$g^{(2)}(\mathbf{r}) = \frac{1}{Vn^2} \left\langle \sum_{i=1}^{N} \frac{\Delta N_i(\mathbf{r})}{\Delta V(\mathbf{r})} \right\rangle . \tag{11.7}$$

For example, in the spherical coordinate system, the small volume $\Delta V(\mathbf{r})$ becomes ($\Delta r \cdot r\Delta\theta \cdot r \sin\theta\Delta\varphi$), so Eq. (11.7) can be written as

$$g^{(2)}(\mathbf{r}) = g^{(2)}(r,\theta,\varphi) = \frac{1}{Vn^2} \left\langle \sum_{i=1}^{N} \frac{\Delta N_i(r,\theta,\varphi)}{r^2 \sin\theta \Delta r \Delta\theta \Delta\varphi} \right\rangle , \tag{11.8}$$

in which θ is the zenithal angle and φ is the azimuthal angle. Although $\Delta V(\mathbf{r})$ increases with r in this method, the judgment concerning whether or not a particle exists in the volume is straightforward.

If there is no anisotropy in the internal structure of a system, then the radial distribution function is used and Eq. (11.8) reduces to the following equation:

$$g(r) = \frac{1}{Vn^2}\left\langle \sum_{i=1}^{N} \frac{\Delta N_i(r)}{4\pi r^2 \Delta r} \right\rangle. \tag{11.9}$$

In this case $g(r)$ can be evaluated by calculating the number of particles existing in the spherical shell, $\Delta N_i(r)$.

Figure 11.8 shows the results of the radial distribution function for the Lennard-Jones system, which were obtained by Monte Carlo simulations. The nondimensional temperature T^* and number density n^* were taken as T^*=1.2 and n^*=0.1, 0.65, and 1.2, which correspond to the gas, liquid, and solid state, respectively. The Δr in Eq. (11.9) was taken as $\Delta r = \sigma/20$, in which σ is the diameter of Lennard-Jones particles. It is seen from Fig. 11.8 that, for n^*=0.1, the correlation increases significantly near the particle, but rapidly approaches unity over this range with increasing r^*. This is the typical characteristic of the radial distribution for a gas state. For the liquid state of n^*=0.65, the curve converges to unity with a periodic oscillation

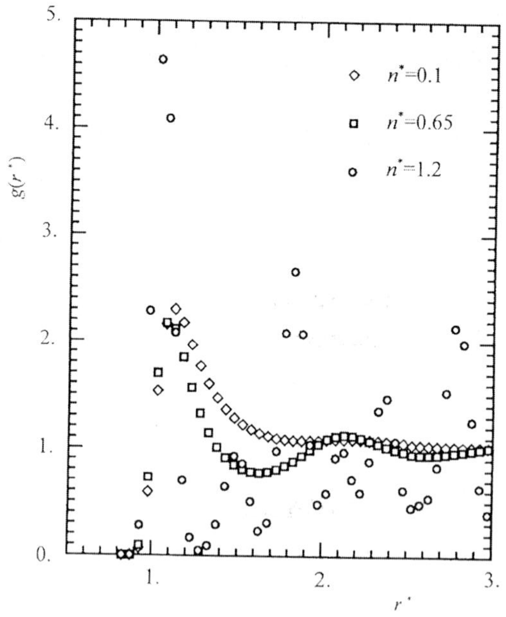

Figure 11.8 Radial distribution function for Lennard-Jones system for T^*=1.2.

around unity. Such convergence with an oscillation comes from the fact that particles cannot freely move to other areas due to the obstacle of the neighboring particles; thus particles have a tendency to stay at their original place to a certain degree. In contrast, for a solid state, particles can seldom move to other areas and so stay at their original place with small oscillational motion. The curve for $n^*=1.2$ in Fig. 11.8 clearly shows this characteristic, that is, there are sharp peaks at particle positions. If the system temperature is absolute zero, where particles do not move, the sharp peaks exhibit the characteristic of the Dirac delta function, with zero width centered at exact lattice positions.

11.6 The Estimate of Errors in the Averaged Values

Results obtained by simulations naturally have statistical errors, so that data should be shown as an average value with an error range. Such an expression of data with an error range is now standard in presenting results obtained by molecular-microsimulations, numerical analysis methods, and experiments.

In molecular-microsimulations, errors are usually classified into a systematic error and a statistical error. A systematic error arises from the limitations of a simulation region and the number of particles, the introduction of the boundary condition, the cutoff distance, etc. This kind of error cannot be removed even by increasing the amount of sampling. The statistical error is due to the finite length of time steps or the finite microscopic states generated in a simulation, and in this case, such errors can be improved by increasing the amount of sampling. We here explain the estimation methods of the statistical errors [1,2]. It should be noted that the influence of the number of particles and the cutoff radius on results has to be checked at least to evaluate the systematic error in simulations.

We consider M sampling values of a certain quantity, $A_1, A_2, \ldots, A_M$, which were obtained by a simulation. The average of these sampling values can be obtained as

$$\langle A \rangle_{run} = \frac{1}{M}\sum_{i=1}^{M} A_i \, . \tag{11.10}$$

Now we evaluate an error included in $\langle A \rangle_{run}$ with the assumption that stochastic variables A_1, $A_2, \ldots, A_M$ are independent of each other and obey a distribution of their population with an average $\langle A \rangle$ and a variance $\sigma^2(A)$. In this case, the central limit theorem says that a stochastic variable $\langle A \rangle_{run}$, which has been made by the arithmetic average of the series of stochastic variables, comes to obey a normal distribution with an average $\langle A \rangle$ and variance $\sigma^2(\langle A \rangle_{run})=\sigma^2(A)/M$ as the number of samples increases such as $M\rightarrow\infty$ [14]. If M is sufficiently large, the following relations can be obtained:

$$\left.\begin{aligned}&\langle A\rangle \approx \langle A\rangle_{run},\\&\sigma^2(A) \approx \langle(\delta A)^2\rangle_{run} = \frac{1}{M}\sum_{i=1}^{M}\left(A_i - \langle A\rangle_{run}\right)^2.\end{aligned}\right\} \tag{11.11}$$

Hence, the average $\langle A\rangle_{run}$ obtained by a simulation is seen to have the following error using the standard deviation:

$$\pm 1.96\sqrt{\langle(\delta A)^2\rangle_{run}/M}\,, \tag{11.12}$$

in which the coefficient 1.96 has appeared by using the reliability of 95 percent; if the reliability of 99 percent is used, the coefficient 1.96 has to be replaced with 2.58. The 95% reliability means that the value of $\langle A\rangle_{run}$ drops into the range from ($\langle A\rangle$ - 1.96$\sigma(A)/M^{1/2}$) to ($\langle A\rangle$ + 1.96$\sigma(A)/M^{1/2}$) with the probability of 95%. The sampled values A_1, A_2, …, A_M are generally not independent of each other, so that the error range expressed in Eq. (11.12) needs to be modified. We show a method of this modification below.

We now decompose a series of data into equal size of blocks in which M_B sampling data are included per block and pick up the first data alone of each block. If M_{corr} is the minimum value of M_B so that such selected data become independent of each other, then Eq. (11.12) can be corrected as

$$\pm 1.96\sqrt{\langle(\delta A)^2\rangle_{run}(M_{corr}/M)}\,. \tag{11.13}$$

Hence, by finding the value of M_{corr} from sampling data, the error expressed in Eq.(11.13) can be evaluated. We show a method of deriving the value of M_{corr} in the following.

If it is assumed that there are N_B blocks composed of M_B sampling data in each block, the equation of $N_B M_B = M$ is satisfied. Then the sub-average for each block may be written as

$$\langle A\rangle_{B_1} = \frac{1}{M_B}\sum_{i=1}^{M_B} A_i,\quad \langle A\rangle_{B_2} = \frac{1}{M_B}\sum_{i=M_B+1}^{2M_B} A_i,\quad \cdots. \tag{11.14}$$

The variance $\sigma^2(\langle A\rangle_B)$ of the sub-averages can be expressed as

$$\sigma^2(\langle A\rangle_B) = \langle(\delta\langle A\rangle_B)^2\rangle = \frac{1}{N_B}\sum_{b=1}^{N_B}(\langle A\rangle_{B_b} - \langle A\rangle_{run})^2\,. \tag{11.15}$$

The correlation among $\langle A\rangle_{B1}$, $\langle A\rangle_{B2}$, … becomes independent with increasing the value of M_B. Thus it is seen from Eq. (11.12) that the variance in Eq. (11.15) is inversely proportional to the value of M_B. That is, Eq. (11.15) can be rewritten as

$$\sigma^2\left(\langle A\rangle_B\right)=\frac{\beta\sigma^2(A)}{M_B} \qquad (M_B\to\infty), \tag{11.16}$$

in which β is a coefficient of proportionality to be evaluated. If there is no correlation among variables A_1, A_2, ..., A_M, then β is equal to unity. Reforming Eq. (11.16) leads to the following equation:

$$\beta=\lim_{M_B\to\infty}\frac{M_B\sigma^2(\langle A\rangle_B)}{\sigma^2(A)}. \tag{11.17}$$

Using Eqs. (11.11) and (11.15), the value of β can be obtained from Eq. (11.17); if the data of $M_B\sigma^2(\langle A\rangle_B)/\sigma^2(A)$ are plotted against the values of $(1/M_B)$ as the abscissa, the value of β can be calculated more straightforwardly. It is clearly seen from the correspondence between Eqs. (11.16) and (11.13) that M_{corr} is equal to β.

11.7 The Ewald Sum (the Treatment of Long-Range Order Potential)

In the case of short-range order potentials such as the Lennard-Jones potential, a significant reduction in computation time can be accomplished by introducing a cutoff distance. In contrast, for long-range order potentials such as the Coulomb potential, the introduction of a cutoff distance may cause significant errors. Hence the interactions with particles in virtual cells far from the central one may need to be taken into account. Here we consider a method useful for colloidal dispersion simulations, known as the Ewald sum.

11.7.1 Interactions between charged particles

We now consider a cubic simulation box with the side length L together with the periodic boundary condition. The system is composed of N charged particles and an arbitrary particle i has a charge q_i and located at a position $\mathbf{r}_i$. Furthermore, the system is assumed to be electrically neutral. With these assumptions, the total energy of the system (or the actual cell), E, due to the Coulomb interactions between particles can be expressed as

$$E=\frac{1}{2}\sum_{\mathbf{n}}{}'\left(\sum_{i=1}^{N}\sum_{j=1}^{N}\frac{q_i q_j}{|\mathbf{r}_{ji}+L\mathbf{n}|}\right), \tag{11.18}$$

in which the CGS unit system is used for simplicity of expressions, $\mathbf{r}_{ji}=\mathbf{r}_j-\mathbf{r}_i$, and $\mathbf{n}=(n_x,n_y,n_z)$; n_x, n_y, and $n_z=0,\pm1,\pm2,\pm3,\ldots$ Note that Eq. (11.18) needs to be divided by $4\pi\varepsilon_0$ (ε_0 the permittivity of free space) to transform it to the expression in SI units. The sum over n means the sum over all replicated cells including the central cell and the prime on this sum indicates that the case of $i=j$ is omitted for $\mathbf{n}=0$, which means the neglect of the interaction with itself.

It is noted that **n**=(0,0,0) corresponds to the central cell and the other cases of **n**≠(0,0,0) correspond to the replicated (or virtual) cells. It is seen from Eq. (11.18) that this equation is not suitable for simulations since the convergence of the sum over **n** is significantly slow. Therefore a modified version of Eq. (11.18) was derived using a well-devised technique to improve its convergence, and this is well known as the Ewald method [15-18].

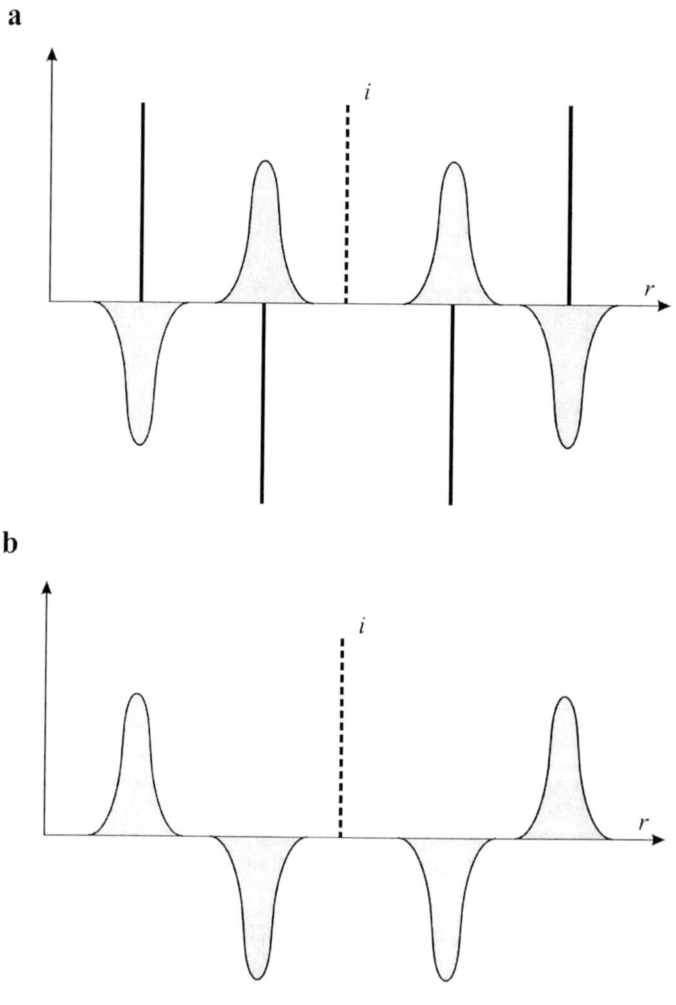

Figure 11.9 Two types of charge distribution used in the Ewald sum: (a) original point charges plus the screening distribution and (b) canceling distribution.

We consider the interaction energy E_i of particle i with all the others including the virtual particles. This expression is written as

$$E_i = q_i \sum_{\mathbf{n}}{}' \sum_{j=1}^{N} \frac{q_j}{|\mathbf{r}_{ji} + L\mathbf{n}|} . \tag{11.19}$$

According to the Ewald method, an arbitrary point charge is regarded as the sum of a point charge with a screening distribution and a canceling distribution. A schematic figure for explaining this concept is shown in Fig. 11.9. We concentrate our attention on particle j. If a normal distribution $\rho_j(\mathbf{r}')$ is used as a screening distribution, the point charge of particle j with the screening effect can be expressed as q_j- $\rho_j(\mathbf{r}')$, in which $\mathbf{r}'$ is the relative position from the point charge of interest (in the above case, from particle j). The canceling distribution is for canceling the screening effect and expressed as $\rho_j(\mathbf{r}')$ itself:

$$\rho_j(\mathbf{r}') = q_j \sigma(\mathbf{r}') = q_j \frac{\kappa^3}{\pi^{3/2}} \exp(-\kappa^2 r'^2) , \tag{11.20}$$

in which κ is an important parameter for determining the convergence speed of the summation in Eq. (11.18), as will be shown. Figure 11.9(a) corresponds to point charges with the screening effects and Fig. 11.9(b) to the canceling distributions. If the screening distribution does not have a large deviation, then the particle of interest will regard the other particles with the screening effects as electrically neutral, with increasing particle-particle separation. Hence, when we calculate the interaction energies between the particle of interest and the other particles with the screening effects, the neighboring particles are sufficient to be considered and virtual particles existing far from the central cell do not need to be taken into account. This is a significant feature of the Ewald sum method.

We here show the final results but the detailed derivation may be found in Appendix A6 because this derivation is not straightforward. Equation (11.18) can be rewritten according to the Ewald sum method as

$$\begin{aligned} E = \frac{1}{2} \sum_{i=1}^{N} \sum_{j=1}^{N} \Bigg\{ & \sum_{\mathbf{n}}{}' q_i q_j \frac{\mathrm{erfc}\left(\kappa |\mathbf{r}_{ji} + L\mathbf{n}|\right)}{|\mathbf{r}_{ji} + L\mathbf{n}|} \\ & + \frac{1}{\pi L^3} \sum_{\mathbf{k}(\neq 0)} q_i q_j \frac{4\pi^2}{k^2} \exp(-k^2/4\kappa^2) \cos(\mathbf{k} \cdot \mathbf{r}_{ji}) \Bigg\} - \frac{\kappa}{\pi^{1/2}} \sum_{i=1}^{N} q_i^2 , \end{aligned} \tag{11.21}$$

in which $\mathbf{k}$ is the reciprocal vector expressed as $\mathbf{k}=2\pi\mathbf{n}/L$, and erfc($x$) is the complementary error function which is defined as

$$\mathrm{erf\,c}(x) = 1 - \frac{2}{\sqrt{\pi}} \int_0^x e^{-t^2} dt . \tag{11.22}$$

The first term of the right-hand side in Eq. (11.21) is the interaction between the point charges and the screened particle charges, the second term is the interaction between the point charges and the canceling distributions, and the third term arises to remove the interaction with their own canceling distributions, which have been taken into account in the second term.

It is seen that the sum of the first term in Eq. (11.21) over **n** will converge quickly as the value of κ becomes large. In other words, the screened potentials are short-ranged for such a case, so that the first term can be calculated just by summing over all particles in the central cell and the nearest neighbor virtual cells. In contrast, the sum of the second term in Eq. (11.21) over **k** converges more slowly for large values of κ. Appropriate values of κ are requested for a fast convergence of the first and second terms in Eq. (11.21) simultaneously. For example, if κ is chosen as $\kappa=5/L$, then 100~200 waves of **k** are necessary to obtain the sufficient convergence of the second term in Eq. (11.21) [19].

11.7.2 Interactions between electric dipoles

The above Ewald sum method for charged particle-particle interactions can relatively straightforwardly be extended to the evaluation of electric dipole-dipole interactions. Such an evaluation may be accomplished by regarding an electric dipole as a pair of the same positive and negative point charges with an infinitesimal separation [16]. We just show the final result, and the detailed derivation may be found in Appendix A6. If the electric dipole moment of particle i is denoted by $\boldsymbol{\mu}_i$, the total interaction energy of the central cell for the electric dipole-dipole interaction, E, can be expressed as

$$
\begin{aligned}
E = \frac{1}{2}\sum_{i=1}^{N}\sum_{j=1}^{N}\Bigg[\sum_{\mathbf{n}}{}' \Big\{ & A\big(|\mathbf{r}_{ji}+L\mathbf{n}|\big)(\boldsymbol{\mu}_j\cdot\boldsymbol{\mu}_i) \\
& - B\big(|\mathbf{r}_{ji}+L\mathbf{n}|\big)\big((\mathbf{r}_{ji}+L\mathbf{n})\cdot\boldsymbol{\mu}_j\big)\big((\mathbf{r}_{ji}+L\mathbf{n})\cdot\boldsymbol{\mu}_i\big)\Big\} \\
& + \frac{1}{\pi L^3}\sum_{\mathbf{k}(\neq 0)} \frac{4\pi^2}{k^2}\exp(-k^2/4\kappa^2)(\mathbf{k}\cdot\boldsymbol{\mu}_j)(\mathbf{k}\cdot\boldsymbol{\mu}_i)\cos(\mathbf{k}\cdot\mathbf{r}_{ji})\Bigg],
\end{aligned}
\tag{11.23}
$$

in which

$$
\left.
\begin{aligned}
A(r) &= \mathrm{erfc}(\kappa r)/r^3 + (2\kappa/\pi^{1/2})\exp(-\kappa^2 r^2)/r^2, \\
B(r) &= 3\,\mathrm{erfc}(\kappa r)/r^5 + (2\kappa/\pi^{1/2})(2\kappa^2 + 3/r^2)\exp(-\kappa^2 r^2)/r^2.
\end{aligned}
\right\}
\tag{11.24}
$$

In these equation, the CGS unit system is used for simplicity.

From the similarity between electric and magnetic dipoles, Eq. (11.23) becomes the interaction energy for a magnetic dipole system by just replacing the electric dipole moment $\boldsymbol{\mu}_i$ with the magnetic one $\mathbf{m}_i$ together with the multiplication by the constant ($\mu_0/4\pi$).

11.8 The Criterion for the Overlap of Spherocylinder Particles

Judging whether or not there are particle overlaps is an indispensable procedure in conducting microsimulations for a colloidal dispersion. Physically unreasonable overlaps cause extraordinarily large repulsive forces between particles (which lead to the divergence of a system), physically unreasonable particle motion (such as penetration through other particles), and so forth. Hence, making computational programs without such unreasonable overlaps is one of the important factors to ensure reliable results with sufficient accuracy. For a spherical particle system, it is quite straightforward to derive the condition of particle overlaps and to develop a program to test for the overlap of particles. Thus we here show an example for a non-spherical particle system composed of spherocylinder particles; a spherocylinder is a typical model for a non-spherical particle, besides a spheroid.

We use a spherocylinder shown in Fig. 15.1: the length of the cylindrical part is denoted by l and the diameter of the hemispheres is denoted by d. The position of the mass center and the axis direction of particle i are denoted by $\mathbf{r}_i$ and $\mathbf{n}_i$, respectively. We now consider the conditions for the overlap between spherocylindrical particles i and j. We draw a straight line which is perpendicular to both particle axes (including the prolongation) and crosses such axis lines at certain points. If this line is defined by the line crossing point P_i (expressed as $\mathbf{r}_i+k_i\mathbf{n}_i$ on the axis line of particle i) and point P_j (expressed as $\mathbf{r}_j+k_j\mathbf{n}_j$ on the axis line of particle j), then this line has to satisfy the following conditions:

$$\left.\begin{aligned} \mathbf{n}_i \cdot \{(\mathbf{r}_i + k_i\mathbf{n}_i) - (\mathbf{r}_j + k_j\mathbf{n}_j)\} = 0 \ , \\ \mathbf{n}_j \cdot \{(\mathbf{r}_i + k_i\mathbf{n}_i) - (\mathbf{r}_j + k_j\mathbf{n}_j)\} = 0 \ . \end{aligned}\right\} \tag{11.25}$$

From these equations, k_i and k_j can readily be solved as

$$\begin{bmatrix} k_i \\ k_j \end{bmatrix} = \frac{1}{1-(\mathbf{n}_i \cdot \mathbf{n}_j)^2} \begin{bmatrix} -1 & \mathbf{n}_i \cdot \mathbf{n}_j \\ -\mathbf{n}_i \cdot \mathbf{n}_j & 1 \end{bmatrix} \begin{bmatrix} \mathbf{n}_i \cdot \mathbf{r}_{ij} \\ \mathbf{n}_j \cdot \mathbf{r}_{ij} \end{bmatrix} . \tag{11.26}$$

This equation holds when $\mathbf{n}_i \cdot \mathbf{n}_j \neq \pm 1$, that is, when the two particles are not parallel. Since the condition of the overlap for the parallel situation is quite straightforwardly derived, we focus our attention on non-parallel situations. When the separation between points P_i and P_j is larger than the diameter d of the hemispheres, particle cannot overlap. Thus the following discussion is limited to the case in which the separation is smaller than the diameter d.

The overlap of two spherocylinders arises in the following three cases: the overlaps between (1) two hemispheres, (2) a hemisphere and a cylindrical part, or (3) two cylindrical parts. Firstly, the condition for the overlap between two cylindrical parts can be straightforwardly derived using Eq. (11.26). That is, the overlap between two cylindrical parts arises if the following conditions are satisfied:

$$\left.\begin{array}{l} |(\mathbf{r}_i + k_i\mathbf{n}_i) - (\mathbf{r}_j + k_j\mathbf{n}_j)| < d, \\ |k_i| < l/2, \quad |k_j| < l/2. \end{array}\right\} \tag{11.27}$$

Secondly, when the conditions of ($|k_i| \le l/2$ and $|k_j| > l/2$) or ($|k_i| > l/2$ and $|k_j| \le l/2$) hold, the overlap between a hemisphere and a cylindrical part may arise. We now consider the former condition of ($|k_i| \le l/2$ and $|k_j| > l/2$), which means a possibility of the overlap between the hemisphere of particle j and the cylindrical part of particle i. If a line which is drawn from the center of the sphere (hemisphere) of particle j to the axis line of particle i perpendicularly is assumed to cross the axis line of particle i at $(\mathbf{r}_i + k_i^s\mathbf{n}_i)$, then k_i^s can be solved as $k_i^s = \mathbf{n}_i \cdot (\mathbf{r}_j^s - \mathbf{r}_i)$, in which $\mathbf{r}_j^s$ is the position of sphere (hemisphere) of particle j (similarly, $\mathbf{r}_i^s$). The overlap between the cylindrical part of particle i and the hemisphere of particle j appears when the following conditions are satisfied with the value of k_i^s:

$$\left.\begin{array}{l} |k_i^s| \le l/2, \\ |(\mathbf{r}_i + k_i^s\mathbf{n}_i) - \mathbf{r}_j^s| < d. \end{array}\right\} \tag{11.28}$$

The hemispheres of particles i and j overlap if the following condition is satisfied:

$$\left.\begin{array}{l} |k_i^s| > l/2, \\ |\mathbf{r}_i^s - \mathbf{r}_j^s| < d. \end{array}\right\} \tag{11.29}$$

Finally, we consider the overlap between two hemispheres under the condition of $|k_i| > l/2$ and $|k_j| > l/2$. For simplicity, we assume that $|k_j|$ is larger than $|k_i|$. In this case, there is a possibility of the overlap between two hemispheres or between the cylindrical part of particle i and the hemisphere of particle j. If k_i^s is calculated, and $|k_i^s|$ is larger than $l/2$, then there is a possibility of the overlap between two hemispheres. If $|k_i^s|$ is smaller than $l/2$, then there is a possibility of the overlap between the cylindrical part of particle i and the hemisphere of particle j, in which the judgment procedure concerning the overlap is the same as before. Hence we consider the former case. The condition of the overlap of the two hemispheres can quite straightforwardly be written as

$$|\mathbf{r}_i^s - \mathbf{r}_j^s| < d. \tag{11.30}$$

The conditions for particles i and j are summarized as follows:

(1) When $|(\mathbf{r}_i + k_i\mathbf{n}_i) - (\mathbf{r}_j + k_j\mathbf{n}_j)| \ge d$, there is no overlap.

(2) When $|(\mathbf{r}_i+k_i\mathbf{n}_i)-(\mathbf{r}_j+k_j\mathbf{n}_j)| < d$, there is a possibility of overlap:

(2)-1 When $|k_i| \le l/2$ and $|k_j| \le l/2$, there is an overlap between cylindrical parts;

(2)-2 When $|k_i| \le l/2$, $|k_j| > l/2$, and $|k_i^s| < l/2$, there is a possibility of the overlap between the cylindrical part of particle i and the hemisphere of particle j:

(2)-2-1 When $|(\mathbf{r}_i+k_i^s\mathbf{n}_i)-\mathbf{r}_j^s| \ge d$, there is no overlap;

(2)-2-2 When $|(\mathbf{r}_i+k_i^s\mathbf{n}_i)-\mathbf{r}_j^s| < d$, there is an overlap;

(2)-3 When $|k_i| \le l/2$, $|k_j| > l/2$, and $|k_i^s| \ge l/2$, there is a possibility of the overlap between hemispheres:

(2)-3-1 When $|\mathbf{r}_i^s-\mathbf{r}_j^s| \ge d$, there is no overlap;

(2)-3-2 When $|\mathbf{r}_i^s-\mathbf{r}_j^s| < d$, there is an overlap;

(2)-4 When $|k_j| > |k_i| > l/2$, and $|k_i^s| < l/2$, there is a possibility of the overlap between the cylindrical part of particle i and the hemisphere of particle j:

(2)-4-1 When $|(\mathbf{r}_i+k_i^s\mathbf{n}_i)-\mathbf{r}_j^s| \ge d$, there is no overlap;

(2)-4-2 When $|(\mathbf{r}_i+k_i^s\mathbf{n}_i)-\mathbf{r}_j^s| < d$, there is an overlap;

(2)-5 When $|k_j| > |k_i| > l/2$, and $|k_i^s| \ge l/2$, there is a possibility of the overlap between hemispheres:

(2)-5-1 When $|\mathbf{r}_i^s-\mathbf{r}_j^s| \ge d$, there is no overlap;

(2)-5-2 When $|\mathbf{r}_i^s-\mathbf{r}_j^s| < d$, there is an overlap.

The procedure of step (2)-2 is for judging the overlap between the cylindrical part of particle i and the hemisphere of particle j, and the similar procedure for the hemisphere of particle i and the cylindrical part of particle j can straightforwardly be shown. Similarly, the judgment for the case of $|k_j| \le l/2$, $|k_i| > l/2$, and $|k_j^s| \ge l/2$ is straightforwardly written from the procedure of step (2)-3. Also, for the case of $|k_i| > |k_j| > l/2$, the procedure similar to steps (2)-4 and (2)-5 can readily be derived.

References

1. J.M. Haile, " Molecular Dynamics Simulation: Elementary Methods," John Wiley & Sons, New York, 1992
2. M.P. Allen and D.J. Tildesley, "Computer Simulation of Liquids," Clarendon Press, Oxford, 1987
3. D.W. Heermann, "Computer Simulation Methods in Theoretical Physics," 2nd Ed., Springer-Verlag, Berlin, 1990
4. D.C. Rapaport, "The Art of Molecular Dynamics Simulation," Cambridge University Press, Cambridge, 1995

5. N. Metropolis, A.W. Rosenbluth, M.N. Rosenbluth, and A. Teller, Equation of State Calculations by Fast Computing Machines, J. Chem. Phys., 21(1953) 1087
6. A.W.Lees and S. F. Edwards, The Computer Study of Transport Processes under Extreme Conditions, J. Phys. C, 5 (1972) 1921
7. D. J. Evans and G. P. Morriss, Non-Newtonian Molecular Dynamics, Comput. Phys. Rep., 1 (1984) 297
8. A. Satoh, A New Outer Boundary Condition for Molecular Dynamics Simulations and Its Application to a Rarefied Gas Flow past a Cylinder (Periodic-Shell Boundary Condition), Advanced Powder Tech., 5 (1994) 105
9. A. Satoh, Periodic-Shell Boundary Condition for Soft Molecular Systems: Molecular Dynamics Simulations of Flow past a Circular Cylinder, Molec. Phys., 92 (1997) 715
10. A. Satoh, Molecular Dynamics Simulations on Internal Structures of Normal Shock Waves in Lennard-Jones Liquids, ASME, J. Fluids Eng., 117 (1995) 97
11. B. Quentrec and C. Brot, New Method for Searching for Neighbors in Molecular Dynamics Computations, J. Comput. Phys., 13 (1975) 430
12. R.W. Hockney and J.W. Eastwood, "Computer Simulation using Particles," McGraw-Hill, New York, 1981
13. L. Verlet, Computer Experiments on Classical Fluids. I. Thermodynamical Properties of Lennard-Jones Molecules, Phys. Rev., 159 (1967) 98
14. A. Papoulis, "Probability, Random Variables, and Stochastic Processes," 3rd Ed., 188-190, McGraw-Hill, Boston, 1991
15. P. Ewald, Die Berechnung Optischer and Elektrostatischer Gitterpotentiale, Ann. Phys., 64 (1921) 253
16. S.W. de Leeuw, J.W. Perram, and E.R. Smith, Simulation of Electrostatic Systems in Periodic Boundary Conditions. I. Lattice Sums and Dielectric Constants, Proc. Roy. Soc. London A, 373 (1980) 27
17. C. Kittel, "Introduction to Solid State Physics," 6th Ed., 606-610, John Wiley & Sons, New York, 1986
18. D.M. Heyes, Electrostatic Potentials and Fields in Infinite Point Charge Lattices, J. Chem. Phys., 74 (1981) 1924
19. L.V. Woodcock and K. Singer, Thermodynamic and Structural Properties of Liquid Ionic Salts obtained by Monte Carlo Computation, Trans. Faraday Soc., 67 (1971) 12

Exercises

11.1 We consider a three-dimensional system composed of 1000 spherical particles with the dimensionless number density $n^*(=nd^3$, d the particle diameter$)=0.1$. If the cell index method with a cutoff radius r_{coff}^* $(=r_{coff}/d)=5$ is used , how many cells can a cubic simulation box be divided into? Also, consider whether or not the cell index method functions effectively for this case.

11.2 We consider the same problem as in the previous problem 11.1 with the same assumptions, but a two-dimensional system is treated for this case. How many cells can a square simulation region be divided into? Also, consider whether or not the cell index method functions effectively for the present two-dimensional system.

11.3 Show that the pair correlation function can be expressed for a two-dimensional system as

$$g^{(2)}(\mathbf{r}) = g^{(2)}(r,\theta,) = \frac{S}{N^2}\left\langle \sum_{i=1}^{N} \frac{\Delta N_i(r,\theta,)}{r\Delta r\Delta\theta} \right\rangle, \tag{11.31}$$

which corresponds to Eq. (11.8) for a three-dimensional system. In Eq. (11.31), S is the area of a simulation region and $\Delta N_i(r,\theta\ )$ is the number of particles existing in a small area at a position (r,θ).

11.4 Similarly to the previous problem 11.3, show that the radial distribution function can be written for a two-dimensional system as

$$g(r) = \frac{S}{N^2}\left\langle \sum_{i=1}^{N} \frac{\Delta N_i(r)}{2\pi r\Delta r} \right\rangle, \tag{11.32}$$

in which $\Delta N_i(r)$ is the number of particles existing in a small shell area with a width Δr at a radius r.

CHAPTER 12

SOME EXAMPLES OF MICROSIMULATIONS

In this chapter we show the results which were obtained by the Stokesian dynamics [1,2] and Brownian dynamics simulations [3] for ferromagnetic colloidal dispersions in order to understand the role of microsimulation methods in investigating physical phenomena. A simple shear flow in the x-direction is considered, as shown in Fig. 8.1, and the magnetic field is applied in the y-direction.

12.1 Stokesian Dynamics Simulations

Even if particles are smaller than micron-order, the translational and rotational Brownian motion may be negligible when the applied magnetic field is strong and also when the magnetic interaction between particles is dominant compared with the Brownian motion. In this section, we describe Stokesian dynamics simulation results obtained under such conditions. The results obtained by the additivity of velocities will be compared with those obtained by the additivity of forces; these approximations have already been explained in Sec. 8.

12.1.1 The dispersion model

As shown in Sec. 3.7.2, a particle is idealized as a spherical particle, with central point dipole,

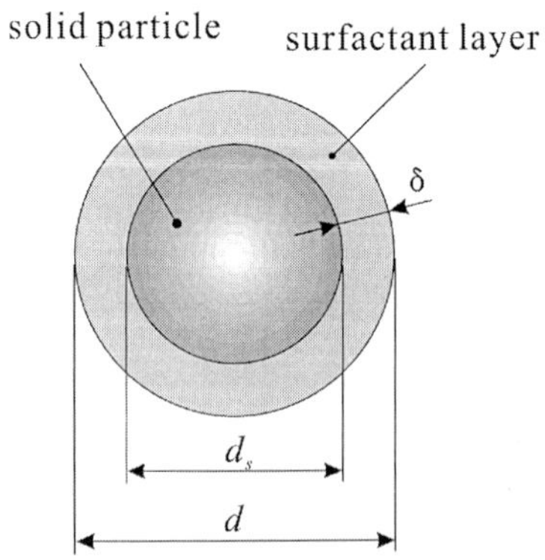

Figure 12.1 Particle model.

coated with a uniform surfactant layer (or a steric layer). A colloidal dispersion is assumed to be composed of such particles with a radius a. The magnetic interaction energy between particles i and j, $u_{ij}^{(m)}$, and the particle-field interaction energy, $u_i^{(H)}$, are written in Eq. (3.57). Repulsive forces arise due to the overlapping of the steric layers, and the expression for the interaction energy for this case, $u_{ij}^{(V)}$, was developed by Rosensweig et al. [4] as

$$u_{ij}^{(V)} = \frac{\pi d_s^{\,2} n_s kT}{2}\left\{2-\left(\frac{r_{ij}}{\delta}\right)\ln\left(\frac{d}{r_{ij}}\right)-\frac{r_{ij}-(d-2\delta)}{\delta}\right\}, \tag{12.1}$$

in which r_{ij} is the distance between particles i and j, n_s is the number of surfactant molecules in a unit area of the particle surface, d_s is the diameter of the solid particle as shown in Fig. 12.1, δ is the thickness of the steric layer, k is Boltzmann's constant, and T is the temperature.

The $u_{ij}^{(m)}$, $u_i^{(H)}$, and $u_{ij}^{(V)}$ are nondimensionalized by the thermal energy kT as

$$u_{ij}^{(m)*} = u_{ij}^{(m)}/kT = \lambda\left(\frac{d_s}{r_{ij}}\right)^3 \{\mathbf{n}_i\cdot\mathbf{n}_j - 3(\mathbf{n}_i\cdot\mathbf{t}_{ij})(\mathbf{n}_j\cdot\mathbf{t}_{ij})\}, \tag{12.2}$$

$$u_i^{(H)*} = u_i^{(H)}/kT = -\xi\,\mathbf{n}_i\cdot\mathbf{h}, \tag{12.3}$$

$$u_{ij}^{(V)*} = u_{ij}^{(V)}/kT = \lambda_V\left\{2-\frac{2r_{ij}/d_s}{t_\delta}\ln\left(\frac{d}{r_{ij}}\right)-2\frac{r_{ij}/d_s-1}{t_\delta}\right\}, \tag{12.4}$$

in which $\mathbf{n}_i$ and $\mathbf{n}_j$ are the unit vectors denoting the direction of the magnetic moment of the particles ($\mathbf{n}_i=\mathbf{m}_i/m$, $m=|\mathbf{m}_i|$), $\mathbf{t}_{ij}=(\mathbf{r}_i-\mathbf{r}_j)/|\mathbf{r}_i-\mathbf{r}_j|$, $\mathbf{h}=\mathbf{H}/H$, and t_δ is the ratio of the thickness of the steric layer δ to the radius of the solid part of the particle, equal to $2\delta/d_s$. The nondimensional parameters, appearing in the above equations, λ, ξ, and λ_V are written as

$$\lambda = \frac{\mu_0 m^2}{4\pi d_s^{\,3} kT}, \quad \xi = \frac{\mu_0 mH}{kT}, \quad \lambda_V = \frac{\pi d_s^{\,2} n_s}{2}. \tag{12.5}$$

It is straightforward to derive the expressions for the forces and torques acting on particle i using the above energy equations. We show here the final result for the magnetic force $\mathbf{F}_{ij}^{(m)}$ exerted on particle i by particle j:

$$\mathbf{F}_{ij}^{(m)} = -\frac{3\mu_0 m^2}{4\pi d^4}\cdot\frac{1}{(r_{ij}/d)^4}\left[-(\mathbf{n}_i\cdot\mathbf{n}_j)\mathbf{t}_{ij}+5(\mathbf{n}_i\cdot\mathbf{t}_{ij})(\mathbf{n}_j\cdot\mathbf{t}_{ij})\mathbf{t}_{ij}-\{(\mathbf{n}_j\cdot\mathbf{t}_{ij})\mathbf{n}_i+(\mathbf{n}_i\cdot\mathbf{t}_{ij})\mathbf{n}_j\}\right]. \tag{12.6}$$

If we take into account the magnetic field induced at the position of particle i by the magnetic dipole moment of particle j, then the torque $\mathbf{T}_{ij}^{(m)}$ acting on particle i due to the magnetic interaction with particle j is expressed as

$$\mathbf{T}_{ij}^{(m)} = -\frac{\mu_0 m^2}{4\pi d^3} \cdot \frac{1}{(r_{ij}/d)^3} \{\mathbf{n}_i \times \mathbf{n}_j - 3(\mathbf{n}_j \cdot \mathbf{t}_{ij})\mathbf{n}_i \times \mathbf{t}_{ij}\}. \tag{12.7}$$

Similarly, the torque $\mathbf{T}_i^{(H)}$ acting on particle i due to the deviation of the magnetic moment from the applied magnetic field direction is written as

$$\mathbf{T}_i^{(H)} = \mu_0 \mathbf{m}_i \times \mathbf{H} = \mu_0 m H \mathbf{n}_i \times \mathbf{h}. \tag{12.8}$$

Finally, the repulsive force $\mathbf{F}_{ij}^{(V)}$ exerted on particle i by the overlap with the steric layer of particle j is expressed as

$$\mathbf{F}_{ij}^{(V)} = kT\lambda_V \frac{1}{\delta} \cdot \mathbf{t}_{ij} \ln\left(\frac{d}{r_{ij}}\right) \quad (d_s \le r_{ij} \le d). \tag{12.9}$$

Since the interaction due to the overlap of the steric layers is dependent only on the particle-particle separation, torques do not arise due to this factor. In addition to the above forces and torques, the van der Waals force may act between particles, but this force is not an important factor in this case since the particles are coated with a steric layer. We therefore do not take into account van der Waals interactions. The force $\mathbf{F}_i$ and torque $\mathbf{T}_i$ (i=1,2,...,N) acting on particle i can now be expressed as

$$\mathbf{F}_i = \sum_{\substack{j=1 \\ (j \ne i)}}^{N} (\mathbf{F}_{ij}^{(m)} + \mathbf{F}_{ij}^{(V)}), \quad \mathbf{T}_i = \sum_{\substack{j=1 \\ (j \ne i)}}^{N} \mathbf{T}_{ij}^{(m)} + \mathbf{T}_i^{(H)}. \tag{12.10}$$

12.1.2 Nondimensional numbers characterizing phenomena

According to the method of nondimensionalizing physical quantities which has been explained in Sec. 8.3, the dimensionless force $\mathbf{F}_i^*$ and torque $\mathbf{T}_i^*$ (i=1,2,...,N) are written as

$$\mathbf{F}_i^* = \sum_{\substack{j=1 \\ (j \ne i)}}^{N} (\mathbf{F}_{ij}^{(m)*} + \mathbf{F}_{ij}^{(V)*}), \quad \mathbf{T}_i^* = \sum_{\substack{j=1 \\ (j \ne i)}}^{N} \mathbf{T}_{ij}^{(m)*} + \mathbf{T}_i^{(H)*}, \tag{12.11}$$

in which

$$\mathbf{F}_{ij}^{(m)*} = -R_m \cdot \frac{8}{r_{ij}^{*4}} [-(\mathbf{n}_i \cdot \mathbf{n}_j)\mathbf{t}_{ij} + 5(\mathbf{n}_i \cdot \mathbf{t}_{ij})(\mathbf{n}_j \cdot \mathbf{t}_{ij})\mathbf{t}_{ij} - \{(\mathbf{n}_j \cdot \mathbf{t}_{ij})\mathbf{n}_i + (\mathbf{n}_i \cdot \mathbf{t}_{ij})\mathbf{n}_j\}], \tag{12.12}$$

$$\mathbf{F}_{ij}^{(V)*} = R_V \mathbf{t}_{ij} \ln\left(\frac{2}{r_{ij}^*}\right) \quad (2/(1+t_\delta) \le r_{ij}^* \le 2), \tag{12.13}$$

$$\mathbf{T}_{ij}^{(m)*} = -R_m \cdot \frac{2}{r_{ij}^{*3}} \{\mathbf{n}_i \times \mathbf{n}_j - 3(\mathbf{n}_j \cdot \mathbf{t}_{ij})\mathbf{n}_i \times \mathbf{t}_{ij}\}, \tag{12.14}$$

$$\mathbf{T}_i^{(H)*} = R_H \mathbf{n}_i \times \mathbf{h}\,. \tag{12.15}$$

The nondimensional numbers, appearing in these equations, R_m, R_H, and R_V arise due to forces and torques being nondimensionalized by the viscous shear force, and expressed as

$$R_m = \frac{\mu_0 m^2}{64\pi^2 \eta a^6 \dot{\gamma}}, \quad R_H = \frac{\mu_0 mH}{8\pi\eta a^3 \dot{\gamma}}, \quad R_V = \frac{kT\lambda_V}{6\pi\eta a^2 \dot{\gamma}\delta}. \tag{12.16}$$

R_m is the ratio of the representative magnetic particle-particle force to the representative hydrodynamic shear force, R_H is the ratio of the representative torque due to an applied field to the hydrodynamic torque, and R_V is the ratio of the representative repulsion due to steric layers to the hydrodynamic shear force.

As pointed out in Sec. 8.3, since the nondimensional shear rate γ^* is always unity, we cannot change the flow rate using this quantity. However, by changing the value of R_m with constant values of R_H/R_m and R_V/R_m, one can generate a simple shear flow with an arbitrary strength of shear rate. The expressions for R_H/R_m and R_V/R_m are written as

$$R_H / R_m = \frac{\pi d^3 H}{m} = \frac{\pi d^3 kT}{\mu_0 m^2} \cdot \frac{\mu_0 mH}{kT} = \frac{(1+t_\delta)^3}{4} \cdot \frac{\xi}{\lambda}, \tag{12.17}$$

$$R_V / R_m = \frac{2\pi d^4 kT\lambda_V}{3\mu_0 m^2 \delta} = \frac{4\pi d^3 kT}{3\mu_0 m^2} \cdot \frac{\lambda_V}{2\delta / d} = \frac{(1+t_\delta)^4}{3t_\delta} \cdot \frac{\lambda_V}{\lambda}. \tag{12.18}$$

It is clear from Eqs. (12.17) and (12.18) that the values of R_H/R_m and R_V/R_m are dependent on the properties of the particles, but independent of the flow properties. In other words, if the magnetic properties of the particles are specified, these values are constant for all of the shear rate. Hence the shear rate $\dot{\gamma}$ can be changed from ∞ to 0 by taking the value of R_m from 0 to ∞ with constant values of R_H/R_m and R_V/R_m.

12.1.3 Parameters for simulations

The following conditions and values were used in conducting the Stokesian dynamics simulations. In the simulations based on the additivity of forces, huge computation time is required because the inverse of the resistance matrix has to be calculated every time step. As shown later, even a small system of N=64 is too large from a computation time point of view. We consider, therefore, a special case of a very strong magnetic field (ξ=∞), which leads to the simplifications, $\boldsymbol{\omega}_i^*$=0 and $\mathbf{n}_i$=$\mathbf{h}$ (i=1,...,N). The number of particles, N, is 64, the volumetric fraction is taken as φ_V =0.15, and the number density, n^*(=na^3), is 0.0358. For this case, the length of the simulation box (cubic), $L^*(=L/a)$, is 12.14. The value of R_V/R_m is chosen as 52.89, which corresponds to the case of λ= 9 where thick chainlike clusters are formed in a strong magnetic field [5]. The values of R_m are taken as R_m=1, 5, 20, 50, and 100 to clarify the influence of the shear rate. The cutoff radius for magnetic particle-particle interactions, r_{coff}^* $(=r_{coff}/a)$, is $L^*/2$.

The values of the time interval Δt^* have to be selected carefully. If the time interval is not much shorter than the characteristic times for the shear flow and for the particle motion due to magnetic interactions, the particles may overlap to an unreasonably large extent, thereby causing large repulsive forces and the divergence of the system. Since the force in the nondimensional system is characterized by R_m, as shown in Eq. (12.12), the characteristic time $\Delta\tau^*$ for the case of the nondimensional force R_m and the length ΔL^* is expressed as $\Delta\tau^*=\Delta L^*/R_m$. If we take ΔL^* as $\Delta L^*=0.1t_\delta$, the characteristic time can be obtained as

$$\Delta\tau^* = \frac{0.1t_\delta}{R_m}. \tag{12.19}$$

The time interval Δt^* has to be much shorter than the characteristic time for the shear flow, so that we adopted the time interval as Δt^*=min(0.001, $\Delta\tau^*$). For example, Δt^*=(0.0001, 0.00002) were used for R_m=(20,100), respectively. The simulations were carried out up to 2,500,000 time steps to obtained averaged values with sufficient accuracy.

The values of the resistance and mobility functions are necessary for the arbitrary particle-particle separations in conducting the simulations. In order to achieve this, we tabulated the results of the resistance and mobility functions before the simulations, which were obtained by the numerical method[6] (also see Appendix A.10). The mobility and resistance function data necessary to evaluate the particle velocities was computed, using an interpolation procedure, from these tabulated results.

12.1.4. Results

Figure 12.2 shows the snapshots of the aggregate structures in equilibrium for R_m=20, in which Fig. 12.2(a) is for the additivity of forces (AF), Fig. 12.2(b) for the additivity of velocities (AV), and Fig. 12.2(c) for the approximation of ignoring hydrodynamic interactions between particles (AIHI). The figure on the left-hand is viewed from an oblique angle, and the figure on the right-hand side from the magnetic field direction. It is seen that particles aggregate to form a wall-like structure for each case, and also that a significant difference in these structures is not qualitatively observed. The wall-like structure arises mainly due to the magnetic interactions between particles, since the viscous shear forces are significantly dominated by the magnetic forces for R_m=20, which may be confirmed by the case of no hydrodynamic interactions, AIHI,

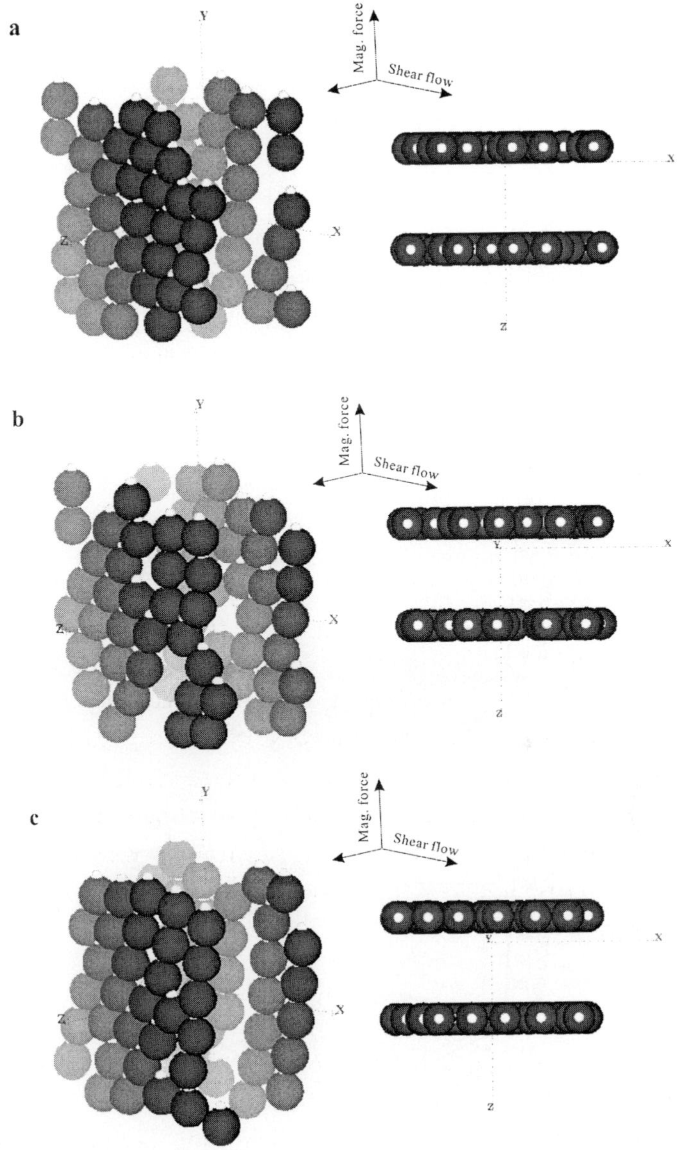

Figure 12.2 Snapshots of aggregate structures for R_m=20: (a) additivity of forces, (b) additivity of velocities, and (c) without hydrodynamic interactions.

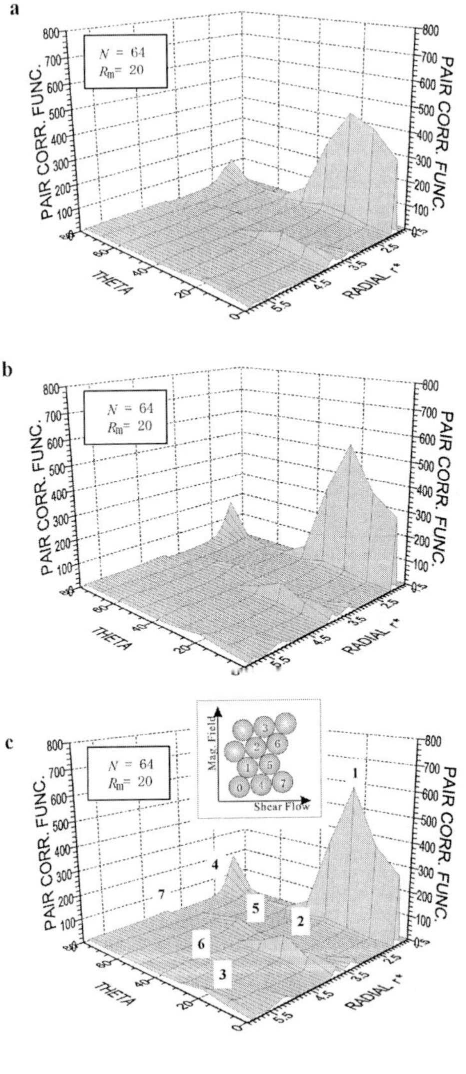

Figure 12.3 Pair correlation functions for R_m =20: (a) additivity of forces, (b) additivity of velocities, and (c) without hydrodynamic interactions.

Fig. 12.2(c). In the small system of N=64, thick chainlike clusters are hardly formed and the predominantly thin chainlike clusters incline towards the shear flow direction. These thin clusters are presumed to aggregate to form wall-like clusters due to the magnetic interactions, as shown in Fig. 12.2. Hence it is expected that thicker wall-like aggregates are formed for a larger system, which will be clearly shown later for N=512.

The results of the pair correlation function are shown in Fig. 12.3 in order to see the quantitative difference in the aggregate structures between AF and AV. Figure 12.3(a) is for

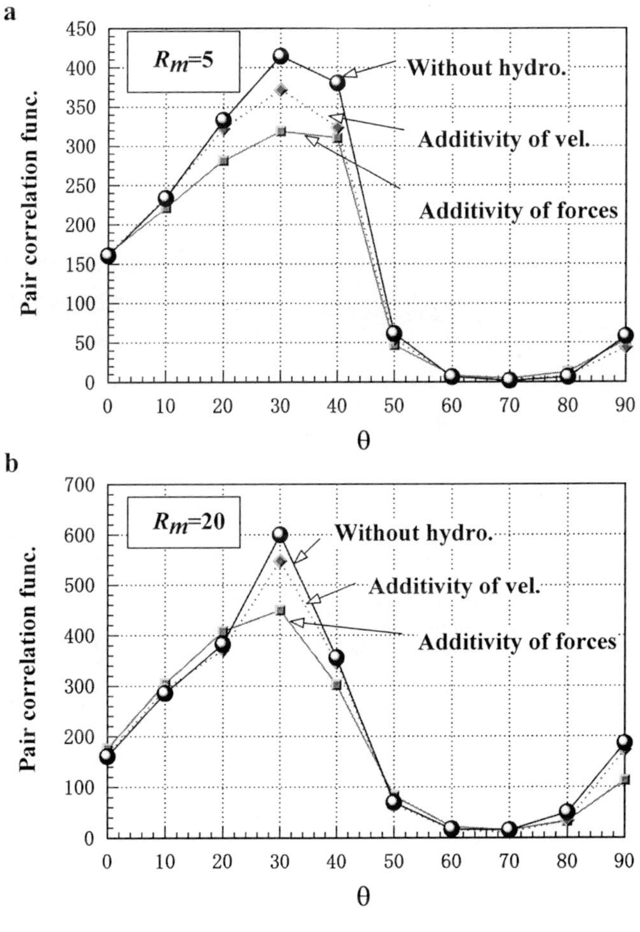

Figure 12.4 Pair correlation functions at r^*=1.99: (a) for R_m =5, and (b) for R_m =20.

the former case, and Fig. 12.3(b) for the latter case. Also, the result for AIHI is shown in Fig. 12.3(c) for comparison. All results were obtained for R_m=20, in which the magnetic interactions dominate the viscous shear forces. The notation of θ in the figures refers to the angle measured from the magnetic field direction to the shear flow direction. In order to clarify the differences more clearly, the correlation function at r^*=1.99 is shown as a function of θ in Fig. 12.4, in which Fig. 12.4(a) is for R_m=5, and Fig. 12.4(b) for R_m=20. We consider the relationship between the aggregate structure and each peak in the correlation function, using Fig. 12.3(c). Since aggregates have a wall-like structure with the thickness of a single particle diameter for R_m=20, it is quite understandable that each peak in the correlation function corresponds to the correlation of particle 0 with the particle of the corresponding number. It is seen from Figs. 12.3 and 12.4(b) that there are quantitative differences among the three cases, although the results obtained by AF and AV are essentially in qualitative agreement with that for AIHI.

Figure 12.5 shows the influence of the shear rate on the averaged viscosities for AF, AV, and AIHI. The maximum and minimum values of the sub-averaged viscosity for 50,000 time steps are used as an error bar for R_m=50 and 100. It is seen that the viscosity increases significantly with the increasing R_m, or, with decreasing shear rate, and this non-Newtonian property almost coincides quantitatively among the three cases. We have already seen that particles aggregate to form a stable wall-like structure in equilibrium when the magnetic forces significantly dominate the viscous shear forces. In this situation, the lubrication effect seldom occurs, so that it is quite understandable that there is no significant difference in the averaged viscosity among these three cases.

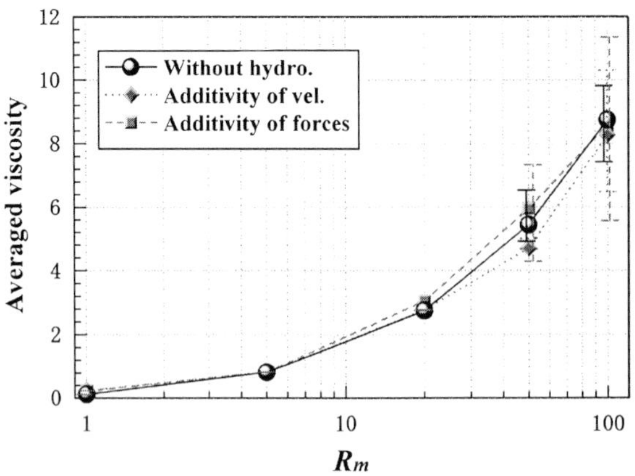

Figure 12.5 Influence of shear rates on averaged viscosities.

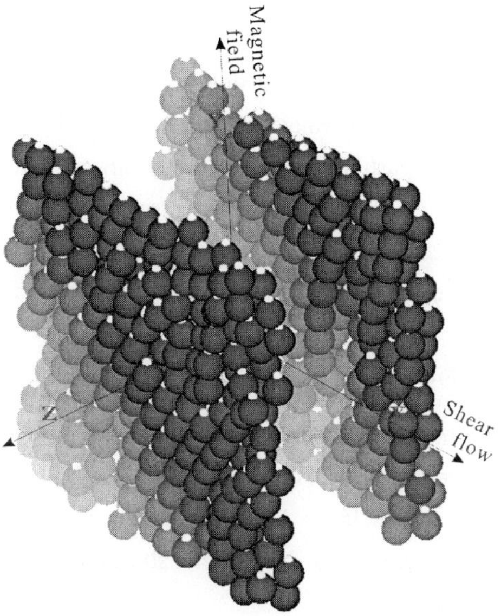

Figure 12.6 Snapshot of aggregate structure for relatively large system, N=512, for R_m =50; the gradiation technique was used to draw the snapshot so that the reader can easily recognize the wall-like clusters along the shear flow direction.

We now consider a snapshot of the aggregate structures in equilibrium for a relatively large system, N=512, shown in Fig. 12.6, which was obtained for AV for R_m=50. It is seen from comparing the result with Fig. 12.2(b) that the wall-like aggregate structure is thicker than that for N=64, and is composed of several layers of particles. For the case of N=512, thick chainlike clusters are expected to be formed along the magnetic field direction [5]. Hence we understand that such thick chainlike clusters associate to form thick wall-like aggregates, shown in Fig. 12.6, under the influences of the applied magnetic field and the shear flow.

Finally we show the comparison of the computation time. All simulations were conducted on a Compaq Alpha workstation with dual CPUs. A total of about 2,500,000 time steps were needed to obtain the results with sufficient accuracy. For the case of N=64, the computation time was 12.5 days for AF, 1.5 hours for AV, and 0.6 hours for AIHI. In addition, for N=512, it was 2.6 days for AV. It is noted that the additivity of velocities has a significant advantage in the computation time compared with the additivity of forces.

12.2 Brownian Dynamics Simulations

In the preceding section, we have considered Stokesian dynamics simulation of ferromagnetic dispersions, in which the particle Brownian motion does not play an important role since the magnetic interactions between particles are sufficiently strong. We here consider Brownian dynamics simulations, which take into account the Brownian motion of particles. However, for simplicity, the translational Brownian motion alone is taken into account, and we neglect the rotational Brownian motion by assuming that a strong magnetic field is applied.

12.2.1 The additivity of velocities for multi-body hydrodynamic interactions

The position of particle i under the influence of the translational Brownian motion, $\mathbf{r}_i$, is written as Eq. (9.106). When the rotational Brownian motion is negligible, the angular velocity $\boldsymbol{\omega}_i$ of particle i can be written as Eq. (8.36). According to the method of nondimensionalization which has been explained in Sec. 9.5, Eq. (9.106) can straightforwardly be nondimensionalized as

$$\begin{aligned}\mathbf{r}_i^*(t^*+\Delta t^*) = \mathbf{r}_i^*(t^*) + \Bigg\{ \mathbf{U}^*(\mathbf{r}_i^*) + \sum_{j=1}^{N} \mathbf{D}_{ij}^{T*}(t^*)\cdot\mathbf{F}_j^{P*}(t^*) + 2\sum_{j=1}^{N} \widetilde{\mathbf{D}}_{ij}^{C*}(t^*)\cdot\mathbf{T}_j^{P*}(t^*) \\ + 2\widetilde{\mathbf{g}}_i'^*(t^*):\mathbf{E}^* + \frac{1}{Pe}\sum_{j=1}^{N}\frac{\partial}{\partial \mathbf{r}_j^*}\cdot\left(\mathbf{D}_{ij}^{T*}(t^*)\right) \Bigg\}\Delta t^* + \Delta\mathbf{r}_i^{B*}(t^*),\end{aligned} \tag{12.20}$$

in which $\widetilde{\mathbf{g}}_i'$ has already been shown in Sec. 6.2. The dimensionless angular velocity can be expressed from Eqs. (8.53) and (9.73) as

$$\begin{aligned}\boldsymbol{\omega}_i^*(t^*) = \boldsymbol{\Omega}^* + \frac{3}{2}\sum_{j=1}^{N}\mathbf{D}_{ij}^{C*}(t^*)\cdot\mathbf{F}_j^{P*}(t^*) + \sum_{j=1}^{N}\mathbf{D}_{ij}^{R*}(t^*)\cdot\mathbf{T}_j^{P*}(t^*) \\ + \widetilde{\mathbf{h}}_i'^*(t^*):\mathbf{E}^*,\end{aligned} \tag{12.21}$$

in which the force $\mathbf{F}_i^{P*}$ and torque $\mathbf{T}_i^{P*}$ are equal to $\mathbf{F}_i^*$ and $\mathbf{T}_i^*$, respectively, shown in Eq. (12.11). Also $\Delta\mathbf{r}_i^{B*}$ is the particle displacement due to the Brownian motion and has the following properties,

$$\left\langle \Delta\mathbf{r}_i^{B*} \right\rangle = 0, \quad \left\langle (\Delta\mathbf{r}_i^{B*})(\Delta\mathbf{r}_j^{B*}) \right\rangle = \frac{2}{Pe}\mathbf{D}_{ij}^{T*}\Delta t^*. \tag{12.22}$$

The nondimensional number Pe is called the Péclet number, which has already been defined in Eq. (9.109), and is the ratio of the representative hydrodynamic shear force to the representative random force. In other words, the influence of the Brownian motion decreases as the value of Pe increases. If the displacement $\Delta\mathbf{r}_i^{B*}$ (i=1,2,…,N) is generated by means of the method explained in Appendix 9 based on the properties in Eq.(12.22), then the Brownian dynamics simulation can be advanced in time.

With the following relation:

$$\sum_{j=1}^{N} \frac{\partial}{\partial \mathbf{r}_j^*} \cdot \left(\mathbf{D}_{ij}^{T*}\right) = \sum_{\substack{j=1 \\ (j\neq i)}}^{N} \frac{\partial}{\partial \mathbf{r}_i^*} \cdot \mathbf{a}_{ii}^* + \sum_{\substack{j=1 \\ (j\neq i)}}^{N} \frac{\partial}{\partial \mathbf{r}_j^*} \cdot \mathbf{a}_{ij}^* , \tag{12.23}$$

the divergence of the diffusion tensor can be evaluated from the data of the mobility tensors $\mathbf{a}_{ii}^*$ and $\mathbf{a}_{ij}^*$. This data is tabulated like the mobility functions, and is referred to and interpolated to evaluate the values necessary during the simulation run.

In addition to the nondimensional numbers R_m, R_H, and R_V, the Péclet number Pe has appeared in Eq. (12.20) as another nondimensional number for characterizing the phenomenon of the Brownian motion in ferromagnetic colloidal dispersions. It has already been pointed out that the value of R_m has to be changed by keeping the values of R_H/R_m and R_V/R_m constant in order to generate a simple shear flow with various strengths of the shear rate. In the present case, the value of $1/(R_m Pe)$ must also be kept constant:

$$\frac{1}{R_m Pe} = \frac{32\pi a^3 kT}{3\mu_0 m^2} = \frac{(1+t_\delta)^3}{3} \cdot \frac{1}{\lambda} . \tag{12.24}$$

The quantity $R_m Pe$ is the nondimensional parameter representing the strength of the magnetic interactions between particles relative to the Brownian motion.

12.2.2 Parameters for simulations

The following values were used in conducting Brownian dynamics simulations. Three values of R_V/R_m were chosen as (158.7, 95.2, 52.89), which were obtained for λ=(3, 5, 9), to compare the present results with the results obtained by the Stokesian dynamics simulations. The values of R_H/R_m, and the number density n^* are the same as shown in Sec. 12.1. In Brownian dynamics simulations, the generation of the random displacement $\Delta\mathbf{r}_i^{B*}$(i=1,2,...,N) is computationally expensive, so that we here consider a relatively large system where the number of particles, N, is 512. For this case, the length of the simulation box (cubic), L^* (=L/a), is 37.13. The cutoff radius for particle-particle interactions, r_{coff}^* (=r_{coff}/a), is 16. The values of $1/(R_m Pe)$ are (0.244, 0.146, 0.0814) for λ=(3, 5, 9), respectively.

The value of the time interval Δt^* has been selected with the following consideration. If we regard the maximum Brownian displacement, $|\ \mathbf{r}_i^{B*}|_{max}$, as

$$\left|\Delta\mathbf{r}_i^{B*}\right|_{max} = 2.5 \times \sqrt{\frac{2}{Pe}\Delta\tau^*} , \tag{12.25}$$

this maximum of the displacement due to the Brownian motion has to be much smaller than the thickness of the steric layer. That is, if we set the following condition:

$$\left|\Delta \mathbf{r}_i^{B*}\right|_{max} = 0.1 t_\delta , \tag{12.26}$$

$\Delta\tau^*$ can be obtained as

$$\Delta\tau^* = 8\times 10^{-4} t_\delta^{\,2} Pe . \tag{12.27}$$

It is desirable that the time interval Δt^* is much shorter than the characteristic time for the shear flow. We have, therefore, adopted the time interval as Δt^*=min(0.001,$\Delta\tau^*$). It is clear from the above discussion that, in Brownian dynamics simulations, large values of the time interval cannot be employed due to the thickness of the steric layer.

It is important to stress the following points before we proceed to the results obtained by the simulations. It is found that the positiveness of the diffusion matrix does not necessarily hold in the present Brownian dynamics simulations. That is, the square root in evaluating L_{ii} in Eq. (A9.19) is not always positive, but sometimes will become imaginary. To circumvent this difficulty, some corrections concerning the components of the diffusion matrix were conducted in such a way that the condition in Eq. (9.85) is satisfied. The details concerning the loss of the positiveness of the diffusion matrix will not be discussed here.

12.2.3 Results

Figure 12.7 shows snapshots of the aggregate structures in equilibrium in a quiescent flow field, which were obtained by the Brownian dynamics simulations using the following initial condition: the particles were placed on face-centered lattice points and the magnetic moments of the particles were set in the direction of the applied magnetic field (y-axis direction). Figure 12.7(a) is for λ=9 and Fig. 12.7(b) is for λ=5. For the case of λ=9, where the effect of Brownian motion is suppressed by the strength of the magnetostatic interaction, the result shown in Fig. 12.7(a) qualitatively agrees well with both the Stokesian dynamics result [1], in which the Brownian effect was not taken into consideration, and also with the Monte Carlo result [5]. For this case, as might be expected, all three simulation methods are able to capture thick chainlike clusters formed along the field direction. However, if we look at the internal structure of the clusters in more detail, some differences between the three methods become clear [3]. That is, the thick chainlike clusters are formed more densely for the case of the Stokesian dynamics simulations, in which the Brownian motion is ignored. In the Monte

a

b

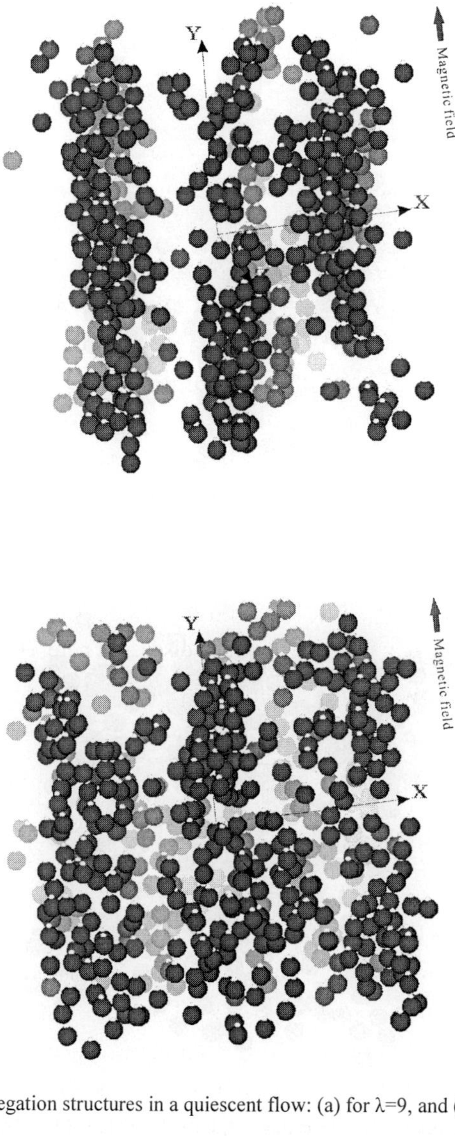

Figure 12.7 Aggregation structures in a quiescent flow: (a) for λ=9, and (b) for λ=5.

Carlo simulations, where hydrodynamic interactions have been ignored, the particle Brownian motion is taken into consideration indirectly through Boltzmann's factor. Hence, we can account for the fact that the thick chains of the Monte Carlo method are not so dense as those of the Stokesian dynamics method.

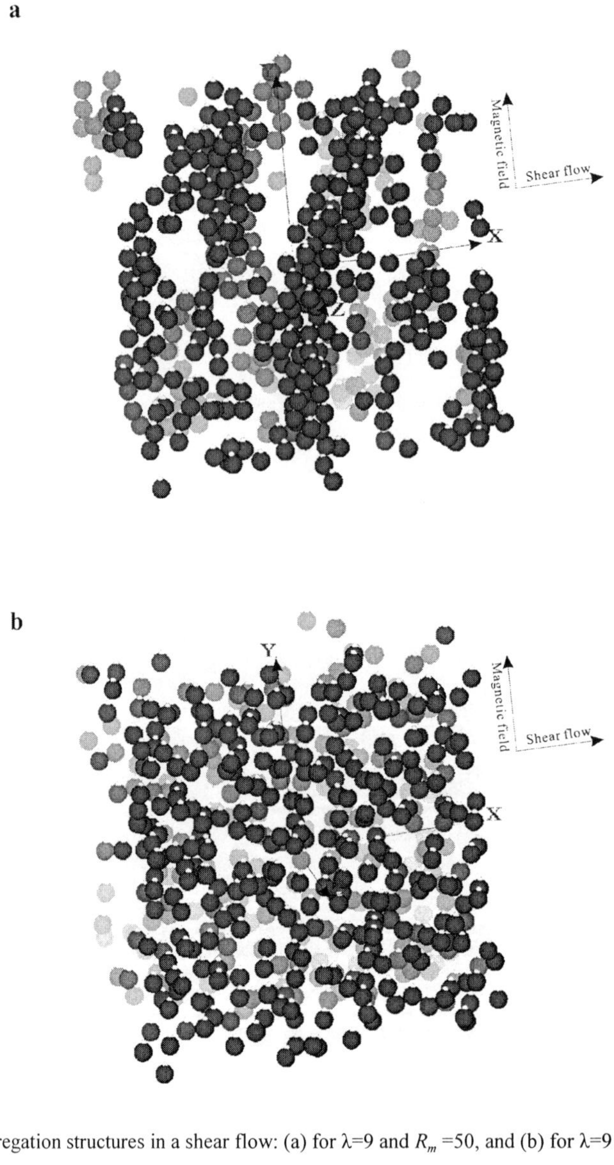

Figure 12.8 Aggregation structures in a shear flow: (a) for λ=9 and R_m =50, and (b) for λ=9 and R_m =1.

Figure 12.8 shows snapshots in the steady state for the Brownian dynamics simulations, which were obtained in a simple shear flow using the aggregate formation in Fig. 12.7(a) as an initial configuration. Figure 12.8(a) is for (λ,R_m)=(9,50) and Fig. 12.8(b) is for (λ,R_m)=(9,1). The values of (λ,R_m)=(9,50) mean that the magnetic particle-particle interaction is dominant compared with both the Brownian effect and the viscous forces due to the simple shear flow. Figure 12.8(a) clearly shows that thick chainlike clusters are formed along the field direction, although they are slightly tilted to the flow direction. Hence the Brownian dynamics method can capture the thick chainlike formation in a simple shear flow as well as in a quiescent flow. On the other hand, significant aggregates are not seen in Fig. 12.8(b). Although the effect of the magnetostatic interaction is larger than the Brownian motion for this case, the viscous shear force dominates the magnetic force because of the large shear rate of R_m=1, so that the magnetic interaction cannot induce an aggregate formation.

References

1. A. Satoh, R.W. Chantrell, G.N. Coverdale, and S. Kamiyama, Stokesian Dynamics Simulations of Ferromagnetic Colloidal Dispersions in a Simple Shear Flow, J. Colloid Inter. Sci., 203 (1988) 233
2. A. Satoh, Comparison of Approximations between Additivity of Velocities and Additivity of Forces for Stokesian Dynamics Methods, J. Colloid Inter. Sci., 243 (2001) 342
3. A. Satoh, R.W. Chantrell, and G. N. Coverdale, Brownian Dynamics Simulations of Ferromagnetic Colloidal Dispersions in a Simple Shear Flow, J. Colloid Inter. Sci., 209 (1999) 44
4. R.E. Rosensweig, "Ferrohydrodynamics," 46-50, Cambridge University Press, Cambridge, 1985
5. A. Satoh, R.W. Chantrell, S. Kamiyama, and G.N. Coverdale, Three-Dimensional Monte Carlo Simulations of Thick Chainlike Clusters Composed of Ferromagnetic Fine Particles, J. Colloid Inter. Sci., 181 (1996) 422
6. S. Kim and R. T. Mifflin, The Resistance and Mobility Functions of Two Equal Spheres in Low-Reynolds-Number Flow, Phys. Fluids, 28 (1985) 2033

Exercises

12.1 The potential energy is related to the force by the following relation:

$$\mathbf{F}_{ij}^{(m)} = -\frac{\partial}{\partial \mathbf{r}_i} u_{ij}^{(m)} = -\frac{\partial}{\partial \mathbf{r}_{ij}} u_{ij}^{(m)} . \qquad (12.28)$$

With Eq. (12.2) for the potential energy, derive the following expression for the force from Eq. (12.28):

$$\mathbf{F}_{ij}^{(m)} = -\frac{3\mu_0}{4\pi}\cdot\frac{1}{r_{ij}^{\ 4}}\left[-(\mathbf{m}_i\cdot\mathbf{m}_j)\frac{\mathbf{r}_{ij}}{r_{ij}} + 5(\mathbf{m}_i\cdot\mathbf{r}_{ij})(\mathbf{m}_j\cdot\mathbf{r}_{ij})\frac{\mathbf{r}_{ij}}{r_{ij}^{\ 3}} \right.$$
$$\left. -\left\{(\mathbf{m}_j\cdot\mathbf{r}_{ij})\mathbf{m}_i + (\mathbf{m}_i\cdot\mathbf{r}_{ij})\mathbf{m}_j\right\}\frac{1}{r_{ij}}\right]. \tag{12.29}$$

12.2 Similarly, show that the expression for the force can be written in Eq. (12.9) for the potential energy in Eq.(12.4).

12.3 By evaluating the magnetic field $\mathbf{H}_{ij}$ induced at the position of particle i by the magnetic moment $\mathbf{m}_j$ of particle j, show that the torque acting on particle i by the induced field can be expressed as follows:

$$\mathbf{T}_{ij}^{(m)} = -\frac{\mu_0}{4\pi}\cdot\frac{1}{r_{ij}^{\ 3}}\left\{\mathbf{m}_i\times\mathbf{m}_j - \frac{3}{r_{ij}^{\ 2}}(\mathbf{m}_j\cdot\mathbf{r}_{ij})\mathbf{m}_i\times\mathbf{r}_{ij}\right\}. \tag{12.30}$$

CHAPTER 13

HIGHER ORDER APPROXIMATIONS OF MULTI-BODY HYDRODYNAMIC INTERACTIONS

In Chapter 6, we have shown the two representative approximations for treating multi-body hydrodynamic interactions, that is, the additivity of forces and the additivity of velocities. In this chapter we explain more accurate approximation methods. If the three-body interaction can be solved analytically, a much more accurate simulation method may be developed; some studies concerning the three-body interaction have been attempted [1,2]. However, in the case of a multi-particle system, a simulation involving the calculation of the three-body interaction would be extraordinarily expensive in terms of computational effect. Therefore we here show some methods which are relatively valuable from a practicable point of view.

13.1 The Durlofsky-Brady-Bossis Method

We consider a simple shear flow problem. As already pointed out in Chapter 4, the flow field can be solved from the Stokes equation in Eq. (4.20) and with the equation of continuity in Eq. (4.6). Another approach is to use the integral representation for the velocity field, which is derived from the concept of the force acting on unit area of the particle surface, rather than the Stokes equation itself. In this method, the flow field is obtained by solving an integral equation, which will be shown later.

If the force acting on a unit area of the surface of a particle is denoted by **f**, the flow field **U**(**r**) at a position **r** with the influence of all dispersed particles can be expressed as [3]

$$\mathbf{U}(\mathbf{r}) = \mathbf{U}_0(\mathbf{r}) - \frac{1}{8\pi\eta}\sum_{\alpha=1}^{N}\int_{S_\alpha}\mathbf{J}(\mathbf{r}-\mathbf{r}')\cdot\mathbf{f}(\mathbf{r}')d\mathbf{r}', \tag{13.1}$$

in which $\mathbf{U}_0$ is the unperturbed flow field of a simple shear flow before particles being dispersed, $\mathbf{r}'$ is the position vector at the surface of each particle ($\mathbf{r}_\alpha$ is the center of particle α), $\mathbf{r}$ is the position vector of an arbitrary point ($r=|\mathbf{r}|$), S_α is the whole surface area of particle α, and **J** is the Oseen tensor shown in Sec 5.2.1 and rewritten as

$$\mathbf{J}(\mathbf{r}) = \frac{1}{r}\left(\mathbf{I} + \frac{\mathbf{rr}}{r^2}\right). \tag{13.2}$$

The force density $\mathbf{f}(\mathbf{r}')$ can be expressed using the stress tensor $\boldsymbol{\tau}$ and the unit vector **n** pointing outwards from the particle surface:

$$\mathbf{f}(\mathbf{r}') = \mathbf{n} \cdot \boldsymbol{\tau} . \tag{13.3}$$

It is seen from the results shown in Sec. 5.2.1 that the integral term in Eq. (13.1) corresponds to the velocity at the position **r** in the liquid induced by the distribution of forces over each particle surface. The solution for the flow field can be obtained by solving the integral equation in Eq. (13.1), but it is possible to use the following approximate approach [4].

If the Oseen tensor is expressed as a Taylor series about the center of each particle, we have the following expression for particle α:

$$\begin{aligned}
\mathbf{J}(\mathbf{r}-\mathbf{r}') &= \mathbf{J}(\mathbf{r}-\mathbf{r}_\alpha) + (\mathbf{r}'-\mathbf{r}_\alpha)\cdot\frac{\partial}{\partial \mathbf{r}'}\mathbf{J}(\mathbf{r}-\mathbf{r}')\Big|_{\mathbf{r}'=\mathbf{r}_\alpha} \\
&\quad + \frac{1}{2}(\mathbf{r}'-\mathbf{r}_\alpha)(\mathbf{r}'-\mathbf{r}_\alpha):\frac{\partial}{\partial \mathbf{r}'}\frac{\partial}{\partial \mathbf{r}'}\mathbf{J}(\mathbf{r}-\mathbf{r}')\Big|_{\mathbf{r}'=\mathbf{r}_\alpha} + \cdots \\
&= \mathbf{J}(\mathbf{r}-\mathbf{r}_\alpha) - (\mathbf{r}'-\mathbf{r}_\alpha)\cdot\frac{\partial}{\partial \mathbf{r}}\mathbf{J}(\mathbf{r}-\mathbf{r}_\alpha) \\
&\quad + \frac{1}{2}(\mathbf{r}'-\mathbf{r}_\alpha)(\mathbf{r}'-\mathbf{r}_\alpha):\frac{\partial}{\partial \mathbf{r}}\frac{\partial}{\partial \mathbf{r}}\mathbf{J}(\mathbf{r}-\mathbf{r}_\alpha) + \cdots .
\end{aligned} \tag{13.4}$$

Substitution of this expression into Eq. (13.1) leads to

$$\begin{aligned}
\mathbf{U}(\mathbf{r}) &= \mathbf{U}_0(\mathbf{r}) - \frac{1}{8\pi\eta}\sum_{\alpha=1}^{N}\int_{S_\alpha}\mathbf{J}(\mathbf{r}-\mathbf{r}_\alpha)\cdot\mathbf{f}(\mathbf{r}')d\mathbf{r}' \\
&\quad + \frac{1}{8\pi\eta}\sum_{\alpha=1}^{N}\int_{S_\alpha}(\mathbf{r}'-\mathbf{r}_\alpha)\cdot\frac{\partial}{\partial \mathbf{r}}\mathbf{J}(\mathbf{r}-\mathbf{r}_\alpha)\cdot\mathbf{f}(\mathbf{r}')d\mathbf{r}' \\
&\quad - \frac{1}{16\pi\eta}\sum_{\alpha=1}^{N}\int_{S_\alpha}(\mathbf{r}'-\mathbf{r}_\alpha)(\mathbf{r}'-\mathbf{r}_\alpha):\frac{\partial}{\partial \mathbf{r}}\frac{\partial}{\partial \mathbf{r}}\mathbf{J}(\mathbf{r}-\mathbf{r}_\alpha)\cdot\mathbf{f}(\mathbf{r}')d\mathbf{r}' + \cdots .
\end{aligned} \tag{13.5}$$

If the force exerted on the ambient fluid by particle α is denoted by $\mathbf{F}_\alpha$, the second term (including the negative sign) on the right-hand side in Eq. (13.5) reduces to the following equation, using Eqs. (4.34) and (13.3):

$$\frac{1}{8\pi\eta}\sum_{\alpha=1}^{N}\mathbf{J}(\mathbf{r}-\mathbf{r}_\alpha)\cdot\mathbf{F}_\alpha . \tag{13.6}$$

This equation is equivalent to the sum of the velocity in Eq. (5.58) induced by the motion of each individual particle in a liquid. Thus it is seen that the third and higher-order terms in Eq. (13.5) are due to the rotational motion of particles and the hydrodynamic interaction. Since the third term in Eq. (13.5) is very similar to Eq. (4.38), the contribution of the torques and the stresslets is given by this term. The other higher-order terms in Eq. (13.5) are due to the higher-order moments, such as the quadrupole moment, and these terms make contributions of correction. By taking account of $(\partial/\partial\mathbf{r})(\partial/\partial\mathbf{r})\mathbf{J}=\nabla^2\mathbf{J}$ and reforming Eq. (13.5), the final expression for $\mathbf{U}(\mathbf{r})$ can be expressed as [4]

$$\mathbf{U}(\mathbf{r}) = \mathbf{U}_0(\mathbf{r}) + \frac{1}{8\pi\eta}\sum_{\alpha=1}^{N}\left(1+\frac{1}{6}a^2\nabla^2\right)\mathbf{J}(\mathbf{r}-\mathbf{r}_\alpha)\cdot\mathbf{F}_\alpha + \sum_{\alpha=1}^{N}\boldsymbol{\varepsilon}\cdot\frac{\mathbf{r}}{r^3}\cdot\mathbf{T}_\alpha$$
$$+\sum_{\alpha=1}^{N}\left(1+\frac{1}{10}a^2\nabla^2\right)\mathbf{L}(\mathbf{r}-\mathbf{r}_\alpha):\mathbf{S}_\alpha + \cdots, \tag{13.7}$$

in which $\mathbf{L}(\mathbf{r})$ is a third-rank tensor and written in a component expression as

$$L_{ijk} = \frac{1}{2}\left(\nabla_k J_{ij} + \nabla_j J_{ik}\right). \tag{13.8}$$

Equation (13.7) is the solution of the flow field which is expressed in terms of the forces, torques, stresslets, etc., exerted by the particles on the fluid. The particle velocity itself $\mathbf{v}_\alpha$ in the flow field given by Eq. (13.7) can be written from the Faxen formula for spherical particles as [5]

$$\mathbf{v}_\alpha - \mathbf{U}_0(\mathbf{r}_\alpha) = \frac{\mathbf{F}_\alpha}{6\pi\eta a} + \left(1+\frac{1}{6}a^2\nabla^2\right)\mathbf{U}'(\mathbf{r}_\alpha), \tag{13.9}$$

$$\boldsymbol{\omega}_\alpha - \boldsymbol{\Omega} = \frac{\mathbf{T}_\alpha}{8\pi\eta a^3} + \frac{1}{2}\boldsymbol{\varepsilon}:\frac{\partial}{\partial\mathbf{r}_\alpha}\mathbf{U}'(\mathbf{r}_\alpha), \tag{13.10}$$

$$-\mathbf{E} = \frac{3\mathbf{S}_\alpha}{20\pi\eta a^3} + \left(1+\frac{1}{10}a^2\nabla^2\right)\mathbf{E}'(\mathbf{r}_\alpha), \tag{13.11}$$

in which $\mathbf{U}'(\mathbf{r}_\alpha)$ is the disturbance velocity field arising at the position $\mathbf{r}_\alpha$ due to the other particles (excluding particle α itself) and $\mathbf{E}'$ is the rate-of-strain tensor of the disturbance flow field, which is expressed as

$$\mathbf{E}' = \frac{1}{2}\left\{\frac{\partial}{\partial\mathbf{r}_\alpha}\mathbf{U}' + \left(\frac{\partial}{\partial\mathbf{r}_\alpha}\mathbf{U}'\right)^t\right\}. \tag{13.12}$$

Also $\mathbf{E}$ and $\boldsymbol{\Omega}$ are the rate-of-strain tensor and rotational velocity vector of the imposed shear flow in the absence of particles. If we take into account only the terms due to the forces, torques and stresslets, and neglect higher-multipole moment terms, then the velocities, angular velocities and rate-of-strain tensor can be related to the forces, torques and stresslets, using the mobility matrix $\mathbf{M}^\infty$, as

$$\begin{bmatrix}\hat{\mathbf{v}}\\ -\mathbf{E}\end{bmatrix} = \frac{1}{\eta}\mathbf{M}^\infty\begin{bmatrix}\hat{\mathbf{F}}\\ \mathbf{S}\end{bmatrix}, \tag{13.13}$$

in which $\hat{\mathbf{v}}$ is the column vector containing the translational and angular velocities of all N particles, similarly, $\hat{\mathbf{F}}$ is the column vector containing the forces and torques, and $\mathbf{S}$ is the column vector containing the stresslets. It is quite clear from Eq. (13.13) that $\hat{\mathbf{F}}$ and $\mathbf{S}$ can be solved as a function of $\hat{\mathbf{v}}$ and $\mathbf{E}$:

the following equation [9]:

$$\nabla^2\varphi(\mathbf{r}) = \kappa^2\varphi(\mathbf{r}) \ . \tag{13.22}$$

This equation can be solved as

$$\varphi(\mathbf{r}) \propto e^{-\kappa r} / r \ . \tag{13.23}$$

It is clearly seen by comparing Eq. (13.23) with Eq. (13.21) that the electrostatic potential with the screening effect in Eq. (13.23) is short-ranged and rapidly approaches zero with increasing distance r, which is in strong contrast with the property of Eq. (13.21). In the potential of Eq. (13.23), the strength of the screening effect is described by the value of the constant κ.

We have now prepared the stage for discussing the screening effect for hydrodynamic interactions. The Stokes equation with a screening effect is written in similar form to Eq. (13.22) as [6,10]

$$\eta(\nabla^2\mathbf{u}(\mathbf{r}) - \kappa^2\mathbf{u}(\mathbf{r})) = \nabla p(\mathbf{r}) - \mathbf{F}(\mathbf{r}') \ . \tag{13.24}$$

The flow velocity $\mathbf{u}(\mathbf{r})$ has to satisfy the equation of continuity:

$$\nabla \cdot \mathbf{u}(\mathbf{r}) = 0 \ . \tag{13.25}$$

In Eq. (13.24), $\mathbf{F}(\mathbf{r}')$ is the force per unit volume at the position $\mathbf{r}'$ exerted by the particle on the fluid. If we assume that a point particle is at the origin, $\mathbf{F}(\mathbf{r}')$ may be expressed as

$$\mathbf{F}(\mathbf{r}') = \mathbf{F}_0\delta(\mathbf{r}) \ . \tag{13.26}$$

The substitution of this equation into Eq. (13.24) leads to the following equation:

$$\eta(\nabla^2 - \kappa^2)\mathbf{u}(\mathbf{r}) = \nabla p(\mathbf{r}) - \mathbf{F}_0\delta(\mathbf{r}) \ . \tag{13.27}$$

The flow velocity $\mathbf{u}(\mathbf{r})$ at the position $\mathbf{r}$ induced by the force $\mathbf{F}_0$ with the screening effect can be obtained by solving Eqs. (13.27) and (13.25). We will now show the derivation of the solution from Eq. (13.27).

If the formulae shown in Appendix A2 are referred to, we can obtain the Fourier transformation of Eq. (13.27) as follows:

$$-\eta(k^2 + \kappa^2)\mathbf{J}(\mathbf{k}) = i\mathbf{k}P(\mathbf{k}) - \frac{1}{(2\pi)^{3/2}}\mathbf{F}_0 \ , \tag{13.28}$$

in which $\mathbf{J}(\mathbf{k})$ and $P(\mathbf{k})$ are the Fourier transforms of $\mathbf{u}(\mathbf{r})$ and $p(\mathbf{r})$, respectively, and i is the imaginary unit. Furthermore, the Fourier transform of Eq. (13.25) leads to the following equation:

$$\mathbf{k} \cdot \mathbf{J}(\mathbf{k}) = 0 \ . \tag{13.29}$$

Hence, by solving $P(\mathbf{k})$ from Eqs. (13.28) and (13.29) and substituting $P(\mathbf{k})$ into Eq. (13.28), $\mathbf{J}(\mathbf{k})$ can be obtained as

$$\mathbf{J}(\mathbf{k}) = \frac{1}{\eta(k^2+\kappa^2)}\left(\mathbf{I}-\frac{\mathbf{kk}}{k^2}\right)\cdot\frac{\mathbf{F}_0}{(2\pi)^{3/2}}\,. \tag{13.30}$$

The Fourier inverse transform of this equation gives the relationship between $\mathbf{u}(\mathbf{r})$ and $\mathbf{F}_0$ as

$$\mathbf{u}(\mathbf{r}) = \frac{1}{4\pi\eta r}\left\{e^{-\kappa r}(\mathbf{I}-\mathbf{ee})+\left(\frac{e^{-\kappa r}}{\kappa r}+\frac{e^{-\kappa r}-1}{\kappa^2 r^2}\right)(\mathbf{I}-3\mathbf{ee})\right\}\cdot\mathbf{F}_0\delta(\mathbf{r})\,. \tag{13.31}$$

Next we apply this result to the velocity $\mathbf{v}_{\alpha(\beta)}$ of particle α at the position $\mathbf{r}_\alpha$ induced by particle β at $\mathbf{r}_\beta$, which exerts a force $\mathbf{F}_\beta$ on the ambient fluid. If the induced velocity $\mathbf{v}_{\alpha(\beta)}$ is written as

$$\mathbf{v}_{\alpha(\beta)} = \frac{1}{\eta}\mathbf{a}_{\alpha\beta}\cdot\mathbf{F}_\beta\,, \tag{13.32}$$

the mobility tensor $\mathbf{a}_{\alpha\beta}$ can be expressed with Eq. (13.31) as

$$\mathbf{a}_{\alpha\beta} = \frac{1}{6\pi a}\cdot\frac{3a}{2r_{\beta\alpha}}\left\{e^{-\kappa_2 r_{\beta\alpha}}(\mathbf{I}-\mathbf{ee})+\left(\frac{e^{-\kappa_2 r_{\beta\alpha}}}{\kappa_2 r_{\beta\alpha}}+\frac{e^{-\kappa_2 r_{\beta\alpha}}-1}{\kappa_2^2 r_{\beta\alpha}^2}\right)(\mathbf{I}-3\mathbf{ee})\right\}, \tag{13.33}$$

in which $\mathbf{e}=(\mathbf{r}_\beta-\mathbf{r}_\alpha)/r_{\beta\alpha}$, $r_{\beta\alpha}=|\mathbf{r}_\beta-\mathbf{r}_\alpha|$, and κ_2 is a constant describing the screening effect. This is the effective mobility tensor between particles α and β with the screening effect. It is clearly seen that Eq. (13.33) becomes equivalent to Eq. (13.19), by neglecting higher-order terms, in the limit of $\kappa_2\to 0$.

The following equation, with the screening effect, for the tensor $\mathbf{a}_{\alpha\alpha}$ may be used [6,10]

$$\mathbf{a}'_{\alpha\alpha} = \frac{1}{6\pi a}\left(\mathbf{I}-\sum_{\beta=1(\neq\alpha)}^{N}\frac{15a^4}{4r^4}e^{-\kappa_1 r_{\beta\alpha}}\mathbf{ee}\right), \tag{13.34}$$

in which κ_1 is another constant describing the screening effect. Similarly, Eq. (13.34) reduces to Eq. (13.20), without higher-order terms, in the limit of $\kappa_1\to 0$.

It was pointed out [6] that, if the two coefficients κ_1 and κ_2 for the screening effect are taken as $\kappa_1=0.25\varphi_V$ and $\kappa_2=8.5\varphi_V+400\varphi_V^7$ for the volumetric fraction $\varphi_V\leq 0.45$, then the simulation results agree well with the experimental results concerning the diffusion coefficient.

Finally, we make some general statements concerning the effective mobility tensors explained above. It is seen that Eqs. (13.33) and (13.34) correspond to Eqs. (13.19) and (13.20) without higher-order tensors, respectively. Hence these effective mobility tensors are applicable to simulations of two particles that are sufficiently far from each other, but not to a nearly-touching case. This means that the lubrication effect cannot be reproduced by these effective mobility tensors. However, from the discussion in Sec. 13.1, it seems to be possible to use the following modified mobility matrix for taking into account the near-field interactions:

$$\mathbf{M} = \mathbf{M}_{SC} + \mathbf{M}_{2B} - \mathbf{M}_{2B}^{\infty}, \tag{13.35}$$

in which $\mathbf{M}_{SC}$ is the mobility matrix composed of the mobility tensors, including the screening effect, expressed in Eqs. (13.33) and (13.34), $\mathbf{M}_{2B}$ is the mobility matrix describing the near-field interactions, and $\mathbf{M}_{2B}{}^{\infty}$ is the canceling mobility tensors for removing the double count of the far-field interaction.

References

1. Mazur and W. van Saarloos, Many-Sphere Hydrodynamic Interactions and Mobilities in a Suspension, Physica A, 115(1982) 21
2. Van Saarloos and P. Mazur, Many-Sphere Hydrodynamic Interactions. II. Mobilities at Finite Frequencies, Physica A, 120 (1983) 77
3. O.A. Ladyzhenskaya, "The Mathematical Theory of Viscous Incompressible Flow," Gordon & Breach, 1963
4. L. Durlofsky, J.F. Brady, and G. Bossis, Dynamics Simulation of Hydrodynamically Interacting Particles, J. Fluid Mech., 180 (1987) 21
5. G.K. Batchlor and J.T. Green, The Hydrodynamic Interaction of Two Small Freely-Moving Spheres in a Linear Flow Field, J. Fluid Mech., 56 (1972) 375
6. I. Snook, W. van Megen, and R.J.A. Tough, Diffusion in Concentrated Hard Sphere Dispersions: Effective Two Particle Mobility Tensors, J. Chem. Phys., 78 (1983) 5825
7. W. van Megen and I. Snook, Brownian-Dynamics Simulation of Concentrated Charge-Stabilized Dispersions, J. Chem. Soc., Faraday Trans. II, 80 (1984) 383
8. W. van Megen and I. Snook, Dynamic Computer Simulation of Concentrated Dispersions, J. Chem. Phys., 88 (1988) 1185
9. R.J. Hunter, "Foundations of Colloid Science," Vol. 1, Clarendon Press, Oxford, 1986
10. S.A. Adelman, Hydrodynamic Screening and Viscous Drag at Finite Concentration, J. Chem. Phys., 68 (1978) 49

Exercises

13.1 If the force exerted by a particle on the ambient fluid is denoted by $\hat{\mathbf{F}}^{hyd}$, Eq. (13.17) can be rewritten as

$$\left.\begin{aligned} \hat{\mathbf{F}}^{hyd} &= \eta(\mathbf{R}_{\hat{F}\hat{v}} \cdot \hat{\mathbf{v}} - \mathbf{R}_{\hat{F}E} : \mathbf{E}), \\ \mathbf{S} &= \eta(\mathbf{R}_{S\hat{v}} \cdot \hat{\mathbf{v}} - \mathbf{R}_{SE} : \mathbf{E}). \end{aligned}\right\} \tag{13.36}$$

When the particle Brownian motion is negligible, the force exerted on particles by the fluid has to be balanced with any non-hydrodynamic force acting on the particle, $\hat{\mathbf{F}}^{P}$, such as magnetic forces. That is,

$$\hat{\mathbf{F}}^{P} - \hat{\mathbf{F}}^{hyd} = 0 . \qquad (13.37)$$

With these equations, show that $\mathbf{S}$ can be written as

$$\mathbf{S} = -\eta \mathbf{R}_{SE} : \mathbf{E} + \eta \mathbf{R}_{S\hat{v}} \cdot \mathbf{R}_{\hat{F}\hat{v}}^{-1} \cdot \mathbf{R}_{\hat{F}E} : \mathbf{E} + \mathbf{R}_{S\hat{v}} \cdot \mathbf{R}_{\hat{F}\hat{v}}^{-1} \cdot \hat{\mathbf{F}}^{P} . \qquad (13.38)$$

If we decompose $\mathbf{S}$ into the components of each particle, for example, $\mathbf{S}_i$ for particle i, then the stress tensor $\boldsymbol{\tau}$ can be evaluated from Eq. (10.10).

13.2 Derive Eq. (13.19) by substituting Eqs. (5.91) and (5.92) into Eq. (5.73) and truncating higher-order terms.

13.3 Show that the Oseen tensor can be obtained from Eq. (13.31) in the limit of $\kappa \to 0$.

CHAPTER 14

OTHER MICROSIMULATION METHODS

In the previous chapters we have described the molecular dynamics, Stokesian dynamics, and Brownian dynamics methods. In the present chapter we focus our attention on other microsimulation methods which may be useful for simulations of colloidal dispersions. First the cluster-based Stokesian dynamics method is explained, which has been developed to reduce significantly the calculation of the inverse of the very large resistance matrix; this calculation is indispensable in the ordinary Stokesian dynamics method based on the approximation of the additivity of forces. Next the dissipative particle dynamics method is discussed, in which a fluid is regarded as being composed of virtual particles or fluid particles, and the flow field is solved by simulating such fluid particles. Finally the lattice Boltzmann method is explained, in which a fluid is regarded as being composed of fluid particles that move from lattice point to lattice point in the simulation region, followed by the treatment of the collisions of particles on all lattice points.

14.1 The Cluster-Based Stokesian Dynamics Method

14.1.1 Theoretical background

It is clear from Eq.(6.14) that the inverse of the resistance matrix must be calculated to obtain the translational and angular velocities in the approximation of the additivity of forces. If we consider a three-dimensional system of N particles, this means that a $6N\times6N$ resistance matrix is treated in simulations. Hence, the simulation method based on this approximation is generally limited to a small system, which has already been pointed out, and it is almost impossible to apply this method to a strongly-interacting system in which particles aggregate to form clusters. To circumvent this drawback from a simulation point of view, the following approximate treatment of the multi-body hydrodynamic interaction has been developed [1].

Strong particle-particle interactions due to the lubrication effect arise when two particles are in a nearly touching case. This effect, therefore, has to be taken into account as accurately as possible while developing an approximate method. In a ferromagnetic colloidal dispersion, particles aggregate to form clusters along the magnetic field direction [2,3]. It is presumed that the dominant factors, for determining the internal structure of aggregates in equilibrium and the process of the cluster formation, are mainly the magnetic force between particles and the hydrodynamic interaction due to the lubrication effect.

The drawback of the above-mentioned ordinary SD method comes from the terms of $\sum \mathbf{A}_{ij}^{*} \cdot \mathbf{v}_{j}^{*}$ and $\sum \widetilde{\mathbf{B}}_{ij}^{*} \cdot \boldsymbol{\omega}_{j}^{*}$ (concerning particles *i* and *j*) in Eq.(8.50) and the similar terms in Eq.(8.51). Due to these terms, the inverse of a $6N*6N$ resistance matrix for velocities and angular velocities has to be calculated. If it is possible to take into account only the important particles which have a more significant influence on the motion of particle *i*, smaller resistance matrices are sufficient in determining the particle motion. This means that a drastic reduction in the computation time is accomplished with a reliable prediction of the particle motion as a first approximation.

We adopt the following approximation for $\sum \mathbf{A}_{ij}^{*} \cdot \mathbf{v}_{j}^{*}$ in Eq. (8.50):

$$\sum_{j=1(\neq i)}^{N} \mathbf{A}_{ij}^{*} \cdot (\mathbf{v}_{j}^{*} - \mathbf{U}^{*}) \approx \sum_{j=1(\neq i)}^{N} {}^{clstr} \mathbf{A}_{ij}^{*} \cdot (\mathbf{v}_{j}^{*} - \mathbf{U}^{*}), \tag{14.1}$$

in which the superscript "clstr" for the summation means that only the interactions with the particles belonging to the same cluster of particle *i* are taken into account. Similarly, the term of $\sum \widetilde{\mathbf{B}}_{ij}^{*} \cdot \boldsymbol{\omega}_{j}^{*}$ and the similar terms in Eq. (8.51) are approximated as

$$\left.\begin{aligned}
\sum_{j=1(\neq i)}^{N} \widetilde{\mathbf{B}}_{ij}^{*} \cdot (\boldsymbol{\omega}_{j}^{*} - \boldsymbol{\Omega}^{*}) &\approx \sum_{j=1(\neq i)}^{N} {}^{clstr} \widetilde{\mathbf{B}}_{ij}^{*} \cdot (\boldsymbol{\omega}_{j}^{*} - \boldsymbol{\Omega}^{*}), \\
\sum_{j=1(\neq i)}^{N} \mathbf{B}_{ij}^{*} \cdot (\mathbf{v}_{j}^{*} - \mathbf{U}^{*}) &\approx \sum_{j=1(\neq i)}^{N} {}^{clstr} \mathbf{B}_{ij}^{*} \cdot (\mathbf{v}_{j}^{*} - \mathbf{U}^{*}), \\
\sum_{j=1(\neq i)}^{N} \mathbf{C}_{ij}^{*} \cdot (\boldsymbol{\omega}_{j}^{*} - \boldsymbol{\Omega}^{*}) &\approx \sum_{j=1(\neq i)}^{N} {}^{clstr} \mathbf{C}_{ij}^{*} \cdot (\boldsymbol{\omega}_{j}^{*} - \boldsymbol{\Omega}^{*}).
\end{aligned}\right\} \tag{14.2}$$

The physical meaning of these approximations is that the particle motion is simulated by means of exactly taking account of the hydrodynamic interactions between particles in the same cluster and evaluating the inverse of such a matrix. Hence, the lubrication effect is exactly treated in each cluster on the level of the approximation of the additivity of forces. It is clear from this approximation that a grand resistance matrix is decomposed into small matrices, so that a drastic reduction in the computation time is expected. We call the simulation method based on the above-mentioned approximation the "cluster-based SD method." It is also noted that the other terms such as $\sum \widetilde{\mathbf{G}}_{ij}^{*} : \mathbf{E}^{*}$ are evaluated in the usual way without the above-mentioned approximation.

The cluster analysis method, which has been explained in Sec. 3.6, may be used for defining the cluster formation of particles. Also, a simple method based on the particle-particle separation may be used for such judgment. In the latter method, particles *i* and *j* are regarded as forming a cluster if the separation r_{ij} between the particles is smaller than a criterion distance r_{clstr}; the neighboring particles in a certain cluster always satisfy the condition of $r_{ij} \leq r_{clstr}$.

14.1.2 The validity of the cluster-based method by simulations

To verify the validity of the cluster-based Stokesian dynamics method, we show the results which were obtained by simulations based on the present method for the ferromagnetic colloidal dispersion explained in Sec. 12.1. The notations which will appear below have completely the same meaning as in Sec. 12.1

To clarify the transient properties from an initial state, the evolution of a sub-averaged viscosity (defined in Eq. (A11.1)) for 500 time steps with time is shown in Fig.14.1 for R_m=20 and 100. These results were obtained by the cluster-based method (CBM) for two cases of $r_{clstr}^*/2$ ($r_{clstr}^* = r_{clstr}/a$)=(1.2, 1.5), by which a cluster is defined. Also, the results for the ordinary method (OM) and for the approximation of ignoring hydrodynamic interactions (AIHI) are shown for comparison. It is seen from Fig. 14.1 that the curves for CBM agree almost completely with those for OM and AIHI in the initial stage. Since the initial configuration of particles was given using simple cubic lattice points, the curves in Fig.14.1(a) are significantly influenced by the initial configuration, and seem to oscillate with time during the initial stage. However, after the initial stage, the sub-averaged viscosity fluctuates randomly around an average value. This feature of the viscosity curves is also relevant to Fig.14.1(b). As the simulation proceeds, the curves for AIHI first start to deviate significantly from those for CBM and OM. This is due to the fact that the lubrication effect, together with the magnetic force, becomes the main factors governing the particle motion after the particles approach each other and aggregate to form clusters. Thus, the simulation based on AIHI cannot capture the particle motion properly in these situations. In contrast, it is seen from Fig. 14.1(b) that the curves for $r_{clstr}^*/2$=1.2 and 1.5 are in very good agreement with that for OM, since CBM takes into account the lubrication effect exactly on the level of the additivity of forces. This good agreement clearly shows that CBM can simulate the particle motion properly in the process of the cluster formation and is highly suitable for the simulations of transient phenomena. The agreement with the curve for OM becomes significant as the value of R_m increases, or, as the influence of magnetostatic interactions increases more significantly than that of viscous shear forces. The curve for $r_{clstr}^*/2$ =1.2 in Fig.14.1(a) gives much better agreement with that for OM than for AIHI, but it starts to deviate from OM at around 13,000 time steps. Since a colloidal dispersion is a multi-particle system, the difference increases rapidly once the trajectory of the particle motion deviates from that of OM, which gives rise to the start of the deviation of the viscosity curve. Even for R_m=20, however, if the criterion for the cluster formation is relaxed as $r_{clstr}^*/2$=1.5, the curve comes to agree well with that for OM, since resistance matrices with more particle interactions are dealt with in determining the particle motion.

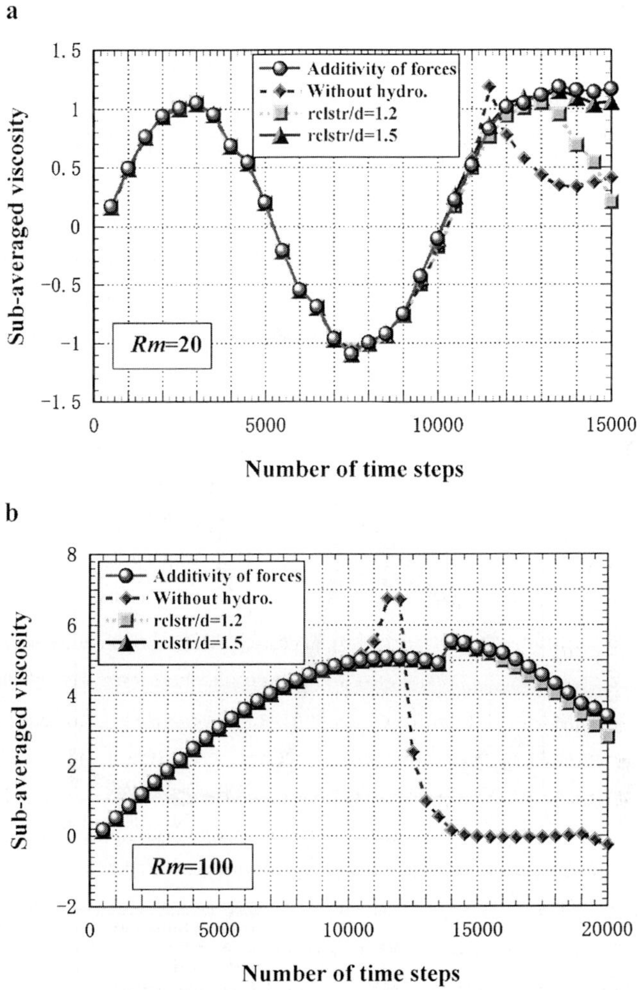

Figure 14.1 Transient properties of sub-averaged viscosities with time steps: (a) for R_m=20 and (b) for R_m =100.

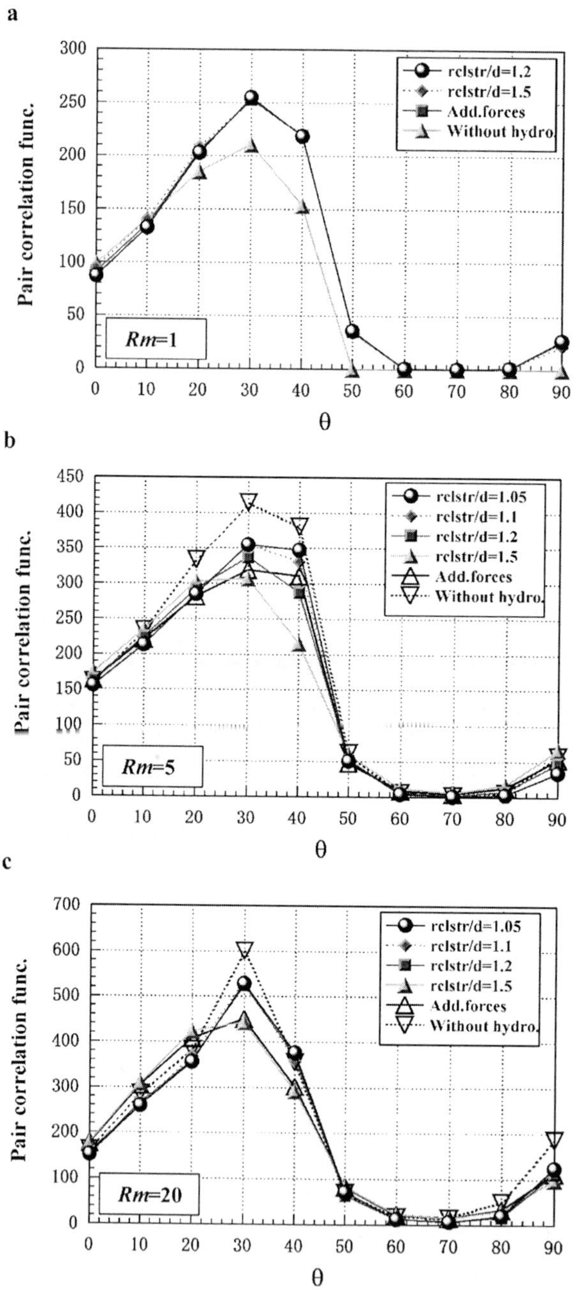

Figure 14.2 Pair correlation functions at r^*=1.99: (a) for R_m =1, (b) for R_m =5, and (c) for R_m =20.

Figure 14.2 shows the pair correlation functions at r^*=1.99 for CBM, in which the results are indicated for R_m=1, 5, and 20; Figs. 14.2(b) and 14.2(c) show the curves for four cases of r_{clstr}^* to compare the influence of values of r_{clstr}^*. In the figures, θ is defined as the angle from the magnetic field direction towards the shear flow direction. It is seen from Fig. 14.2(c) that the curves for $r_{clstr}^*/2$ =1.05, 1.1, and 1.2 almost agree well with each other, and have characteristics intermediate between OM and AIHI. The curve for $r_{clstr}^*/2$=1.5 is in good agreement with that for OM, since the resistance matrices for large clusters are treated in this case. These results clearly show that CBM can simulate properly the particle motion in equilibrium as well as in the transient state. It is seen from Fig. 14.2(a) that, for the case of R_m=1, both curves of $r_{clstr}^*/2$=1.2 and 1.5 agree very well with that for OM, which is in significant contrast with the result for AIHI. The fact that the cluster-based SD method is very useful, even in the situation where the magnetic forces and viscous shear forces are of the same order, suggests that the cluster-based SD method is also useful for usual colloidal dispersions. This is expected to a certain degree from experimental results that, even in a colloidal dispersion composed of rigid spherical particles without magnetic properties, particles form configurations similar to clusters due to the shear forces. These clusters are not due to the interparticle forces such as magnetostatic interactions, but are in cluster-like formation under the circumstances of the shear flow. Although the curves for the different values of r_{clstr}^* are not in good agreement with each other for the case of R_m=5, as shown in Fig. 14.2(b), there is a tendency for the curves to approach that for OM with increasing values of r_{clstr}^* ; the value at θ=40° for $r_{clstr}^*/2$ =1.5 seems to be too small, but the reason for this is currently not clear.

14.2 The Dissipative Particle Dynamics Method

14.2.1 Dissipative particles

Solutions of the flow field for a three-particle system (or for a more-particle system) may be necessary to develop a simulation method with a more accurate multi-body hydrodynamic interaction in the case of a colloidal dispersion. However, it is extremely difficult to solve the flow field even for a three-particle system. Thus, for a more complicated system such as with spherocylinder particles, obtaining analytical solutions of the flow field seems to be almost impossible. Now we change our way of thinking about treating the multi-body hydrodynamic interactions among particles. If both the colloidal particles and the solvent molecules are simulated simultaneously, then we can obtain the solutions of the flow field and the particle motion simultaneously. However, as pointed out in Sec.9, such simulations are seldom practicable since the characteristic times of the motion of particles and molecules are significantly different. To circumvent this difficulty, we stand on a mesoscopic level rather than on a microscopic level. On a mesoscopic level, molecules themselves are not treated, but rather clusters or groups of molecules are dealt with, and therefore these virtual or fluid particles are simulated to obtain a flow field. In this approach a system is regarded as being composed of model fluid particles. These particles then interact with each other dissipatively, exchange momenta, and move randomly like Brownian particles. We call this virtual fluid particle a "dissipative particle."

In a flow problem of colloidal dispersions, the motion of the colloidal particles is governed by the interaction with other colloidal particles, dissipative particles, and an applied field such as a magnetic field. Hence, a simulation method based on dissipative particle dynamics does not require the analytical solution of the flow field for a two- or three-particle system. Multi-body hydrodynamic interactions among colloidal particles are automatically reproduced through the interactions with dissipative particles. This is the most appealing feature of the

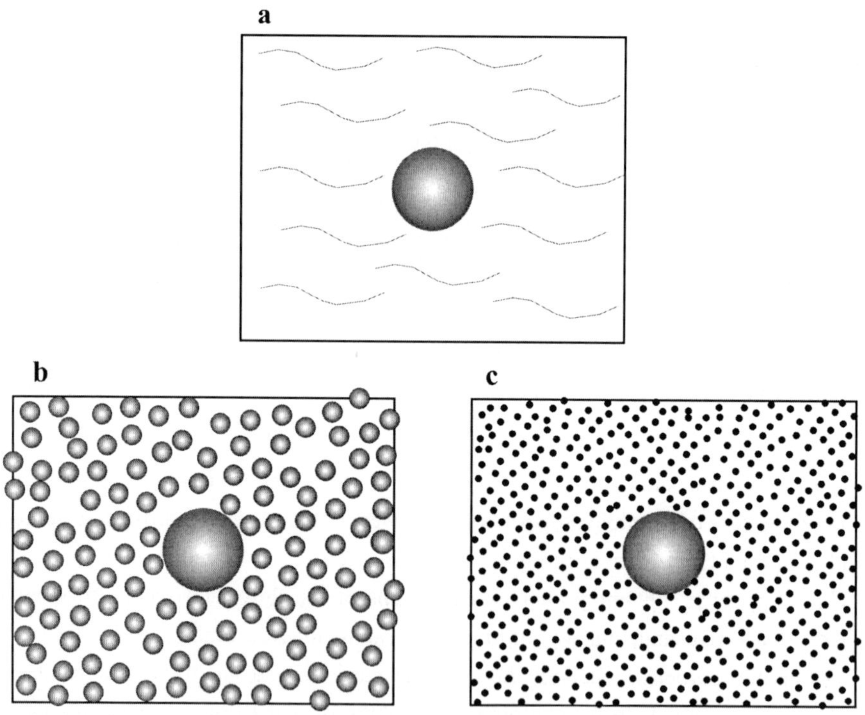

Figure 14.3 Various models of a fluid: (a) a fluid is modeled as a continuum medium (macroscopic model); (b) a fluid is considered to be composed of dissipative particles (mesoscoic model); (c) a fluid is considered to be composed of molecules (microscopic model).

dissipative particle dynamics method. This method is straightforwardly applicable to a colloidal dispersion composed of non-spherical particles, so that it seems to be a significant and promising method from a colloid simulation point of view.

Figure 14.3 shows schematically the classification of the modeling of a fluid. Figure 14.3(a) shows a macroscopic model in which a fluid is regarded as a continuum, Fig. 14.3(c) shows a microscopic model in which a fluid is considered to be composed of molecules, and Fig. 14.3(b) shows a mesoscopic model in which a fluid is regarded as being composed of dissipative particles. To obtain a flow field, the first model uses the Navier-Stokes equation, the second model uses the molecular dynamics method, and the last model uses the dissipative particle dynamics method, which is explained in detail below.

In the following sections, we discuss the theoretical background of the dissipative particle dynamics method, where the system of interest is assumed to be composed of dissipative particles alone, i.e. not including colloidal particles. Unless otherwise specified, we will call dissipative particles just particles in this chapter for convenience.

14.2.2 Dynamics of dissipative particles

The equation of motion for particles has to obey some physical constraints in order that the solution obtained by the dissipative particle dynamics method agrees with that obtained by the Navier-Stokes equation. The conservation of the total momentum of a system and the Galilean invariance must be satisfied by the equation of motion as a minimum request.

We concentrate our attention on particle i and consider the forces acting on this particle. The following three kinds of forces may be physically reasonable as forces acting on particle i: a repulsive conservative force $\mathbf{F}_{ij}^{C}$ exerted by the other particles, a dissipative force $\mathbf{F}_{ij}^{D}$ providing a viscous drag to the system, and a random or stochastic force $\mathbf{F}_{ij}^{R}$ inducing the thermal motion of particles. With these forces, the equation of motion of particle i can be written as [4,5]

$$m\frac{d\mathbf{v}_i}{dt} = \sum_{j(\neq i)}\mathbf{F}_{ij}^{C} + \sum_{j(\neq i)}\mathbf{F}_{ij}^{D} + \sum_{j(\neq i)}\mathbf{F}_{ij}^{R}, \qquad (14.3)$$

in which m is the mass of particle i, $\mathbf{v}_i$ is the velocity, and, concerning the subscripts, for example, $\mathbf{F}_{ij}^{C}$ is the force acting on particle i by particle j.

Now we have to embody specific forms of the above-mentioned forces. It may be reasonable to assume that the conservative force $\mathbf{F}_{ij}^{C}$ depends only on the relative position $\mathbf{r}_{ij}$ ($=\mathbf{r}_i-\mathbf{r}_j$), and not on the particle velocities. An explicit expression for this force will be shown later. Since the Galilean invariance has to be satisfied, the dissipative force $\mathbf{F}_{ij}^{D}$ and the random force $\mathbf{F}_{ij}^{R}$ should not be dependent on the position $\mathbf{r}_i$ and velocity $\mathbf{v}_i$ themselves, but should be functions of the relative position $\mathbf{r}_{ij}$ and relative velocity $\mathbf{v}_{ij}$ ($=\mathbf{v}_i-\mathbf{v}_j$). Additionally, it may be reasonable to assume that the random force $\mathbf{F}_{ij}^R$ does not depend on the relative velocity but the relative position $\mathbf{r}_{ij}$ alone. Furthermore, we have to take into account the isotropy of the particle motion and the decrease in the magnitude of forces with the particle-particle separation. The following expressions for $\mathbf{F}_{ij}^D$ and $\mathbf{F}_{ij}^R$ satisfy these physical requirements [6,7]:

$$\mathbf{F}_{ij}^D = -\gamma w_D(r_{ij})(\mathbf{e}_{ij}\cdot\mathbf{v}_{ij})\mathbf{e}_{ij}, \tag{14.4}$$

$$\mathbf{F}_{ij}^R = \sigma w_R(r_{ij})\mathbf{e}_{ij}\zeta_{ij}, \tag{14.5}$$

in which $r_{ij}=|\,\mathbf{r}_{ij}|$, and $\mathbf{e}_{ij}$ is the unit vector denoting the direction of particle i from particle j, expressed as $\mathbf{e}_{ij}=\mathbf{r}_{ij}/r_{ij}$. Also ζ_{ij} is a random variable inducing the random motion of particles and has to satisfy the following stochastic properties:

$$\langle\zeta_{ij}\rangle = 0, \quad \langle\zeta_{ij}(t)\zeta_{i'j'}(t')\rangle = (\delta_{ii'}\delta_{jj'}+\delta_{ij'}\delta_{ji'})\delta(t-t'). \tag{14.6}$$

This variable satisfies the characteristic of the symmetry $\zeta_{ij}=\zeta_{ji}$, which ensures that the total momentum of the system is conserved. The $w_D(r_{ij})$ and $w_R(r_{ij})$ are weight functions to reproduce the decrease in forces with particle-particle separation, and γ and σ are constants specifying the magnitude of forces. These constants can be related to the system temperature and friction coefficient, which will be shown later. It is seen from the expression for $\mathbf{F}_{ij}^D$ that this force acts in such a way as to relax the relative motion of particles i and j. On the other hand, $\mathbf{F}_{ij}^R$ acts as if inducing the thermal motion of particles, but this force satisfies the action-reaction formula, so that the conservation of the total momentum of the system is ensured.

The substitution of Eqs. (14.4) and (14.5) into Eq. (14.3) leads to the following equation:

$$m\frac{d\mathbf{v}_i}{dt} = \sum_{j(\neq i)}\mathbf{F}_{ij}^C(\mathbf{r}_{ij}) - \sum_{j(\neq i)}\gamma w_D(r_{ij})(\mathbf{e}_{ij}\cdot\mathbf{v}_{ij})\mathbf{e}_{ij} + \sum_{j(\neq i)}\sigma w_R(r_{ij})\mathbf{e}_{ij}\zeta_{ij}. \tag{14.7}$$

If this equation is integrated with respect to time over a small time interval from t to $t+\Delta t$, then the finite difference equations governing the particle motion in simulations can be obtained as

$$\Delta\mathbf{r}_i = \mathbf{v}_i\Delta t, \tag{14.8}$$

$$\Delta\mathbf{v}_i = \frac{1}{m}\left(\sum_{j(\neq i)}\mathbf{F}_{ij}^C(\mathbf{r}_{ij}) - \sum_{j(\neq i)}\gamma w_D(r_{ij})(\mathbf{e}_{ij}\cdot\mathbf{v}_{ij})\mathbf{e}_{ij}\right)\Delta t + \frac{1}{m}\sum_{j(\neq i)}\sigma w_R(r_{ij})\mathbf{e}_{ij}\Delta W_{ij}, \tag{14.9}$$

in which

$$\Delta W_{ij} = \int_{t}^{t+\Delta t} \zeta_{ij} d\tau . \tag{14.10}$$

This ΔW_{ij} has to satisfy the following stochastic properties, which arise from Eq. (14.6):

$$\left.\begin{aligned} &\langle \Delta W_{ij} \rangle = 0, \\ &\langle \Delta W_{ij} \Delta W_{i'j'} \rangle = (\delta_{ii'}\delta_{jj'} + \delta_{ij'}\delta_{ji'})\Delta t . \end{aligned}\right\} \tag{14.11}$$

If a new stochastic variable θ_{ij} is introduced from the definition $\Delta W_{ij} = \theta_{ij} (\Delta t)^{1/2}$, the third term in Eq. (14.9) can be written as

$$\frac{1}{m} \sum_{j(\neq i)} \sigma w_R(r_{ij}) \mathbf{e}_{ij} \theta_{ij} \sqrt{\Delta t} , \tag{14.12}$$

in which θ_{ij} has to satisfy the following stochastic properties:

$$\left.\begin{aligned} &\langle \theta_{ij} \rangle = 0, \\ &\langle \theta_{ij} \theta_{i'j'} \rangle = (\delta_{ii'}\delta_{jj'} + \delta_{ij'}\delta_{ji'}) . \end{aligned}\right\} \tag{14.13}$$

It is seen from Eq. (14.13) that the stochastic variable θ_{ij} is sampled from a uniform or normal distribution with zero average value and unit variance.

The equation of motion in Eq. (14.7) satisfies the conservation law of the total momentum of the system, but not the energy conservation law. The equation of motion has to be modified to satisfy the conservation of the system energy, but this subject is not dealt with in this book. Thus if the reader is interested in the modified version of the equation of motion, some appropriate references [8,9] may be referred to.

The coefficients γ and σ, and the weight functions $w_D(r_{ij})$ and $w_R(r_{ij})$ cannot be determined independently, because there are some relations among these quantities. The derivation of these relations will be shown in Sec. 14.2.4.

14.2.3 The Fokker-Planck equation

We use the notation $\mathbf{r}_i$ for the position vector of particle i and $\mathbf{v}_i$ for the velocity vector. Also, for simplicity of expression, the vector $\mathbf{v}$ is used for describing the velocity vectors of all the particles (that is, $\mathbf{v}$ represents $\mathbf{v}_1$, $\mathbf{v}_2$, $\mathbf{v}_3$, …), and similarly $\mathbf{r}$ is for the position vectors of all the particles (that is, $\mathbf{r}$ represents $\mathbf{r}_1$, $\mathbf{r}_2$, $\mathbf{r}_3$, …). If the probability that a particle position and velocity are found within the range from $(\mathbf{r},\mathbf{v})$ to $(\mathbf{r}+\Delta\mathbf{r}, \mathbf{v}+\Delta\mathbf{v})$ is denoted by $W(\mathbf{r},\mathbf{v},t)d\mathbf{r}d\mathbf{v}$, then the probability density function $W(\mathbf{r},\mathbf{v},t)$ satisfies the following stochastic formula, i.e. the Chapman-Kolmogorov equation:

$$W(\mathbf{r},\mathbf{v},t) = \iint W(\mathbf{r}-\Delta\mathbf{r},\mathbf{v}-\Delta\mathbf{v},t-\Delta t)\Psi(\mathbf{r}-\Delta\mathbf{r},\mathbf{v}-\Delta\mathbf{v};\Delta\mathbf{r},\Delta\mathbf{v})d(\Delta\mathbf{r})d(\Delta\mathbf{v}), \quad (14.14)$$

in which $\Psi(\mathbf{r}\text{-}\Delta\mathbf{r},\mathbf{v}\text{-}\Delta\mathbf{v};\ \Delta\mathbf{r},\Delta\mathbf{v})$ is the transition probability for the state $(\mathbf{r}\text{-}\Delta\mathbf{r},\mathbf{v}\text{-}\Delta\mathbf{v})$ transferring to another state $(\mathbf{r},\mathbf{v})$ during a time interval Δt, and this does not depend on the time point t. If the velocity does not change appreciably during the small time interval Δt, then Ψ can be written as

$$\Psi(\mathbf{r}-\Delta\mathbf{r},\mathbf{v}-\Delta\mathbf{v};\Delta\mathbf{r},\Delta\mathbf{v}) = \psi(\mathbf{r}-\Delta\mathbf{r},\mathbf{v}-\Delta\mathbf{v};\Delta\mathbf{v})\delta(\Delta\mathbf{r}-\mathbf{v}\Delta t), \quad (14.15)$$

in which ψ is the transition probability and is independent of $\Delta\mathbf{r}$. The substitution of Eq. (14.15) into Eq. (14.14) leads to the following different form of the Chapman-Kolmogorov equation:

$$W(\mathbf{r}+\mathbf{v}\Delta t,\mathbf{v},t+\Delta t) = \int W(\mathbf{r},\mathbf{v}-\Delta\mathbf{v},t)\psi(\mathbf{r},\mathbf{v}-\Delta\mathbf{v};\Delta\mathbf{v})d(\Delta\mathbf{v}). \quad (14.16)$$

We now derive the Fokker-Planck equation in phase space (or the Chandrasekhar equation) by reforming Eq. (14.16).

If ψ is a function of $\Delta\mathbf{v}$ which has a sharp peak at $\Delta\mathbf{v}=0$, Eq. (14.16) can be expanded in a Taylor series as

$$W(\mathbf{r}+\mathbf{v}\Delta t,\mathbf{v},t+\Delta t) = W(\mathbf{r},\mathbf{v},t) + \Delta t\frac{\partial W}{\partial t} + \sum_i (\mathbf{v}_i\Delta t)\cdot\frac{\partial W}{\partial \mathbf{r}_i} + \mathrm{O}((\Delta t)^2,\Delta v\Delta t,(\Delta v)^2). \quad (14.17)$$

Similarly,

$$\begin{aligned}\psi(\mathbf{r},\mathbf{v}-\Delta\mathbf{v};\Delta\mathbf{v}) &= \psi(\mathbf{r},\mathbf{v};\Delta\mathbf{v}) - \sum_i \Delta\mathbf{v}_i\cdot\frac{\partial\psi}{\partial\mathbf{v}_i} \\ &\quad + \frac{1}{2}\sum_i\sum_j \Delta\mathbf{v}_i\Delta\mathbf{v}_j : \frac{\partial}{\partial\mathbf{v}_j}\frac{\partial}{\partial\mathbf{v}_i}\psi + \mathrm{O}((\Delta v)^3),\end{aligned} \quad (14.18)$$

$$W(\mathbf{r},\mathbf{v}-\Delta\mathbf{v},t) = W(\mathbf{r},\mathbf{v},t) - \sum_i \Delta\mathbf{v}_i\cdot\frac{\partial W}{\partial\mathbf{v}_i} + \frac{1}{2}\sum_i\sum_j \Delta\mathbf{v}_i\Delta\mathbf{v}_j : \frac{\partial}{\partial\mathbf{v}_j}\frac{\partial}{\partial\mathbf{v}_i}W + \mathrm{O}((\Delta v)^3). \quad (14.19)$$

The transition probability ψ has the following normalization condition:

$$\int\psi(\mathbf{r},\mathbf{v};\Delta\mathbf{v})d(\Delta\mathbf{v}) = 1. \quad (14.20)$$

The substitution of Eqs. (14.17) to (14.19) into Eq. (14.16) leads to the following expression:

$$\frac{\partial W}{\partial t}+\sum_i \mathbf{v}_i\cdot\frac{\partial W}{\partial \mathbf{r}_i}+O((\Delta t),(\Delta v),\frac{(\Delta v)^2}{\Delta t})$$
$$=\frac{1}{\Delta t}\int\left\{-W\sum_i \Delta\mathbf{v}_i\cdot\frac{\partial \psi}{\partial \mathbf{v}_i}-\psi\sum_i \Delta\mathbf{v}_i\cdot\frac{\partial W}{\partial \mathbf{v}_i}+\frac{1}{2}W\sum_i\sum_j \Delta\mathbf{v}_i\Delta\mathbf{v}_j:\frac{\partial}{\partial \mathbf{v}_j}\frac{\partial}{\partial \mathbf{v}_i}\psi\right.$$
$$\left.+\frac{1}{2}\psi\sum_i\sum_j \Delta\mathbf{v}_i\Delta\mathbf{v}_j:\frac{\partial}{\partial \mathbf{v}_j}\frac{\partial}{\partial \mathbf{v}_i}W+\left(\sum_i \Delta\mathbf{v}_i\cdot\frac{\partial \psi}{\partial \mathbf{v}_i}\right)\left(\sum_j \Delta\mathbf{v}_j\cdot\frac{\partial W}{\partial \mathbf{v}_j}\right)+O((\Delta v)^3)\right\}d(\Delta\mathbf{v}).$$
(14.21)

If some vector operations are used to transform the terms on the right-hand side (see Exercises at the end of this chapter), then Eq. (14.21) can be reformed as

$$\frac{\partial W}{\partial t}+\sum_i \mathbf{v}_i\cdot\frac{\partial W}{\partial \mathbf{r}_i}=-\sum_i\frac{\partial}{\partial \mathbf{v}_i}\cdot\left(W\frac{\langle\Delta\mathbf{v}_i\rangle}{\Delta t}\right)+\frac{1}{2}\sum_i\sum_j\frac{\partial^2}{\partial \mathbf{v}_j\partial \mathbf{v}_i}:\left(W\frac{\langle\Delta\mathbf{v}_i\Delta\mathbf{v}_j\rangle}{\Delta t}\right), \tag{14.22}$$

in which

$$\left.\begin{aligned}\langle\Delta\mathbf{v}_i\rangle&=\int\Delta\mathbf{v}_i\psi d(\Delta\mathbf{v}),\\ \langle\Delta\mathbf{v}_i\Delta\mathbf{v}_j\rangle&=\int(\Delta\mathbf{v}_i\Delta\mathbf{v}_j)\psi d(\Delta\mathbf{v}).\end{aligned}\right\} \tag{14.23}$$

In Eq. (14.22) the terms of denoting the order of the neglected higher-order terms have been omitted. Equation (14.22) is the Chandrasekhar equation in its general form.

Next we show the Chandrasekhar equation for the present dissipative particle dynamics method. This is accomplished by evaluating $\langle\Delta\mathbf{v}_i\rangle$ and $\langle\Delta\mathbf{v}_i\Delta\mathbf{v}_j\rangle$ using Eq. (14.9). $\langle\Delta\mathbf{v}_i\rangle$ can be straightforwardly evaluated and expressed as

$$\langle\Delta\mathbf{v}_i\rangle=\frac{1}{m}\left\{\sum_{j(\neq i)}\mathbf{F}_{ij}^C(\mathbf{r}_{ij})-\sum_{j(\neq i)}\gamma w_D(r_{ij})(\mathbf{e}_{ij}\cdot\mathbf{v}_{ij})\mathbf{e}_{ij}\right\}\Delta t. \tag{14.24}$$

Similarly, $\langle\Delta\mathbf{v}_i\Delta\mathbf{v}_j\rangle$ can be evaluated and written as

$$\left.\begin{aligned}\langle\Delta\mathbf{v}_i\Delta\mathbf{v}_j\rangle&=\frac{1}{m^2}\sigma^2 {w_R}^2(r_{ij})\mathbf{e}_{ij}\mathbf{e}_{ji}\Delta t\qquad(i\neq j),\\ \langle\Delta\mathbf{v}_i\Delta\mathbf{v}_i\rangle&=\frac{1}{m^2}\sum_{j(\neq i)}\sigma^2 {w_R}^2(r_{ij})\mathbf{e}_{ij}\mathbf{e}_{ij}\Delta t.\end{aligned}\right\} \tag{14.25}$$

The substitution of Eqs (14.24) and (14.25) into Eq. (14.22) leads to the following Chandrasekhar equation (or the Fokker-Planck equation in phase space) for the dissipative particle dynamics method:

$$\frac{\partial W}{\partial t}+\sum_{i}\mathbf{v}_i\cdot\frac{\partial W}{\partial \mathbf{r}_i}+\sum_{i}\sum_{\substack{j\\(i\neq j)}}\frac{\mathbf{F}_{ij}^C}{m}\cdot\frac{\partial W}{\partial \mathbf{v}_i}=\sum_{i}\sum_{\substack{j\\(i\neq j)}}\mathbf{e}_{ij}\cdot\frac{\partial}{\partial \mathbf{v}_i}\left\{\frac{1}{m}\gamma w_D(r_{ij})(\mathbf{e}_{ij}\cdot\mathbf{v}_{ij})W\right\}$$
$$+\frac{1}{2}\sum_{i}\sum_{\substack{j\\(i\neq j)}}\frac{1}{m^2}\sigma^2 w_R{}^2(r_{ij})\mathbf{e}_{ij}\cdot\frac{\partial}{\partial \mathbf{v}_i}\left(\mathbf{e}_{ij}\cdot\frac{\partial}{\partial \mathbf{v}_i}-\mathbf{e}_{ij}\cdot\frac{\partial}{\partial \mathbf{v}_j}\right)W. \tag{14.26}$$

14.2.4 The equation of motion of dissipative particles

If a system composed of dissipative particles is in equilibrium, then the equilibrium distribution W_{eq} becomes the canonical distribution for an ensemble which is specified by a given particle number N, volume V, and temperature T. With the notation of U for the potential energy, W_{eq} is written as

$$W_{eq}=\frac{1}{Z}\exp\left\{-\frac{1}{kT}\left(\sum_{i=1}^{N}m\frac{v_i^2}{2}+U\right)\right\}, \tag{14.27}$$

in which Z is the partition function, and U is related to the conservative force $\mathbf{F}_{ij}^C$ by

$$\mathbf{F}_{ij}^C=-\frac{\partial U}{\partial \mathbf{r}_{ij}}. \tag{14.28}$$

The equilibrium distribution W_{eq} must satisfy the Fokker-Planck equation in phase space in Eq. (14.26). Since the left-hand side in Eq. (14.26) vanishes for the substitution of W_{eq}, the right-hand side must also become zero. This is accomplished by the following requirements:

$$\left.\begin{aligned} w_D(r_{ij}) &= w_R^2(r_{ij}),\\ \sigma^2 &= 2\gamma kT, \end{aligned}\right\} \tag{14.29}$$

in which k is Boltzmann's constant. The second equation is the fluctuation-dissipation theorem for the dissipative particle dynamics.

We now have to determine the explicit expression for the conservative force $\mathbf{F}_{ij}^C$ and the weight function $w_R(r_{ij})$. $\mathbf{F}_{ij}^C$ is a repulsive force for preventing unphysical excessive overlaps between particles, and $w_R(r_{ij})$ has to be set so that interparticle forces decrease with increasing particle-particle separations. These requirements are satisfied by the following expressions:

$$\mathbf{F}_{ij}^C=\alpha w_R(r_{ij})\mathbf{e}_{ij}, \tag{14.30}$$

$$w_R(r_{ij}) = \begin{cases} 1 - \dfrac{r_{ij}}{r_c} & \text{for} \quad r_{ij} \le r_c, \\ 0 & \text{for} \quad r_{ij} > r_c, \end{cases} \tag{14.31}$$

in which α is a constant representing the magnitude of the repulsive forces. By substituting these equations into Eq. (14.9) with considering Eq. (14.12), the final expression for the equation of motion of the dissipative particle dynamics method can be written as

$$\Delta \mathbf{r}_i = \mathbf{v}_i \Delta t, \tag{14.32}$$

$$\begin{aligned} \Delta \mathbf{v}_i = \frac{\alpha}{m} \sum_{j(\neq i)} w_R(r_{ij}) \mathbf{e}_{ij} \Delta t - \frac{\gamma}{m} \sum_{j(\neq i)} w_R^2(r_{ij})(\mathbf{e}_{ij} \cdot \mathbf{v}_{ij}) \mathbf{e}_{ij} \Delta t \\ + \frac{(2\gamma kT)^{1/2}}{m} \sum_{j(\neq i)} w_R(r_{ij}) \mathbf{e}_{ij} \theta_{ij} \sqrt{\Delta t}, \end{aligned} \tag{14.33}$$

in which the expression for $w_R(r_{ij})$ has already been shown as Eq. (14.31). The stochastic variable θ_{ij} must obey the stochastic properties shown in Eq. (14.13) and is sampled from a uniform or normal distribution with zero average and unit variance, as already pointed out. The dissipative particle dynamics method uses Eqs. (14.32) and (14.33) to simulate the motion of the dissipative particles.

Lastly, we show an example of the nondimensionalization method used for actual simulations. To nondimensionalize each quantity, the following representative values are used: $(kT/m)^{1/2}$ for velocities, r_c for distances, $r_c(m/kT)^{1/2}$ for time, $(1/r_c^3)$ for number densities, etc. With these representative values, Eqs. (14.32) and (14.33) can be nondimensionalized as

$$\Delta \mathbf{r}_i^* = \mathbf{v}_i^* \Delta t^*, \tag{14.34}$$

$$\begin{aligned} \Delta \mathbf{v}_i^* = \alpha^* \sum_{j(\neq i)} w_R(r_{ij}^*) \mathbf{e}_{ij} \Delta t^* - \gamma^* \sum_{j(\neq i)} w_R^2(r_{ij}^*)(\mathbf{e}_{ij} \cdot \mathbf{v}_{ij}^*) \mathbf{e}_{ij} \Delta t^* \\ + (2\gamma^*)^{1/2} \sum_{j(\neq i)} w_R(r_{ij}^*) \mathbf{e}_{ij} \theta_{ij} \sqrt{\Delta t^*}, \end{aligned} \tag{14.35}$$

in which

$$w_R(r_{ij}^*) = \begin{cases} 1 - r_{ij}^* & \text{for} \quad r_{ij}^* \le 1, \\ 0 & \text{for} \quad r_{ij}^* > 0, \end{cases} \tag{14.36}$$

$$\alpha^* = \alpha \frac{r_c}{kT}, \quad \gamma^* = \gamma \frac{r_c}{(mkT)^{1/2}}. \tag{14.37}$$

In these equations, the quantities with the superscript * are dimensionless. Equations (14.34) and (14.35) clearly show that one can start a dissipative particle simulation if appropriate values of the nondimensional parameters α^* and γ^* are adopted, and also the time interval Δt^*, the number density n^* $(=nr_c^3)$, and the particle number N are properly specified. Results obtained by the simulations should not depend on the nondimensional quantities which have been introduced in embodying the equation of motion; one has to be, therefore, careful in setting appropriate values for these values in conducting simulations. For example, since time is nondimensionalized by the representative value based on the mean velocity $\bar{v}$ $(\approx kT/m)^{1/2}$) and the radius r_c, the magnitude of the dimensionless time interval Δt^* may have to be taken sufficiently smaller that unity [6,10,11].

14.2.5 Transport coefficients

Since it is very difficult to derive theoretically the transport coefficients such as viscosity from the equation of motion of dissipative particle dynamics, we only show the final expression for the viscosity, without its derivation process [10]. The dissipative part of the viscosity, η^D, is expressed as

$$\eta^D = \frac{\gamma n^2 \langle r^2 \rangle_w \tilde{w}}{2d(d+2)}, \tag{14.38}$$

in which

$$\left.\begin{aligned} \langle r^2 \rangle_w &= \frac{1}{\tilde{w}} \int r^2 w(r) d\mathbf{r}, \\ \tilde{w} &= \int w(r) d\mathbf{r}. \end{aligned}\right\} \tag{14.39}$$

In this equation $w(r)$ has to be taken as $w(r)=w_D(r)$. Equation (14.38) is seemingly different from the expression which was derived by Hoogerbrugge and Koelman [4], but we can see that Eq. (14.38) reduces to their expression $\eta^D = mn\omega \langle r^2 \rangle / \{2d(d+2)\Delta t\}$ because $n\tilde{w}=1$ and $\omega = \gamma \Delta t/m$ in their theory. In Eq. (14.38), $d=3$ for a three-dimensional system.

Using nondimensional values explained above, the viscosity is nondimensionalized as

$$\eta^{D*} = \frac{\eta^D}{(mkT)^{1/2}/r_c^2} = \frac{\gamma^* n^{*2} \langle r^{*2} \rangle_w \tilde{w}^*}{2d(d+2)}. \tag{14.40}$$

Results of the viscosity predicted by simulations do not necessarily agree with the theoretical values in Eq. (14.40), but depend on the values of α^*, γ^*, and Δt^*. Thus when a more accurate value of the viscosity is necessary to calculate the Reynolds number, one has to evaluate the values of the viscosity from another independent simulation. This may be accomplished using the following Green-Kubo expression for the shear viscosity:

$$\eta_{yx} = \frac{1}{kTV}\int_0^\infty \langle J_{yx}(t)J_{yx}(0)\rangle dt\,, \tag{14.41}$$

in which

$$J_{yx}(t) = \sum_{i=1}^{N}\{mv_{iy}(t)v_{ix}(t) + y_i(t)F_{ix}(t)\} = \sum_{i=1}^{N} mv_{iy}(t)v_{ix}(t) + \sum_{\substack{i=1\\(i<j)}}^{N}\sum_{j=1}^{N} y_{ij}(t)F_{ijx}(t)\,. \tag{14.42}$$

In this equation F_{ijx} is the x-component of $\mathbf{F}_{ij}$, and $\mathbf{F}_{ij}$ is taken as $\mathbf{F}_{ij}=\mathbf{F}_{ij}^{C}+\mathbf{F}_{ij}^{D}$ in the dissipative particle dynamics method. Since the expression in Eq. (14.41) is valid in thermodynamic equilibrium, the symmetric property of $\eta_{yx}=\eta_{xy}=\eta_{yz}=\eta_{zy}=\eta_{zx}=\eta_{xz}$ holds.

According to a similar nondimensionalization to Eq. (14.40), η_{yx} can be written in nondimensional form as

$$\eta_{yx}^{*} = \frac{\eta_{yx}}{(mkT)^{1/2}/r_c^{\,2}} = \frac{1}{V^*}\int_0^\infty \langle J_{yx}^{*}(t^*)J_{yx}^{*}(0)\rangle dt^*\,, \tag{14.43}$$

in which

$$J_{yx}^{*}(t^*) = \sum_{i=1}^{N} v_{iy}^{*}(t^*)v_{ix}^{*}(t^*) + \sum_{\substack{i=1\\(i<j)}}^{N}\sum_{j=1}^{N} y_{ij}^{*}(t^*)F_{ijx}^{*}(t^*)\,. \tag{14.44}$$

When the viscosity is evaluated in a simple shear flow, the following expression can be used:

$$\eta_{yx} = -\frac{1}{V\dot{\gamma}}\langle J_{yx}\rangle_{ne}\,, \tag{14.45}$$

in which V is the system volume, $\dot{\gamma}$ is the shear rate, and $\langle J_{yx}\rangle_{ne}$ is the time average of J_{yx} under a shear flow situation. The expression in Eq. (14.45) is the result from the theory for the nonequilibrium molecular dynamics method.

Similarly, Eq. (14.45) can be nondimensionalized as

$$\eta_{yx}^{*}(t^*) = -\frac{1}{V^*\dot{\gamma}^*}\left\langle \sum_{i=1}^{N} v_{iy}^{*}v_{ix}^{*} + \sum_{\substack{i=1\\(i<j)}}^{N}\sum_{j=1}^{N} y_{ij}^{*}F_{ijx}^{*}\right\rangle_{ne}\,, \tag{14.46}$$

in which the shear rate is nondimensionalized by $(kT/m)^{1/2}/r_c$.

14.2.6 Derivation of transport equations

As a final discussion concerning the dissipative particle dynamics method, the equation of continuity and the momentum equation of the fluid are derived using the equation of motion of the dissipative particle dynamics.

If an arbitrary physical quantity $A(\mathbf{r},\mathbf{v})$ is not dependent on time explicitly, the time average $\langle A \rangle$ can be expressed, using the probability density function W which satisfies Eq. (14.26), as

$$\langle A \rangle = \iint A W(\mathbf{r},\mathbf{v},t) d\mathbf{r} d\mathbf{v} \, , \tag{14.47}$$

in which

$$\iint W(\mathbf{r},\mathbf{v},t) d\mathbf{r} d\mathbf{v} = 1 \, . \tag{14.48}$$

Hence, the time variation of $\langle A \rangle$ can be expressed, from Eq. (14.26), as

$$\begin{aligned} \frac{\partial}{\partial t}\langle A \rangle = \iint A \frac{\partial W}{\partial t} d\mathbf{r} d\mathbf{v} = \iint A \Bigg[& -\sum_i \mathbf{v}_i \cdot \frac{\partial W}{\partial \mathbf{r}_i} - \sum_i \sum_{\substack{j \\ (i \neq j)}} \frac{\mathbf{F}_{ij}^C}{m} \cdot \frac{\partial W}{\partial \mathbf{v}_i} \\ & + \sum_i \sum_{\substack{j \\ (i \neq j)}} \mathbf{e}_{ij} \cdot \frac{\partial}{\partial \mathbf{v}_i} \left\{ \frac{1}{m} \gamma w_D(r_{ij})(\mathbf{e}_{ij} \cdot \mathbf{v}_{ij}) W \right\} \\ & + \frac{1}{2} \sum_i \sum_{\substack{j \\ (i \neq j)}} \frac{1}{m^2} \sigma^2 w_R^{\,2}(r_{ij}) \mathbf{e}_{ij} \cdot \frac{\partial}{\partial \mathbf{v}_i} \left(\mathbf{e}_{ij} \cdot \frac{\partial}{\partial \mathbf{v}_i} - \mathbf{e}_{ij} \cdot \frac{\partial}{\partial \mathbf{v}_j} \right) W \Bigg] d\mathbf{r} d\mathbf{v} \, . \end{aligned} \tag{14.49}$$

The first term can be reformed as

$$\begin{aligned} -\iint A \sum_i \mathbf{v}_i \cdot \frac{\partial W}{\partial \mathbf{r}_i} d\mathbf{r} d\mathbf{v} &= -\iint \sum_i \mathbf{v}_i \cdot \left\{ \frac{\partial}{\partial \mathbf{r}_i}(AW) - W \frac{\partial A}{\partial \mathbf{r}_i} \right\} d\mathbf{r} d\mathbf{v} \\ &= \iint \left(\sum_i \mathbf{v}_i \cdot \frac{\partial A}{\partial \mathbf{r}_i} \right) W d\mathbf{r} d\mathbf{v}, \end{aligned} \tag{14.50}$$

in which the fact that W vanishes at the system boundaries has been taken into account in deriving the second expression from the first expression on the right-hand side. Similarly, the remainder terms in Eq. (14.49) can be reformed and then Eq. (14.49) reduces to

$$\begin{aligned} \frac{\partial}{\partial t}\langle A \rangle = \Bigg\langle & \sum_i \mathbf{v}_i \cdot \frac{\partial A}{\partial \mathbf{r}_i} + \sum_i \sum_{\substack{j \\ (i \neq j)}} \frac{\mathbf{F}_{ij}^C}{m} \cdot \frac{\partial A}{\partial \mathbf{v}_i} - \sum_i \sum_{\substack{j \\ (i \neq j)}} \frac{\gamma}{m} w_D(r_{ij})(\mathbf{e}_{ij} \cdot \mathbf{v}_{ij}) \left(\mathbf{e}_{ij} \cdot \frac{\partial A}{\partial \mathbf{v}_i} \right) \\ & + \frac{1}{2} \sum_i \sum_{\substack{j \\ (i \neq j)}} \frac{1}{m^2} \sigma^2 w_R^{\,2}(r_{ij}) \left(\mathbf{e}_{ij} \cdot \frac{\partial}{\partial \mathbf{v}_i} \right) \left(\mathbf{e}_{ij} \cdot \frac{\partial}{\partial \mathbf{v}_i} - \mathbf{e}_{ij} \cdot \frac{\partial}{\partial \mathbf{v}_j} \right) A \Bigg\rangle . \end{aligned} \tag{14.51}$$

Next we derive the equation of continuity and the momentum equation of the fluid using Eq. (14.51). If A is defined by the following equation,

$$A = \sum_{i=1}^{N} m\delta(\mathbf{r}-\mathbf{r}_i), \tag{14.52}$$

then the average $\langle A \rangle$ which is evaluated from Eq. (14.47) is equal to a local density ρ(**r**). That is,

$$\langle A \rangle = \left\langle \sum_{i=1}^{N} m\delta(\mathbf{r}-\mathbf{r}_i) \right\rangle = \rho(\mathbf{r}). \tag{14.53}$$

If A is taken as

$$A = \sum_{i=1}^{N} m\mathbf{v}_i\delta(\mathbf{r}-\mathbf{r}_i), \tag{14.54}$$

then the average $\langle A \rangle$ is now

$$\langle A \rangle = \rho(\mathbf{r})\mathbf{u}(\mathbf{r}), \tag{14.55}$$

in which $\mathbf{u}(\mathbf{r})$ is a macroscopic fluid velocity at a position $\mathbf{r}$.

Substitution of Eq. (14.52) into Eq. (14.51) leads to the following expression:

$$\frac{\partial}{\partial t}\rho = \left\langle \sum_{i=1}^{N} \mathbf{v}_i \cdot \frac{\partial}{\partial \mathbf{r}_i}(m\delta(\mathbf{r}-\mathbf{r}_i)) \right\rangle = -\left\langle \sum_{i=1}^{N} \mathbf{v}_i \cdot \frac{\partial}{\partial \mathbf{r}}(m\delta(\mathbf{r}-\mathbf{r}_i)) \right\rangle = -\frac{\partial}{\partial \mathbf{r}} \cdot (\rho\mathbf{u}). \tag{14.56}$$

This equation can be reformed as

$$\frac{\partial \rho}{\partial t} + \frac{\partial}{\partial \mathbf{r}} \cdot (\rho\mathbf{u}) = 0. \tag{14.57}$$

This is no other than the equation of continuity.

Next the substitution of Eq. (14.54) into Eq. (14.51) leads to the following equation:

$$\begin{aligned}
\frac{\partial}{\partial t}(\rho\mathbf{u}) = \Bigg\langle & \sum_i m\mathbf{v}_i \cdot \frac{\partial}{\partial \mathbf{r}_i}(\mathbf{v}_i\delta(\mathbf{r}-\mathbf{r}_i)) + \sum_i \sum_{j\,(i\neq j)} \mathbf{F}_{ij}^C \cdot \frac{\partial}{\partial \mathbf{v}_i}(\mathbf{v}_i\delta(\mathbf{r}-\mathbf{r}_i)) \\
& - \sum_i \sum_{j\,(i\neq j)} \gamma w_D(r_{ij})(\mathbf{e}_{ij}\cdot\mathbf{v}_{ij})\left\{\mathbf{e}_{ij}\cdot\frac{\partial}{\partial \mathbf{v}_i}(\mathbf{v}_i\delta(\mathbf{r}-\mathbf{r}_i))\right\} \\
& + \frac{1}{2}\sum_i \sum_{j\,(i\neq j)} \frac{1}{m}\sigma^2 w_R{}^2(r_{ij})\left(\mathbf{e}_{ij}\cdot\frac{\partial}{\partial \mathbf{v}_i}\right)\mathbf{e}_{ij}\cdot\left\{\frac{\partial}{\partial \mathbf{v}_i}(\mathbf{v}_i\delta(\mathbf{r}-\mathbf{r}_i)) - \frac{\partial}{\partial \mathbf{v}_j}(\mathbf{v}_j\delta(\mathbf{r}-\mathbf{r}_j))\right\}\Bigg\rangle.
\end{aligned} \tag{14.58}$$

By taking into account the following relation:

$$\frac{\partial}{\partial \mathbf{v}_i}(\mathbf{v}_i\delta(\mathbf{r}-\mathbf{r}_i)) - \frac{\partial}{\partial \mathbf{v}_j}(\mathbf{v}_j\delta(\mathbf{r}-\mathbf{r}_j)) = (\delta(\mathbf{r}-\mathbf{r}_i) - \delta(\mathbf{r}-\mathbf{r}_j))\mathbf{I}, \tag{14.59}$$

it is seen that the last term in Eq. (14.58) vanishes. Thus, by taking into account Eq. (14.4), and conducting a similar reformation, Eq. (14.58) can be simplified as

$$\frac{\partial}{\partial t}(\rho\mathbf{u}) = \left\langle -\frac{\partial}{\partial \mathbf{r}}\cdot\left\{\sum_i m\mathbf{v}_i\mathbf{v}_i\delta(\mathbf{r}-\mathbf{r}_i)\right\} + \sum_i \sum_{\substack{j \\ (i\neq j)}} (\mathbf{F}_{ij}^C + \mathbf{F}_{ij}^D)\delta(\mathbf{r}-\mathbf{r}_i) \right\rangle, \tag{14.60}$$

The first term in this equation can be reformed as

$$\begin{aligned} \left\langle -\frac{\partial}{\partial \mathbf{r}}\cdot\left\{\sum_i m\mathbf{v}_i\mathbf{v}_i\delta(\mathbf{r}-\mathbf{r}_i)\right\}\right\rangle &= -\frac{\partial}{\partial \mathbf{r}}\cdot(\rho\mathbf{u}\mathbf{u}) \\ &\quad -\frac{\partial}{\partial \mathbf{r}}\cdot\left\{\sum_i m\langle(\mathbf{v}_i-\mathbf{u})(\mathbf{v}_i-\mathbf{u})\delta(\mathbf{r}-\mathbf{r}_i)\rangle\right\}, \end{aligned} \tag{14.61}$$

in which the following relation has been used.

$$\sum_i m\langle(\mathbf{v}_i-\mathbf{u})(\mathbf{v}_i-\mathbf{u})\delta(\mathbf{r}-\mathbf{r}_i)\rangle = \sum_i m\langle\mathbf{v}_i\mathbf{v}_i\delta(\mathbf{r}-\mathbf{r}_i)\rangle - \rho\mathbf{u}\mathbf{u}\,. \tag{14.62}$$

In reforming the second term in Eq. (14.60), the following general Taylor expansion is necessary:

$$\begin{aligned} F(\mathbf{r}+\mathbf{R}) &= F(\mathbf{r}) + \mathbf{R}\cdot\frac{\partial}{\partial \mathbf{r}}\int_0^1 F(\mathbf{r}+\xi\mathbf{R})d\xi \\ &= F(\mathbf{r}) + \mathbf{R}\cdot\frac{\partial}{\partial \mathbf{r}}F(\mathbf{r}) + \frac{1}{2!}\mathbf{R}\mathbf{R}:\frac{\partial}{\partial \mathbf{r}}\frac{\partial}{\partial \mathbf{r}}F(\mathbf{r}) + \cdots. \end{aligned} \tag{14.63}$$

By taking account of $\mathbf{F}_{ij}^D = F_{ij}^D\mathbf{e}_{ij}$, the following equation is obtained:

$$\begin{aligned} \sum_i \sum_{\substack{j \\ (i\neq j)}} \mathbf{F}_{ij}^D\delta(\mathbf{r}-\mathbf{r}_i) &= \frac{1}{2}\sum_i \sum_{\substack{j \\ (i\neq j)}} \{\mathbf{F}_{ij}^D\delta(\mathbf{r}-\mathbf{r}_i) + \mathbf{F}_{ji}^D\delta(\mathbf{r}-\mathbf{r}_j)\} \\ &= \frac{1}{2}\sum_i \sum_{\substack{j \\ (i\neq j)}} F_{ij}^D\mathbf{e}_{ij}\{\delta(\mathbf{r}-\mathbf{r}_i) - \delta(\mathbf{r}-\mathbf{r}_j)\}. \end{aligned} \tag{14.64}$$

If the following relation concerning the Dirac delta function is taken into account,

$$\delta(\mathbf{r}-\mathbf{r}_j) = \delta(\mathbf{r}-\mathbf{r}_i+\mathbf{r}_{ij}) = \delta(\mathbf{r}-\mathbf{r}_i) + \frac{\partial}{\partial \mathbf{r}}\cdot\left\{\mathbf{e}_{ij}\int_0^{r_{ij}}\delta(\mathbf{r}-\mathbf{r}_i+\xi'\mathbf{e}_{ij})d\xi'\right\}, \tag{14.65}$$

then Eq. (14.64) can be reformed further as

$$\sum_i \sum_{\substack{j \\ (i\neq j)}} \mathbf{F}_{ij}^D\delta(\mathbf{r}-\mathbf{r}_i) = -\frac{1}{2}\sum_i \sum_{\substack{j \\ (i\neq j)}} \frac{\partial}{\partial \mathbf{r}}\cdot\left\{F_{ij}^D\mathbf{e}_{ij}\mathbf{e}_{ij}\int_0^{r_{ij}}\delta(\mathbf{r}-\mathbf{r}_i+\xi\mathbf{e}_{ij})d\xi\right\}. \tag{16.66}$$

By conducting a similar reformation concerning $\mathbf{F}_{ij}^{C}$, Eq. (14.60) can finally be written as

$$\frac{\partial}{\partial t}(\rho\mathbf{u}) = -\frac{\partial}{\partial \mathbf{r}}\cdot(\rho\mathbf{u}\mathbf{u}) + \frac{\partial}{\partial \mathbf{r}}\cdot(\boldsymbol{\tau}^{K} + \boldsymbol{\tau}^{U}), \tag{14.67}$$

in which

$$\left.\begin{aligned} \boldsymbol{\tau}^{K} &= -\left\langle \sum_{i} m(\mathbf{v}_i - \mathbf{u})(\mathbf{v}_i - \mathbf{u})\delta(\mathbf{r} - \mathbf{r}_i) \right\rangle, \\ \boldsymbol{\tau}^{U} &= -\frac{1}{2}\left\langle \sum_{i}\sum_{\substack{j \\ (i \neq j)}} (F_{ij}^{C} + F_{ij}^{D})\mathbf{e}_{ij}\mathbf{e}_{ij} \int_{0}^{r_{ij}} \delta(\mathbf{r} - \mathbf{r}_i + \xi\mathbf{e}_{ij})d\xi \right\rangle. \end{aligned}\right\} \tag{14.68}$$

If the equation of continuity in Eq. (14.57) is taken into consideration, Eq. (14.67) reduces to the momentum equation of the fluid:

$$\rho\left(\frac{\partial \mathbf{u}}{\partial t} + \mathbf{u}\cdot\frac{\partial \mathbf{u}}{\partial \mathbf{r}}\right) = \frac{\partial}{\partial \mathbf{r}}\cdot(\boldsymbol{\tau}^{K} + \boldsymbol{\tau}^{U}), \tag{14.69}$$

in which $\boldsymbol{\tau}^{K}$ and $\boldsymbol{\tau}^{U}$ are stress tensors due to the particle momenta and the forces acting between particles, respectively. It has now been shown that the equation of motion of the dissipative particle dynamics method gives rise to the momentum equation of the fluid as shown in Eq. (14.69). In Exercise 14.4 at the end of this chapter, the virial equation of state is derived by evaluating directly $\boldsymbol{\tau}^{K}$ and $\boldsymbol{\tau}^{U}$ in Eq. (14.68).

14.3 The Lattice Boltzmann Method

As in the dissipative particle dynamics method, solvent molecules are modeled as clusters of molecules or as fluid particles in the lattice Boltzmann method. However, these fluid particles cannot freely move to an arbitrary point in the simulation region. They can only transfer from site to site in an adopted lattice. The motion of fluid particles is governed by the discretized Boltzmann equation, which is explained below. Thus the treatment of particle-particle collisions at each site is a key factor for successful application of this method to various flow problems. In this section, we just outline this method from a colloid simulation point of view. If the reader is interested in detailed discussions of the lattice Boltzmann method, typical textbooks should be referred to [13-15]. As in the previous section, unless otherwise specified, we will call these fluid particles just particles in this section for convenience.

14.3.1 Lattice models and particle velocities

In the lattice Boltzmann method, particles transfer from site to site in an adopted lattice with the momentum exchanges at each site according to the collision dynamics. Hence, we first show some typical lattice models which are generally used in simulations based on this method.

For two-dimensional flow problems, a square lattice is widely used, and is shown in Fig. 14.4 (a). The site velocity vectors along the lines connecting neighboring sites are denoted by $\mathbf{c}_\alpha$, and for the case of Fig. 14.4(a) $\mathbf{c}_\alpha$ is taken as $\mathbf{c}_1$, $\mathbf{c}_2$, ..., $\mathbf{c}_8$, and $\mathbf{c}_0$ for the rest particle. Hence a particle is allowed to transfer to its eight neighboring sites or to stay at its original site. Several magnitudes of velocities, or several speeds, for particle movement can be assigned to each particle in the lattice Boltzmann method since particles are restricted to move to only sites of a regular lattice. If one speed alone is assigned to every particle, then this is called the one-speed model. If two possible speeds are given to each particle as in Fig. 14.4(a), then this is called the two-speed model. In the one-speed model, all particles are moved with the same speed from site to site of the adopted lattice. In the two-speed model, particles are divided into two groups according to the speeds: for example, the first group has one speed $|\mathbf{c}_1|=|\mathbf{c}_2|=|\mathbf{c}_3|=|\mathbf{c}_4|$ and the second group has a different speed $|\mathbf{c}_5|=|\mathbf{c}_6|=|\mathbf{c}_7|=|\mathbf{c}_8|$ (and additionally $|\mathbf{c}_0|=0$ for rest particles) for the two-dimensional square lattice model in Fig. 14.4(a).

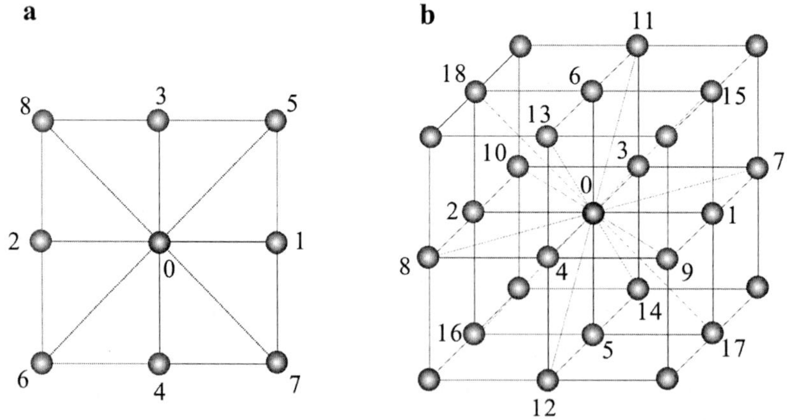

Figure 14.4 Lattice models and discrete velocity vectors. Figure 14.4(a) is for a two-dimensional square lattice and for the two-speed model: 0 for rest particles; 1,2,3,4 for particles with one speed; and 5,6,7,8 for particles with another speed. Figure 14.4(b) is for a cubic lattice and for the two-speed model: 0 for rest particles; 1,2, ..., 6 for particles with a speed; and 7,8, ..., 18 for particles with another speed.

For a three-dimensional system, a regular cubic lattice shown in Fig. 14.4(b) is widely used. In the two-speed model, the following three sets of discrete speeds are assigned to particles including rest particles, as shown in Fig. 14.4(b):

$$|\mathbf{c}_\alpha| = \begin{cases} 0 & \text{for} \quad \alpha = 0 \quad \text{(rest particles)}, \\ c_\mathrm{I} & \text{for} \quad \alpha = 1,2,\cdots,6 \quad \text{(group I)}, \\ c_\mathrm{II} & \text{for} \quad \alpha = 7,8,\cdots,18 \quad \text{(group II)}, \end{cases} \tag{14.70}$$

in which c_I and c_II are constants representing the particle speeds

14.3.2 The lattice Boltzmann equation

In the lattice Boltzmann method, the population density at each site of a lattice is solved at discrete time t. Hence this method is completely different from a simulation technique in which the particle velocities and positions are solved by the equation of motion of particles. We use the notation $f_\alpha(\mathbf{r},t)$ for the population density or the single-particle distribution function, at a site $\mathbf{r}$ of a lattice and at time t, in which the subscript α means quantities related to a discrete velocity vector $\mathbf{c}_\alpha$. The macroscopic state of a fluid system is, therefore, characterized by the single-particle distribution function. With the notation of the number density $n(\mathbf{r},t)$, the mass density $\rho(\mathbf{r},t)$ and the momentum density $\rho(\mathbf{r},t)\mathbf{u}(\mathbf{r},t)$ at the site $\mathbf{r}$ can be defined, respectively, as

$$\rho(\mathbf{r},t) = mn(\mathbf{r},t) = m\sum_\alpha f_\alpha(\mathbf{r},t), \quad \rho(\mathbf{r},t)\mathbf{u}(\mathbf{r},t) = m\sum_\alpha f_\alpha(\mathbf{r},t)\mathbf{c}_\alpha(\mathbf{r},t), \tag{14.71}$$

in which $\mathbf{u}$ is the macroscopic fluid velocity at the site $\mathbf{r}$. The number density n is given by ρ/m (m is the mass of fluid particles). The single-particle distribution changes through particle-particle collisions at sites, so that the time evolution of this distribution is governed by the discretized analogue of the Boltzmann equation [13-15]:

$$f_\alpha(\mathbf{r}+\mathbf{c}_\alpha\Delta t, t+\Delta t) = f_\alpha(\mathbf{r},t) + \Delta f_\alpha^{coll}(\mathbf{r},t)\Delta t\,, \tag{14.72}$$

in which Δt is a time interval, and the last term on the right-hand side is due to the particle-particle collisions. Various models for the collision term have been presented to date. Since we discuss this method from a colloid simulation point of view (the Reynolds number is small for flow problems of colloids), we concentrate our attention on the lattice BGK (Bhatnagar-Gross-Krook) collision model, which is approximately valid unless a fluid system is significantly in nonequilibrium. In this model, the collision term Δf_α^{coll} is expressed as

$$\Delta f_\alpha^{\ coll}(\mathbf{r},t) = \frac{1}{\tau}\left(f_\alpha^{\ (eq)}(\mathbf{r},t) - f_\alpha(\mathbf{r},t)\right), \tag{14.73}$$

in which τ is a constant for all sites, and $f_\alpha^{(eq)}(\mathbf{r},t)$ is a local equilibrium distribution function. This collision model describes the relaxation of a nonequilibrium distribution towards the equilibrium distribution $f_\alpha^{(eq)}$. The local relaxation time is given by τ in the lattice BGK model. Equation (14.73) is very similar to the expression for the BGK collision model in rarefied gas dynamics [16].

The equilibrium distribution function should be chosen in such a way that the lattice BGK collision term satisfies the conservation of mass locally. That is,

$$\sum_\alpha \Delta f_\alpha^{\ coll} = \frac{1}{\tau}\sum_\alpha \left(f_\alpha^{\ (eq)}(\mathbf{r},t) - f_\alpha(\mathbf{r},t)\right) = 0\,. \tag{14.74}$$

In addition, the momentum of all speed components together has to be conserved locally:

$$m\sum_\alpha \mathbf{c}_\alpha \Delta f_\alpha^{\ coll} = \frac{m}{\tau}\sum_\alpha \mathbf{c}_\alpha \left(f_\alpha^{\ (eq)}(\mathbf{r},t) - f_\alpha(\mathbf{r},t)\right) = 0\,. \tag{14.75}$$

The explicit expression for $f_\alpha^{(eq)}$ depends on the types of the lattice and speed models employed in simulations. Flows in colloidal dispersions occur at low Reynolds number, so that in this case, for example, the following equilibrium distribution can be used for a two-dimensional square lattice and two-speed model shown in Fig. 14.4(a) [17-19]:

$$f_\alpha^{\ (eq)} = n w_\alpha^{(8)}\left(1 + \frac{3\mathbf{c}_\alpha \cdot \mathbf{u}}{c^2}\right) \qquad \text{for} \quad \alpha = 0,1,\cdots,8\,, \tag{14.76}$$

in which n and $\mathbf{u}$ are the local number density and local flow velocity, respectively, as defined before and c is the lattice speed $\Delta x/\Delta t$ (Δx is a lattice constant). Also $w_\alpha^{(8)}$ is a weighting factor, and $w_\alpha^{(8)}$ =4/9 for α=0, 1/9 for α=1,2,3, 4 (group I) and 1/36 for α=5,6,7,8 (group II) in Fig. 14.4(a).

For the cubic lattice shown in Fig. 14.4(b), the following equilibrium distribution can be used [17,18,20]:

$$f_\alpha^{\ (eq)} = n w_\alpha^{(18)}\left(1 + \frac{3\mathbf{c}_\alpha \cdot \mathbf{u}}{c^2}\right) \qquad \text{for} \quad \alpha = 0,1,\cdots,18\,, \tag{14.77}$$

in which $w_\alpha^{(18)}$ is a weighting factor, and $w_\alpha^{(18)}$ =1/3 for α=0, 1/18 for α=1,2, …, 6 (group I) and 1/36 for α=7,8, …, 18 (group II) in Fig. 14.4(b).

If the dimensionless relaxation time $\tau/\Delta t$ is denoted by τ^*, the kinematic viscosity ν can be obtained from the Eq. (14.72) with Eq. (14.73) as [18],

$$\nu = \frac{1}{3}\left(\tau^* - \frac{1}{2}\right)c^2 \Delta t\,. \tag{14.78}$$

It is seen from this equation that τ^* has to be taken as $\tau^*>1/2$ to ensure positivity of the kinematic viscosity.

Finally, we show the main part of a typical lattice Boltzmann algorithm based on the lattice BGK model:

(1) Calculate n and $\mathbf{u}$ at each site using Eq. (14.71)
(2) Evaluate $f_\alpha^{(eq)}$ at each site using (14.76), (14.77) or a similar equation
(3) Compute the collision terms at each site using Eq. (14.73)
(4) Evaluate f_α at the next time step at each site using (14.72)

We have just outlined the essence of the lattice Boltzmann method, but omitted the explanation of the boundary conditions at the interfaces between the lattice simulation region and solid walls such as colloidal particles. Some references should be referred to for this subject [13,18,21,22]. The above-mentioned lattice Boltzmann method based on the lattice BGK collision model ensures the conservation of the total momentum of the system but not the conservation of the total system energy. To conserve the total system energy, a modified method including the energy equation must be used [23,24]. Also, we have omitted a discussion concerning the derivation of the macroscopic conservation equations, such as the equation of continuity and the Navier-Stokes equation, from the lattice Boltzmann equation. Some references should be referred to for these derivation procedures [13-15].

References

1. A. Satoh, Development of Effective Stokesian Dynamics Method for Ferromagnetic Colloidal Dispersions (Cluster-Based Stokesian Dynamics Method), J. Colloid Inter. Sci., 255 (2002) 98
2. A. Satoh, R.W. Chantrell, S. Kamiyama, and G.N. Coverdale, Three-Dimensional Monte Carlo Simulations to Thick Chainlike Clusters Composed of Ferromagnetic Fine Particles, J. Colloid Inter. Sci., 181 (1996) 422
3. A. Satoh, R.W. Chantrell, G.N. Coverdale, and S. Kamiyama, Stokesian Dynamics Simulations of Ferromagnetic Colloidal Dispersions in a Simple Shear Flow, J. Colloid Inter. Sci., 203 (1998) 233
4. P.J. Hoogerbrugge and J.M.V. A. Koelman, Simulating Microscopic Hydrodynamic Phenomena with Dissipative Particle Dynamics, Europhys. Lett., 19 (1992) 155
5. J.M.V. A. Koelman and P.J. Hoogerbrugge, Dynamic Simulations of Hard-Sphere Suspensions Under Steady Shear, Europhys. Lett., 21 (1993) 363

6. P. Espanol and P. Warren, Statistical Mechanics of Dissipative Particle Dynamics, Europhys. Lett., 30 (1995) 191
7. P. Espanol, Hydrodynamics from Dissipative Particle Dynamics, Phys. Rev. E, 52 (1995) 1734
8. P. Espanol, Dissipative Particle Dynamics with Energy Conservation, Europhys. Lett., 40 (1997) 631
9. J.B. Avalos and A.D. Mackie, Dissipative Particle Dynamics with Energy Conservation, Europhys. Lett., 40 (1997) 141
10. C.A. Marsh, G. Backx, and M.H. Ernst, Static and Dynamic Properties of Dissipative Particle Dynamics, Phys. Rev. E, 56 (1997) 1676
11. G. Besold, I. Vattulainen, M. Karttunen, and J.M. Polson, Towards Better Integrators for Dissipative Particle Dynamics Simulations, Phys. Rev. E, 62 (2000) R7611
12. M.P. Allen and D.J. Tildesley, "Computer Simulation of Liquids," Clarendon Press, Oxford, 1987
13. D. H. Rothman and S. Zaleski, "Lattice-Gas Cellar Automata," Cambridge Univ. Press, Cambridge, 1997
14. J.-P. Rivet and J.P. Boon, "Lattice Gas Hydrodynamics," Cambridge Univ. Press, Cambridge, 2001
15. S. Succi, "The Lattice Boltzmann Equation," Clarendon Press, Oxford, 2001
16. W.G. Vincenti and C.H. Kruger, Jr., "Introduction to Physical Gas Dynamics," John Wiley & Sons, New York, 1965
17. X. He and L.-S., Luo, Theory of the Lattice Boltzmann Method: From the Boltzmann Equation to the Lattice Boltzmann Equation, Phys. Rev. E, 56 (1997) 6811
18. R. Mei, W. Shyy, D. Yu, and L.-S. Luo, Lattice Boltzmann Method for 3-D Flows with Curved Bounday, J. Comput. Phys., 161 (2000) 680
19. X. He, S. Chen, and R. Zhang, A Lattice Boltzmann Scheme for Incompressible Multiphase Flow and Its Application in Simulation of Rayleigh-Taylor Instability, J. Comput. Phys., 152 (1999) 642
20. A.J.C. Ladd, Short-Time Motion f Colloidal Particles: Numerical Simulation via a Fluctuating Lattice-Boltzmann Equation, Phys. Rev. Lett., 70 (1993) 1339
21. A.J.C. Ladd, Numerical Simulations of Particulate Suspensions via a Discretized Boltzmann Equation. Part 1. Theoretical Foundation, J. Fluid Mech., 271 (1994) 285
22. O. Filippova and D. Hänel, A Novel Lattice BGK Approach for Low Mach Number Combustion, J. Comput. Phys., 158 (2000) 139
23. B.J. Palmer and D.R. Rector, Lattice Boltzmann Algorithm for Simulating Thermal Flow in Compressible Fluids, J. Comput. Phys., 161 (2000) 1

24. B. J. Palmer and D. R. Rector, Lattice-Boltzmann Algorithm for Simulating Thermal Two-Phase Flow, Phys. Rev. E, 61 (2000) 5295

Exercises

14.1 Derive the following equation before reforming Eq. (14.21):

$$\frac{\partial^2}{\partial \mathbf{v}_j \partial \mathbf{v}_i} : (W\psi \Delta \mathbf{v}_i \Delta \mathbf{v}_j) = \left(\Delta \mathbf{v}_i \cdot \frac{\partial \psi}{\partial \mathbf{v}_i} \right)\left(\Delta \mathbf{v}_j \cdot \frac{\partial W}{\partial \mathbf{v}_j} \right) + \left(\Delta \mathbf{v}_i \cdot \frac{\partial W}{\partial \mathbf{v}_i} \right)\left(\Delta \mathbf{v}_j \cdot \frac{\partial \psi}{\partial \mathbf{v}_j} \right)$$
$$+ W \Delta \mathbf{v}_i \Delta \mathbf{v}_j : \frac{\partial}{\partial \mathbf{v}_j} \frac{\partial}{\partial \mathbf{v}_i} \psi + \psi \Delta \mathbf{v}_i \Delta \mathbf{v}_j : \frac{\partial}{\partial \mathbf{v}_j} \frac{\partial}{\partial \mathbf{v}_i} W. \tag{14.79}$$

Using the above equation and the following relation,

$$W \frac{\partial \psi}{\partial \mathbf{v}_i} + \psi \frac{\partial W}{\partial \mathbf{v}_i} = \frac{\partial}{\partial \mathbf{v}_i} (W\psi), \tag{14.80}$$

verify that Eq. (14.21) can be transformed to Eq. (14.22).

14.2 Before deriving Eq. (14.25), prove that the following relation holds:

$$\Delta \mathbf{v}_i \Delta \mathbf{v}_j = \frac{1}{m^2} \left\{ \sum_k \mathbf{F}_{ik}^C - \sum_k \gamma w_D(r_{ik})(\mathbf{e}_{ik} \cdot \mathbf{v}_{ik})\mathbf{e}_{ik} \right\}\left\{ \sum_l \sigma w_R(r_{jl})\mathbf{e}_{jl} \Delta W_{jl} \right\} \Delta t$$
$$+ \frac{1}{m^2} \left\{ \sum_l \mathbf{F}_{jl}^C - \sum_l \gamma w_D(r_{jl})(\mathbf{e}_{jl} \cdot \mathbf{v}_{jl})\mathbf{e}_{jl} \right\}\left\{ \sum_k \sigma w_R(r_{ik})\mathbf{e}_{ik} \Delta W_{ik} \right\} \Delta t \tag{14.81}$$
$$+ \frac{1}{m^2} \sum_k \sum_l \sigma^2 w_R(r_{ik}) w_R(r_{jl}) \mathbf{e}_{ik} \mathbf{e}_{jl} \Delta W_{ik} \Delta W_{jl} .$$

With this relation and the stochastic properties in Eq. (14.11), derive Eq. (14.25).

14.3 By the method of integration by parts, show that the second and third terms on the right-hand side in Eq. (14.49) can be reformed, respectively, as

$$-\iint \sum_i \sum_{j\,(i\neq j)} \frac{\mathbf{F}_{ij}^C}{m} \cdot \left\{ \frac{\partial}{\partial \mathbf{v}_i}(AW) - W \frac{\partial A}{\partial \mathbf{v}_i} \right\} d\mathbf{r} d\mathbf{v} = \iint \left\{ \sum_i \sum_{j\,(i\neq j)} \frac{\mathbf{F}_{ij}^C}{m} \cdot \frac{\partial A}{\partial \mathbf{v}_i} \right\} W d\mathbf{r} d\mathbf{v}, \tag{14.82}$$

particle-particle forces can be obtained as

$$P^U = -\frac{1}{3}(\tau_{xx}^U + \tau_{yy}^U + \tau_{zz}^U) = \frac{2\pi n^2}{3}\int_0^\infty F_{12}^C r_{12}{}^3 g(r_{12}) dr_{12} \, . \tag{14.95}$$

Hence, the pressure P in equilibrium is obtained from Eqs. (14.92) and (14.95) as

$$P = P^K + P^U = nkT + \frac{2\pi n^2}{3}\int_0^\infty F_{12}^C r_{12}{}^3 g(r_{12}) dr_{12} \, . \tag{14.96}$$

This is no other than the virial equation of state which has been shown in Eq. (10.5).

CHAPTER 15

THEORETICAL ANALYSIS OF THE ORIENTATIONAL DISTRIBUTION OF SPHEROCYLINDER PARTICLES WITH BROWNIAN MOTION

In the present chapter, we leave the discussions concerning molecular-microsimulation methods to show the theoretical analysis method of the orientational distribution of ferromagnetic spherocylinder particles exhibiting rotational Brownian motion in a simple shear flow and a uniform magnetic field. Since simulations which take into account multi-body hydrodynamic interactions are very difficult for a rodlike particle system, the friction terms alone are taken into account in many cases of simulations. If a dispersion is sufficiently dilute, interactions between particles are negligible, so that the behavior of particles in a flow field may be solved analytically. The analytical solution, even in such a limiting case, is very important from the viewpoint of verifying the validity of simulation results. We here explain a theoretical method, based on the concept of the orientational distribution, to investigate the behavior of ferromagnetic spherocylinder particles in a simple shear flow [1].

15.1 The Particle Model

As shown in Fig.15.1, a ferromagnetic rod-like particle is idealized as a spherocylinder with the magnetic charges $\pm q$ in the center of each hemisphere. The interaction energy U between such a particle and a uniform applied magnetic field $\mathbf{H}$, and the torque $\mathbf{T}^m$ acting on the particle due to the field are written, respectively, as

$$U = -\mu_0 \mathbf{m} \cdot \mathbf{H}, \quad \mathbf{T}^m = \mu_0 \mathbf{m} \times \mathbf{H}, \tag{15.1}$$

in which μ_0 is the permeability of free space. The magnetic moment $\mathbf{m}$ has the magnitude of the product of the magnetic charge q and the length of the cylinder l, and its direction is given by the unit vector $\mathbf{e}$ along the cylinder axis. It is noted that the interaction with an applied field in Eq. (15.1) is equivalent to a model in which the magnetic dipole moment $\mathbf{m}$ is in the center of the cylinder; in the general spherocylinder model of a non-dilute dispersion, particle-particle interactions are based on magnetic charge interactions. We here consider a dilute colloidal dispersion composed of such particles, so that the interactions among particles are assumed to be negligible.

15.2 Rotational Motion of a Particle in a Simple Shear Flow

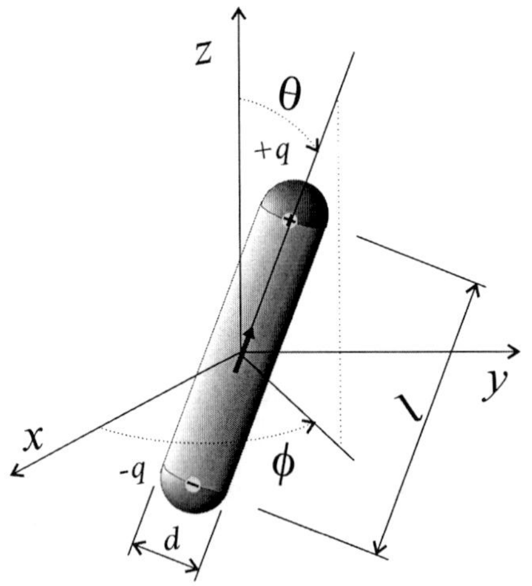

Figure 15.1 Particle model and system of coordinates.

In this section, we derive an equation which gives the direction of a particle under the assumption that the particle moves with the local flow velocity of a simple shear flow. There are three kinds of torques acting on the particle: $\mathbf{T}^m$ due to the magnetic force, $\mathbf{T}^{Br}$ due to the rotational Brownian motion, and $\mathbf{T}^{fl}$ due to the shear flow.

The torque due to the rotational Brownian motion is written using an orientational distribution function Ψ as [2]

$$\mathbf{T}^{Br} = -kT\mathbf{e}\times\frac{\partial}{\partial \mathbf{e}}(\ln\Psi), \tag{15.2}$$

in which k is Boltzmann's constant, T is the absolute temperature of the fluid, and Ψ will be defined in the next section.

In Chapter 5 we saw that, for axisymmetric particles such as spheroids and spherocylinders, $\mathbf{T}^{fl}$ is expressed in terms of the angular velocity vector $\boldsymbol{\Omega}$ and the rate-of-strain tensor $\mathbf{E}$ of a simple shear flow:

$$\mathbf{T}^{fl} = \eta_s\left\{X^C\mathbf{ee} + Y^C(\mathbf{I}-\mathbf{ee})\right\}\cdot(\boldsymbol{\Omega}-\boldsymbol{\omega}) - \eta_s Y^H(\boldsymbol{\varepsilon}\cdot\mathbf{ee}):\mathbf{E}, \tag{15.3}$$

in which $\boldsymbol{\omega}$ is the angular velocity vector of the particle, η_s is the viscosity of the base liquid, $\mathbf{e}$ is the unit vector of the long particle axis as defined previously, $\boldsymbol{\varepsilon}$ is called Eddington's epsilon (a three-rank tensor), $\mathbf{I}$ is the unit tensor, and X^C,Y^C and Y^H are resistance functions that are dependent only on the particle shape.

Since the inertia term is negligible for usual colloidal dispersions, the governing equation for the rotational motion of particle can be derived from the balance of torques as

$$\mathbf{T}^{fl}+\mathbf{T}^{Br}+\mathbf{T}^{m}=0. \tag{15.4}$$

If Eqs. (15.1)~(15.3) are substituted into Eq.(15.4), we get the following equation:

$$\begin{aligned}\eta_s\left\{X^C\mathbf{ee}+Y^C(\mathbf{I}-\mathbf{ee})\right\}\cdot(\boldsymbol{\Omega}-\boldsymbol{\omega})-\eta_s Y^H(\boldsymbol{\varepsilon}\cdot\mathbf{ee}):\mathbf{E}\\ -kT\mathbf{e}\times\frac{\partial}{\partial\mathbf{e}}(\ln\Psi)+\mu_0\mathbf{m}\times\mathbf{H}=0.\end{aligned} \tag{15.5}$$

If we multiply both sides of Eq. (15.5) by (×**e**) and solve it for (**ω**×**e**), then the equation giving the change in the particle direction $\dot{\mathbf{e}}$ is obtained as

$$\begin{aligned}\dot{\mathbf{e}}=\boldsymbol{\omega}\times\mathbf{e}=\boldsymbol{\Omega}\times\mathbf{e}+\frac{Y^H}{Y^C}\left\{\mathbf{E}\cdot\mathbf{e}-(\mathbf{E}:\mathbf{ee})\mathbf{e}\right\}-\frac{kT}{\eta_s Y^C}\frac{\partial}{\partial\mathbf{e}}(\ln\Psi)\\ -\frac{\mu_0 mH}{\eta_s Y^C}\left\{(\mathbf{e}\cdot\mathbf{h})\mathbf{e}-\mathbf{h}\right\}.\end{aligned} \tag{15.6}$$

In the derivation of Eq. (15.6), the following relations have been used:

$$\left.\begin{aligned}&\mathbf{e}\times\left\{(\mathbf{ee})\cdot\boldsymbol{\Omega}\right\}=\mathbf{e}\times\left\{(\mathbf{ee})\cdot\boldsymbol{\omega}\right\}=0,\\ &\mathbf{e}\times\left\{(\boldsymbol{\varepsilon}\cdot\mathbf{ee}):\mathbf{E}\right\}=\mathbf{E}\cdot\mathbf{e}-(\mathbf{E}:\mathbf{ee})\mathbf{e},\\ &\mathbf{e}\times\left[\mathbf{e}\times\frac{\partial}{\partial\mathbf{e}}\{\ln(\Psi)\}\right]=-\frac{\partial}{\partial\mathbf{e}}\{\ln(\Psi)\},\\ &\mathbf{e}\times(\mathbf{m}\times\mathbf{H})=mH\left\{(\mathbf{e}\cdot\mathbf{h})\mathbf{e}-\mathbf{h}\right\},\end{aligned}\right\} \tag{15.7}$$

in which $m=|\mathbf{m}|$, $H=|\mathbf{H}|$, and $\mathbf{h}=\mathbf{H}/H$.

15.3 The Basic Equation of the Orientational Distribution Function

Although the motion of particle in a simple shear flow is dynamic, we describe this phenomenon from a probability point of view using the orientational distribution function.

As shown in Fig.15.1, if the direction of the particle is described by the zenithal angle θ and the azimuthal angle φ, then the orientational distribution function Ψ is defined such that the probability of the particle being found in the range (θ,φ) to (θ+*d*θ,φ+*d*φ) is written as

$$\Psi(\theta,\varphi,t)\sin\theta d\theta d\varphi. \tag{15.8}$$

From the normalization condition, the following equation has to be satisfied:

$$\int_0^{2\pi}\int_0^{\pi}\Psi(\theta,\varphi,t)\sin\theta d\theta d\varphi=1. \tag{15.9}$$

Thus, the average value $\langle G\rangle$ of an arbitrary quantity *G*, which is dependent on the particle direction, is expressed as

$$\langle G\rangle = \int_0^{2\pi}\int_0^{\pi} G(\theta,\varphi)\Psi(\theta,\varphi,t)\sin\theta d\theta d\varphi \,. \tag{15.10}$$

The average viscosity will be evaluated later using this definition.

The orientational distribution function has to satisfy the equation of continuity:

$$\frac{\partial\Psi}{\partial t} = -\frac{\partial}{\partial\mathbf{e}}\cdot(\dot{\mathbf{e}}\Psi) = -\Psi\frac{\partial}{\partial\mathbf{e}}\cdot\dot{\mathbf{e}} - \dot{\mathbf{e}}\cdot\frac{\partial\Psi}{\partial\mathbf{e}}\,. \tag{15.11}$$

If Eq. (15.6) is substituted into Eq. (15.11), we obtain the following expression:

$$\begin{aligned}\frac{\partial\Psi}{\partial t} &= 3\frac{Y^H}{Y^C}(\mathbf{E}:\mathbf{ee})\Psi - \left\{\mathbf{\Omega}\times\mathbf{e} + \frac{Y^H}{Y^C}(\mathbf{E}\cdot\mathbf{e})\right\}\cdot\frac{\partial\Psi}{\partial\mathbf{e}} \\ &\quad + \frac{kT}{\eta_s Y^C}\left(\frac{\partial}{\partial\mathbf{e}}\cdot\frac{\partial}{\partial\mathbf{e}}\right)\Psi + \frac{\mu_0 mH}{\eta_s Y^C}\left\{2(\mathbf{e}\cdot\mathbf{h})\Psi - \mathbf{h}\cdot\frac{\partial\Psi}{\partial\mathbf{e}}\right\}.\end{aligned} \tag{15.12}$$

In the derivation of Eq. (15.12), the following relations have been used:

$$\begin{aligned}&\frac{\partial}{\partial\mathbf{e}}\cdot(\mathbf{E}\cdot\mathbf{e}) = -\mathbf{E}:\mathbf{ee}, \quad \frac{\partial}{\partial\mathbf{e}}\cdot\{(\mathbf{E}:\mathbf{ee})\mathbf{e}\} = 2\mathbf{E}:\mathbf{ee}, \\ &\frac{\partial}{\partial\mathbf{e}}\cdot\{(\mathbf{e}\cdot\mathbf{h})\mathbf{e} - \mathbf{h}\} = 2\mathbf{e}\cdot\mathbf{h}.\end{aligned} \tag{15.13}$$

It is noted that D_r is equal to $kT/\eta_s Y^C$ and called the rotational diffusion coefficient. For an axisymmetric particle with a large aspect ratio such as r_p $(=l/d)>>1$, Y^H/Y^C is nearly equal to unity [3]. Thus Eq. (15.12) can be simplified as

$$\begin{aligned}\frac{\partial\Psi}{\partial t} &= 3(\mathbf{E}:\mathbf{ee})\Psi - (\mathbf{\Omega}\times\mathbf{e} + \mathbf{E}\cdot\mathbf{e})\cdot\frac{\partial\Psi}{\partial\mathbf{e}} + D_r\left(\frac{\partial}{\partial\mathbf{e}}\cdot\frac{\partial}{\partial\mathbf{e}}\right)\Psi \\ &\quad + \frac{\mu_0 mH}{\eta_s Y^C}\left\{2(\mathbf{e}\cdot\mathbf{h})\Psi - \mathbf{h}\cdot\frac{\partial\Psi}{\partial\mathbf{e}}\right\}.\end{aligned} \tag{15.14}$$

We now concentrate our attention on the special case of a simple shear flow in the x-axis direction. If the shear rate is denoted by $\dot{\gamma}$, then the flow velocity **U**, the angular velocity of the fluid **Ω**, and the rate-of-strain tensor **E** are written as

$$\mathbf{U} = \dot{\gamma}y\boldsymbol{\delta}_x, \quad \mathbf{\Omega} = -\frac{\dot{\gamma}}{2}\boldsymbol{\delta}_z, \quad \mathbf{E} = \frac{\dot{\gamma}}{2}\begin{bmatrix}0 & 1 & 0\\ 1 & 0 & 0\\ 0 & 0 & 0\end{bmatrix}, \tag{15.15}$$

in which $\boldsymbol{\delta}_x, \boldsymbol{\delta}_y, \boldsymbol{\delta}_z$ are the fundamental vectors. Equation (15.15) is valid before the particles are dispersed. Furthermore, the magnetic field is assumed to be applied in the *y*-axis direction; that is, $\mathbf{h}=\boldsymbol{\delta}_y$. If these relations are substituted into Eq. (15.14), an equation for Ψ is obtained as

$$\Lambda(\Psi) - Pe\Omega_s(\Psi) + \xi\Omega_{my}(\Psi) = 0, \tag{15.16}$$

in which the operators Λ, Ω_s, and Ω_{my} are defined as follows:

$$\left.\begin{aligned}
\Lambda(\Psi) &= \frac{1}{S}\cdot\frac{\partial}{\partial\theta}\left(S\frac{\partial\Psi}{\partial\theta}\right) + \frac{1}{S^2}\cdot\frac{\partial^2\Psi}{\partial\varphi^2}, \\
\Omega_s(\Psi) &= \frac{sc}{S}\cdot\frac{\partial}{\partial\theta}\left(S^2C\Psi\right) - \frac{\partial}{\partial\varphi}\left(s^2\Psi\right), \\
\Omega_{my}(\Psi) &= 2Ss\Psi - Cs\frac{\partial\Psi}{\partial\theta} - \frac{c}{S}\cdot\frac{\partial\Psi}{\partial\varphi}.
\end{aligned}\right\} \tag{15.17}$$

In Eq. (15.17), the simplified notations of $S=\sin\theta$, $C=\cos\theta$, $s=\sin\varphi$, $c=\cos\varphi$, $s_m=\sin m\varphi$, and $c_m=\cos m\varphi$ have been used. The nondimensional numbers in Eq. (15.16) are as follows:

$$Pe = \dot{\gamma}/D_r, \quad \xi = \mu_0 mH/kT. \tag{15.18}$$

Pe is called the Péclet number, as already defined before, and is the ratio of the representative hydrodynamic shear force to the representative Brownian force. Also, ξ is the ratio of the representative magnetic force to the representative rotational Brownian force.

Finally we show some useful results that will be useful for the approximate solution by Galerkin's method in the next section. We apply the operators shown in Eq. (15.17) to spherical harmonics $P_n^m s_m$ and $P_n^m c_m$, and these results are reformed using the relations of Legendre's polynomial functions. Then, the results are expressed using spherical harmonics themselves. The results for the operators Λ and Ω_s have already been obtained as [2]

$$\Lambda(P_n^m c_m) = -n(n+1)P_n^m c_m, \quad \Lambda(P_n^m s_m) = -n(n+1)P_n^m s_m, \tag{15.19}$$

$$\begin{aligned}
\Omega_s(P_n^m c_m) = &-\sum_{i=-1}^{1} a_{n,n+2i}^{m,m-2} P_{n+2i}^{m-2} s_{m-2} - a_{n,n}^{m,m} P_n^m s_m \\
&-\sum_{i=-1}^{1} a_{n,n+2i}^{m,m+2} P_{n+2i}^{m+2} s_{m+2} \qquad (m \geq 0, m \leq n),
\end{aligned} \tag{15.20}$$

$$\begin{aligned}
\Omega_s(P_n^m s_m) = &\sum_{i=-1}^{1} a_{n,n+2i}^{m,m-2} P_{n+2i}^{m-2} c_{m-2} + a_{n,n}^{m,m} P_n^m c_m \\
&+\sum_{i=-1}^{1} a_{n,n+2i}^{m,m+2} P_{n+2i}^{m+2} c_{m+2} \qquad (m > 0, m \leq n).
\end{aligned} \tag{15.21}$$

If the following relations are taken into account for the operator Ω_{my}:

$$\left.\begin{aligned}
SP_n^m &= (P_{n+1}^{m+1} - P_{n-1}^{m+1})/(2n+1),\\
SP_n^m &= \frac{(n+m)(n+m-1)}{(2n+1)} P_{n-1}^{m-1} - \frac{(n-m+1)(n-m+2)}{(2n+1)} P_{n+1}^{m-1},\\
CP_n^m &= \frac{(n-m+1)}{(2n+1)} P_{n+1}^{m} + \frac{(n+m)}{(2n+1)} P_{n-1}^{m},
\end{aligned}\right\} \tag{15.22}$$

the final results can be obtained as

$$\begin{aligned}
\Omega_{my}(P_n^m c_m) &= \sum_{i=0}^{1} b_{n,n-1+2i}^{m,m-1} P_{n-1+2i}^{m-1} s_{m-1}\\
&+ \sum_{i=0}^{1} b_{n,n-1+2i}^{m,m+1} P_{n-1+2i}^{m+1} s_{m+1} \qquad (m \geq 0, m \leq n),
\end{aligned} \tag{15.23}$$

$$\begin{aligned}
\Omega_{my}(P_n^m s_m) &= -\sum_{i=0}^{1} b_{n,n-1+2i}^{m,m-1} P_{n-1+2i}^{m-1} c_{m-1}\\
&- \sum_{i=0}^{1} b_{n,n-1+2i}^{m,m+1} P_{n-1+2i}^{m+1} c_{m+1} \qquad (m > 0, m \leq n),
\end{aligned} \tag{15.24}$$

in which $a_{n,n-2}^{m,m-2}$, …, $a_{n,n+2}^{m,m+2}$, $b_{n,n-1}^{m,m-1}$, …, $b_{n,n+1}^{m,m+1}$ are as follows:

$$\left.\begin{aligned}
a_{n,n-2}^{m,m-2} &= \frac{(n-2)(n+m)(n+m-1)(n+m-2)(n+m-3)}{4(2n+1)(2n-1)}(1-\delta_{m0}),\\
a_{n,n}^{m,m-2} &= \frac{3(n+m)(n+m-1)(n-m+1)(n-m+2)}{4(2n-1)(2n+3)}(1-\delta_{m0}),\\
a_{n,n+2}^{m,m-2} &= -\frac{(n+3)(n-m+1)(n-m+2)(n-m+3)(n-m+4)}{4(2n+1)(2n+3)}(1-\delta_{m0}),\\
a_{n,n}^{m,m} &= -\frac{m}{2}, \quad a_{n,n-2}^{m,m+2} = -\frac{(n-2)}{4(2n+1)(2n-1)}(1+\delta_{m0}),\\
a_{n,n}^{m,m+2} &= -\frac{3}{4(2n-1)(2n+3)}(1+\delta_{m0}), \quad a_{n,n+2}^{m,m+2} = \frac{(n+3)}{4(2n+1)(2n+3)}(1+\delta_{m0}),
\end{aligned}\right\} \tag{15.25}$$

$$\left.\begin{aligned}
b_{n,n-1}^{m,m-1} &= \frac{(n+m)(n+m-1)(n-1)}{2(2n+1)}(1-\delta_{m0}),\\
b_{n,n+1}^{m,m-1} &= \frac{(n-m+1)(n-m+2)(n+2)}{2(2n+1)}(1-\delta_{m0}),\\
b_{n,n-1}^{m,m+1} &= \frac{(n-1)}{2(2n+1)}(1+\delta_{m0}),\\
b_{n,n+1}^{m,m+1} &= \frac{(n+2)}{2(2n+1)}(1+\delta_{m0}).
\end{aligned}\right\} \tag{15.26}$$

In this equation, δ_{m0} is Kronecker's delta.

15.4 Solution by Means of Galerkin's Method

Although Eq. (15.16) is generally solved by the perturbation method under the assumption of $Pe<<1$, we here adopt Galerkin's method in which such a limitation is unnecessary. If we expand Ψ in terms of spherical harmonics $P_n^m c_m$ and $P_n^m s_m$, the M-th order approximation $\Psi^{(M)}$ to Ψ can be expressed as

$$\Psi^{(M)} = \frac{1}{4\pi}\left(\sum_{m=0}^{M}\sum_{n=m}^{M} A_n^m P_n^m c_m + \sum_{m=1}^{M}\sum_{n=m}^{M} B_n^m P_n^m s_m\right), \tag{15.27}$$

in which A_n^m and B_n^m are coefficients to be determined by Galerkin's method. For the present special case of the simple shear flow, the distribution function has to be unchanged by the replacement of (θ,φ) to $(\pi-\theta,\varphi)$, so that n has to be even if m is even and n has to be odd if m is odd. The substitution of Eq. (15.27) into Eq. (15.16) with consideration of Eqs. (15.19)~(15.21), (15.23), and (15.24) leads to the following expression:

$$\begin{aligned}
&-\frac{1}{4\pi}\left\{\sum_{m=0}^{M}\sum_{n=m}^{M} n(n+1)A_n^m P_n^m c_m + \sum_{m=1}^{M}\sum_{n=m}^{M} n(n+1)B_n^m P_n^m s_m\right\}\\
&-\frac{Pe}{4\pi}\left\{\sum_{m=0}^{M}\sum_{n=m}^{M} A_n^m\left(-\sum_{i=-1}^{1} a_{n,n+2i}^{m,m-2} P_{n+2i}^{m-2} s_{m-2} - a_{n,n}^{m,m} P_n^m s_m - \sum_{i=-1}^{1} a_{n,n+2i}^{m,m+2} P_{n+2i}^{m+2} s_{m+2}\right)\right.\\
&\left.+\sum_{m=1}^{M}\sum_{n=m}^{M} B_n^m\left(\sum_{i=-1}^{1} a_{n,n+2i}^{m,m-2} P_{n+2i}^{m-2} c_{m-2} + a_{n,n}^{m,m} P_n^m c_m + \sum_{i=-1}^{1} a_{n,n+2i}^{m,m+2} P_{n+2i}^{m+2} c_{m+2}\right)\right\}\\
&+\frac{\xi}{4\pi}\left\{\sum_{m=0}^{M}\sum_{n=m}^{M} A_n^m\left(\sum_{i=0}^{1} b_{n,n-1+2i}^{m,m-1} P_{n-1+2i}^{m-1} s_{m-1} + \sum_{i=0}^{1} b_{n,n-1+2i}^{m,m+1} P_{n-1+2i}^{m+1} s_{m+1}\right)\right.\\
&\left.+\sum_{m=1}^{M}\sum_{n=m}^{M} B_n^m\left(-\sum_{i=0}^{1} b_{n,n-1+2i}^{m,m-1} P_{n-1+2i}^{m-1} c_{m-1} - \sum_{i=0}^{1} b_{n,n-1+2i}^{m,m+1} P_{n-1+2i}^{m+1} c_{m+1}\right)\right\}\\
&= R^{(M)}.
\end{aligned} \tag{15.28}$$

Since $\Psi^{(M)}$ is not the exact solution, the residual function $R^{(M)}$ has arisen on the right-hand side of Eq. (15.28). According to Galerkin's method, the weighted mean value of the residual function is forced to be zero using spherical harmonics themselves as a weighting function, so that

$$\int_0^{2\pi}\int_0^{\pi} R^{(M)} P_q^p c_p S d\theta d\varphi = \int_0^{2\pi}\int_0^{\pi} R^{(M)} P_q^p s_p S d\theta d\varphi = 0 . \tag{15.29}$$

With the following orthogonality condition of spherical functions:

$$\int_0^{2\pi}\int_0^{\pi} P_n^m \begin{Bmatrix} c_m \\ s_m \end{Bmatrix} P_{n'}^{m'} \begin{Bmatrix} c_{m'} \\ s_{m'} \end{Bmatrix} S d\theta d\varphi = \frac{2\pi(n+m)!}{(2n+1)(n-m)!}\delta_{nn'}\delta_{mm'}(1\pm\delta_{m0}), \tag{15.30}$$

the requirement in Eq. (15.29) leads to the following equations to be satisfied by the coefficients, A_n^m (m=0,1,...,M; n=m,m+1,...,M) and B_n^m (m=1,2,...,M; n=m,m+1,...,M):

$$\begin{aligned} n(n+1)A_n^m &+ Pe\left(\sum_{i=-1}^{1} a_{n-2i,n}^{m+2,m} B_{n-2i}^{m+2} + a_{n,n}^{m,m} B_n^m + \sum_{i=-1}^{1} a_{n-2i,n}^{m-2,m} B_{n-2i}^{m-2}\right) \\ &+ \xi\left(\sum_{i=0}^{1} b_{n+1-2i,n}^{m+1,m} B_{n+1-2i}^{m+1} + \sum_{i=0}^{1} b_{n+1-2i,n}^{m-1,m} B_{n+1-2i}^{m-1}\right) = 0 \\ &(m = 0,1,\cdots,M; n = m, m+1,\cdots,M), \end{aligned} \tag{15.31}$$

$$\begin{aligned} n(n+1)B_n^m &- Pe\left(\sum_{i=-1}^{1} a_{n-2i,n}^{m+2,m} A_{n-2i}^{m+2} + a_{n,n}^{m,m} A_n^m + \sum_{i=-1}^{1} a_{n-2i,n}^{m-2,m} A_{n-2i}^{m-2}\right) \\ &- \xi\left(\sum_{i=0}^{1} b_{n+1-2i,n}^{m+1,m} A_{n+1-2i}^{m+1} + \sum_{i=0}^{1} b_{n+1-2i,n}^{m-1,m} A_{n+1-2i}^{m-1}\right) = 0 \\ &(m = 1,2,\cdots,M; n = m, m+1,\cdots,M). \end{aligned} \tag{15.32}$$

It is noted that the terms A_n^m and B_n^m which are not used in the definition of $\Psi^{(M)}$ have to be removed from Eqs. (5.31) and (5.32). If the system is in equilibrium with Pe=ξ=0, the particle points in every direction with equal probability, so that A_0^0 has to be unity. Now we have the same number of conditions of the unknown quantities, so that we may say that the M-th level of approximation to Ψ has in principle been obtained.

It is straightforward to solve numerically the algebraic equations of Eqs. (5.31) and (5.32). Figure 15.2 shows the orientational distribution function for three different cases of the magnetic field strength, (a) ξ=1, (b) ξ=10, and (c) ξ=70; all results are for the Péclet number Pe=10. It is noted that a part of the whole distribution, for θ=30°~90°, is shown because of the symmetrical properties of Ψ. The result shown in Fig. 15.2 is for a case where the shear flow is dominant compared with the rotational Brownian motion. For the case of ξ=1, the shear force also dominates the magnetic force, so that a particle has a tendency to incline in the flow direction (φ=0°). As the magnetic field increases, such as ξ=10 and 70, the particle has a tendency to point to the magnetic field direction more significantly.

References

1. A. Satoh, Rheological Properties and Orientational Distributions of Dilute Ferromagnetic Spherocylinder Particle Dispersions (Approximate Solutions by Means of Galerkin's

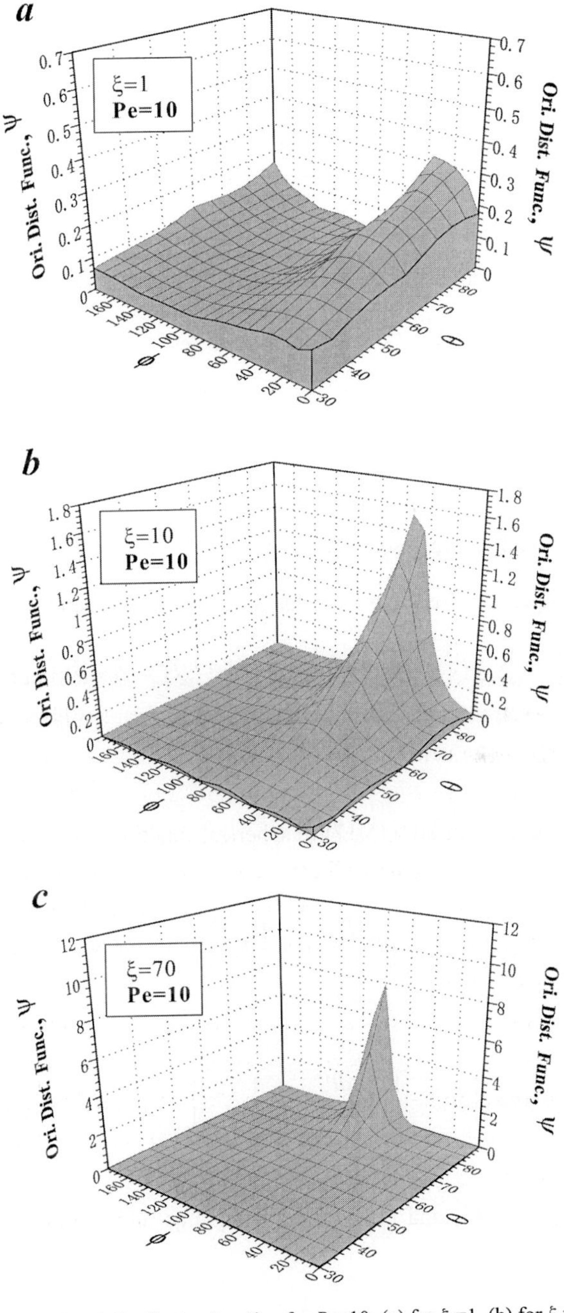

Figure 15.2 Orientational distribution function for Pe=10: (a) for ξ =1, (b) for ξ =10, and (c) for ξ =70.

Method), J. Colloid Inter. Sci., 234 (2001) 425

2. R.B. Bird, R.C. Armstrong, and O. Hassager, "Dynamics of Polymeric Liquid, Vol.l, Fluids Mechanics," John Wiley & Sons, New York, 1977.
3. Brenner, H., J. Multiphase Flow, 1 (1974) 195

Exercises

15.1 With the following relations for the simple shear flow expressed in Eq.(15.15):

$$\left.\begin{aligned}
&3\mathbf{E}:\mathbf{ee} = 3\dot{\gamma}S^2 sc,\quad \mathbf{\Omega}\times\mathbf{e} = \frac{\dot{\gamma}}{2}(Ss\boldsymbol{\delta}_x - Sc\boldsymbol{\delta}_y),\quad \mathbf{E}\cdot\mathbf{e} = \frac{\dot{\gamma}}{2}(Ss\boldsymbol{\delta}_x + Sc\boldsymbol{\delta}_y),\\
&\frac{\partial}{\partial\mathbf{e}} = \boldsymbol{\delta}_\theta\frac{\partial}{\partial\theta} + \boldsymbol{\delta}_\varphi\frac{\partial}{S\partial\varphi} = (Cc\boldsymbol{\delta}_x + Cs\boldsymbol{\delta}_y - S\boldsymbol{\delta}_z)\frac{\partial}{\partial\theta} + (-s\boldsymbol{\delta}_x + c\boldsymbol{\delta}_y)\frac{\partial}{S\partial\varphi},\\
&\mathbf{e}\cdot\mathbf{h} = Ss,\quad \mathbf{h}\cdot\frac{\partial\Psi}{\partial\mathbf{e}} = Cs\frac{\partial\Psi}{\partial\theta} + \frac{c}{S}\frac{\partial\Psi}{\partial\varphi},
\end{aligned}\right\}\tag{15.33}$$

show that Eq. (15.14) can be expressed as Eq. (15.16).

15.2 The following equations are some formulae for associated Legendre functions:

$$\left.\begin{aligned}
&\frac{dP_n^m}{d\theta} = -P_n^{m+1} + m\frac{C}{S}P_n^m,\quad \frac{dP_n^m}{d\theta} + m\frac{C}{S}P_n^m - (n+m)(n-m+1)P_n^{m-1} = 0,\\
&(n-m)P_{n+1}^{m+1} - (2n+1)CP_n^{m+1} + (n+m+1)P_{n-1}^{m+1} = 0,\\
&(n-m+2)P_{n+1}^{m-1} - (2n+1)CP_n^{m-1} + (n+m-1)P_{n-1}^{m-1} = 0.
\end{aligned}\right\}\tag{15.34}$$

If we apply the operator Ω_{my} in Eq. (15.17) to spherical functions, the following equations can be obtained:

$$\Omega_{my}(P_n^m c_m) = s_{m+1}\left\{SP_n^m - \frac{1}{2}\left(C\frac{dP_n^m}{d\theta} - m\frac{1}{S}P_n^m\right)\right\} - s_{m-1}\left\{SP_n^m - \frac{1}{2}\left(C\frac{dP_n^m}{d\theta} + m\frac{1}{S}P_n^m\right)\right\},\tag{15.35}$$

$$\Omega_{my}(P_n^m s_m) = -c_{m+1}\left\{SP_n^m - \frac{1}{2}\left(C\frac{dP_n^m}{d\theta} - m\frac{1}{S}P_n^m\right)\right\} + c_{m-1}\left\{SP_n^m - \frac{1}{2}\left(C\frac{dP_n^m}{d\theta} + m\frac{1}{S}P_n^m\right)\right\}.\tag{15.36}$$

By starting from Eqs. (15.35) and (15.36) and using Eqs. (15.34) and (15.22), derive Eqs. (15.23) and (15.24).

APPENDICES

A1. VECTORS AND TENSORS

Useful expressions for vectors and tensors are briefly summarized in the following [1]. Boldface-type small alphabet are used as vector quantities, boldface-type capital alphabet as second-rank tensors, and boldface-type Greek characters as higher-rank tensors. Also, the Cartesian coordinate system is used.

If the fundamental vectors for the Cartesian coordinate system are denoted by $\boldsymbol{\delta}_1$, $\boldsymbol{\delta}_2$, and $\boldsymbol{\delta}_3$, an arbitrary vector $\mathbf{a}=(a_1, a_2, a_3)$ is written as

$$\mathbf{a} = a_1\boldsymbol{\delta}_1 + a_2\boldsymbol{\delta}_2 + a_3\boldsymbol{\delta}_3 . \tag{A1.1}$$

The dyadic product of **a** and **b** is defined as

$$\mathbf{ab} = \begin{bmatrix} a_1b_1 & a_1b_2 & a_1b_3 \\ a_2b_1 & a_2b_2 & a_2b_3 \\ a_3b_1 & a_3b_2 & a_3b_3 \end{bmatrix} = \sum_{i=1}^{N}\sum_{j=1}^{N} a_i b_j \boldsymbol{\delta}_i \boldsymbol{\delta}_j , \tag{A1.2}$$

which is a second-rank tensor. From the correspondence to Eq. (A1.1), the ij-component of the tensor **ab** is a_ib_j. The expression for $\boldsymbol{\delta}_i\boldsymbol{\delta}_j$ in Eq. (A1.2) is written as follows, for example, $\boldsymbol{\delta}_1\boldsymbol{\delta}_2$:

$$\boldsymbol{\delta}_1\boldsymbol{\delta}_2 = \begin{bmatrix} 0 & 1 & 0 \\ 0 & 0 & 0 \\ 0 & 0 & 0 \end{bmatrix} . \tag{A1.3}$$

This clearly shows that a_ib_j is the ij-component of the tensor **ab**. A general second-rank tensor **T** is described in component form as

$$\mathbf{T} = \sum_{i=1}^{3}\sum_{j=1}^{3} T_{ij}\boldsymbol{\delta}_i\boldsymbol{\delta}_j . \tag{A1.4}$$

The product of second-rank tensors **S** and **T** may be expressed in two different ways:

$$\mathbf{S}\cdot\mathbf{T} = \sum_{i=1}^{3}\sum_{j=1}^{3}\left(\sum_{k=1}^{3} S_{ik}T_{kj}\right)\boldsymbol{\delta}_i\boldsymbol{\delta}_j , \tag{A1.5}$$

$$\mathbf{S}:\mathbf{T} = \sum_{i=1}^{3}\sum_{j=1}^{3} S_{ij}T_{ji} , \tag{A1.6}$$

in which **S·T** is a second-rank tensor and **S:T** is a scalar quantity. The product of a vector **a** and a tensor **T** is expressed as

$$\mathbf{T}\cdot\mathbf{a}=\sum_{i=1}^{3}\left(\sum_{j=1}^{3}T_{ij}a_j\right)\boldsymbol{\delta}_i, \tag{A1.7}$$

$$\mathbf{a}\cdot\mathbf{T}=\sum_{i=1}^{3}\left(\sum_{j=1}^{3}a_jT_{ji}\right)\boldsymbol{\delta}_i. \tag{A1.8}$$

It is seen from these equations that $\mathbf{T}\cdot\mathbf{a}$ is not equal to $\mathbf{a}\cdot\mathbf{T}$ unless $\mathbf{T}$ is a symmetric tensor.

Useful expressions for the products of vectors and tensors are summarized as follows:

$$\left.\begin{aligned}
&(\mathbf{ab})\cdot\mathbf{c}=\mathbf{a}(\mathbf{b}\cdot\mathbf{c}),\\
&\mathbf{a}\cdot(\mathbf{bc})=(\mathbf{a}\cdot\mathbf{b})\mathbf{c},\\
&\mathbf{ab}:\mathbf{cd}=\mathbf{ac}:\mathbf{bd}=(\mathbf{a}\cdot\mathbf{d})(\mathbf{b}\cdot\mathbf{c}),\\
&\mathbf{T}:\mathbf{ab}=(\mathbf{T}\cdot\mathbf{a})\cdot\mathbf{b},\\
&\mathbf{ab}:\mathbf{T}=\mathbf{a}\cdot(\mathbf{b}\cdot\mathbf{T}),\\
&\mathbf{a}(\mathbf{b}\cdot\mathbf{T})=(\mathbf{ab})\cdot\mathbf{T},\\
&(\mathbf{T}\cdot\mathbf{a})\mathbf{b}=\mathbf{T}\cdot(\mathbf{ab}),\\
&(\mathbf{T}\cdot\mathbf{a})(\mathbf{b}\cdot\mathbf{S})=\mathbf{T}\cdot(\mathbf{ab})\cdot\mathbf{S}.
\end{aligned}\right\} \tag{A1.9}$$

Using the following alternate tensor or Eddington's epsilon $\boldsymbol{\varepsilon}$ (ε_{ijk} for ijk-component), which is a third-rank tensor:

$$\varepsilon_{ijk}=\left\{\begin{array}{ll}1 & \text{for}(i,j,k)=(1,2,3),(2,3,1),(3,1,2),\\ -1 & \text{for}(i,j,k)=(3,2,1),(2,1,3),(1,3,2),\\ 0 & \text{for the other cases,}\end{array}\right\} \tag{A1.10}$$

the vector product $\mathbf{a}\times\mathbf{b}$ of $\mathbf{a}$ and $\mathbf{b}$ can be written as

$$\mathbf{a}\times\mathbf{b}=\begin{bmatrix}\boldsymbol{\delta}_1 & \boldsymbol{\delta}_2 & \boldsymbol{\delta}_3\\ a_1 & a_2 & a_3\\ b_1 & b_2 & b_3\end{bmatrix}=\sum_{i=1}^{3}\left(\sum_{j=1}^{3}\sum_{k=1}^{3}\varepsilon_{ijk}a_jb_k\right)\boldsymbol{\delta}_i=\boldsymbol{\varepsilon}:\mathbf{ba}=-\boldsymbol{\varepsilon}:\mathbf{ab}. \tag{A1.11}$$

By extending Eq. (A1.6), the product of a third-rank tensor $\boldsymbol{\sigma}$ and a second-rank tensor $\mathbf{T}$ can be defined as

$$\boldsymbol{\sigma}:\mathbf{T}=\sum_{i=1}^{3}\left(\sum_{j=1}^{3}\sum_{k=1}^{3}\sigma_{ijk}T_{kj}\right)\boldsymbol{\delta}_i, \tag{A1.12}$$

$$\mathbf{T}:\boldsymbol{\sigma}=\sum_{i=1}^{3}\left(\sum_{j=1}^{3}\sum_{k=1}^{3}T_{jk}\sigma_{kji}\right)\boldsymbol{\delta}_i. \tag{A1.13}$$

Similarly to Eq. (A1.6), the product of third-rank tensors $\boldsymbol{\sigma}$ and $\boldsymbol{\tau}$ is expressed as

$$\boldsymbol{\sigma} \vdots \boldsymbol{\tau} = \sum_{i=1}^{3}\sum_{j=1}^{3}\sum_{k=1}^{3} \sigma_{ijk}\tau_{kji} \,. \tag{A1.14}$$

Next we show differential formulae for vectors and tensors which are functions of the position **r** (=(x_1,x_2,x_3)). The vector differential operator ∇, known as the nabla, is defined as

$$\nabla = \sum_{i=1}^{3} \boldsymbol{\delta}_i \frac{\partial}{\partial x_i}, \tag{A1.15}$$

in which (x_1,x_2,x_3) is written as (x,y,z). Basic formulae concerning the nabla operator are summarized as follows:

$$\nabla \cdot \mathbf{a} = \sum_{i=1}^{3} \frac{\partial a_i}{\partial x_i}, \qquad \nabla \mathbf{a} = \sum_{i=1}^{3}\sum_{j=1}^{3} \frac{\partial a_j}{\partial x_i} \boldsymbol{\delta}_i \boldsymbol{\delta}_j \,, \tag{A1.16}$$

$$\begin{aligned} \nabla \times \mathbf{a} &= \begin{bmatrix} \boldsymbol{\delta}_1 & \boldsymbol{\delta}_2 & \boldsymbol{\delta}_3 \\ \dfrac{\partial}{\partial x_1} & \dfrac{\partial}{\partial x_2} & \dfrac{\partial}{\partial x_3} \\ a_1 & a_2 & a_3 \end{bmatrix} = \sum_{i=1}^{3}\sum_{j=1}^{3}\sum_{k=1}^{3} \varepsilon_{ijk} \boldsymbol{\delta}_i \frac{\partial}{\partial x_j} a_k \\ &= \boldsymbol{\delta}_1 \left(\frac{\partial a_3}{\partial x_2} - \frac{\partial a_2}{\partial x_3} \right) + \boldsymbol{\delta}_2 \left(\frac{\partial a_1}{\partial x_3} - \frac{\partial a_3}{\partial x_1} \right) + \boldsymbol{\delta}_3 \left(\frac{\partial a_2}{\partial x_1} - \frac{\partial a_1}{\partial x_2} \right), \end{aligned} \tag{A1.17}$$

$$\nabla \cdot \mathbf{T} = \sum_{i=1}^{3} \left(\sum_{j=1}^{3} \frac{\partial T_{ji}}{\partial x_j} \right) \boldsymbol{\delta}_i \,. \tag{A1.18}$$

Additionally,

$$\left. \begin{aligned} &\nabla \cdot (\mathbf{ab}) = \mathbf{a} \cdot \nabla \mathbf{b} + \mathbf{b}(\nabla \cdot \mathbf{a}), \\ &\mathbf{ab} : \nabla \mathbf{c} = \mathbf{a} \cdot (\mathbf{b} \cdot \nabla)\mathbf{c}, \end{aligned} \right\} \tag{A1.19}$$

$$\nabla \cdot \nabla \mathbf{v} = \sum_{k=1}^{3} \boldsymbol{\delta}_k \left(\sum_{i=1}^{3} \frac{\partial^2}{\partial x_i^2} v_k \right). \tag{A1.20}$$

Expanding an arbitrary vector **a**(**r**) in a Taylor series about the origin leads to the following equation:

$$\mathbf{a}(\mathbf{r}) = \mathbf{a}_0 + \mathbf{r} \cdot \nabla \mathbf{a} + \frac{1}{2!}(\mathbf{rr} : \nabla\nabla \mathbf{a}) + \frac{1}{3!}(\mathbf{rrr} \vdots \nabla\nabla\nabla \mathbf{a}) + \cdots, \tag{A1.21}$$

in which $\mathbf{a}_0$ is the value of **a** at the origin, and ∇ **a** and $\nabla\ \nabla$ **a** have to be evaluated at **r**=0 after the differential operators have been applied.

Finally we show the divergence theorem and the Stokes theorem for a second-rank tensor **T** which is a function of the position **r**. That is,

$$\int_V (\nabla \cdot \mathbf{T}) dV = \int_S (\mathbf{n} \cdot \mathbf{T}) dS , \qquad \text{(A1.22)}$$

$$\int_S \mathbf{n} \cdot (\nabla \times \mathbf{T}) dS = \oint_C (\mathbf{t} \cdot \mathbf{T}) dC , \qquad \text{(A1.23)}$$

in which **n** is the unit normal vector directed outwards from a surface element *dS* on the surface *S* of the volume *V*, and **t** is the unit tangential vector along the boundary line *C* of the surface *S*.

References

1. R.B. Bird, R.C. Armstrong, and O. Hassager, “Dynamics of Polymeric Liquids, Vol.1, Fluid Mechanics,”Appendix A, John Wiley & Sons, New York, 1977

A2. THE DIRAC DELTA FUNCTION AND FOURIER INTEGRALS

The Dirac delta function δ is defined in a general way by the following equations:

$$\left.\begin{aligned} &\int_{-\infty}^{\infty}\delta(x-a)\Phi(x)dx = \Phi(a), \\ &\int_{-\infty}^{\infty}\delta(x)dx = 1, \end{aligned}\right\} \tag{A2.1}$$

in which $\Phi(x)$ is an arbitrary function which is continuous at $x=a$. It is clear from Eqs. (A2.1) that the Dirac delta function has a special meaning as an integrand. We explain the characteristics of the Dirac delta function using the square function $\Delta(x)$ shown in Fig. A2.1. This square function $\Delta(x)$ can be defined as

$$\Delta(x) = \begin{cases} \dfrac{1}{\varepsilon} & (-\varepsilon/2 \le x \le \varepsilon/2), \\ 0 & (x < -\varepsilon/2, x > \varepsilon/2). \end{cases} \tag{A2.2}$$

The square function is symmetric about the origin and an even function with an area of unity. This function satisfies the relations expressed in Eqs. (A2.1) in the limit of $\varepsilon \to 0$, so that the limiting function becomes the Dirac delta function.

If a three-dimensional system is considered, Eqs. (A2.1) can be replaced with

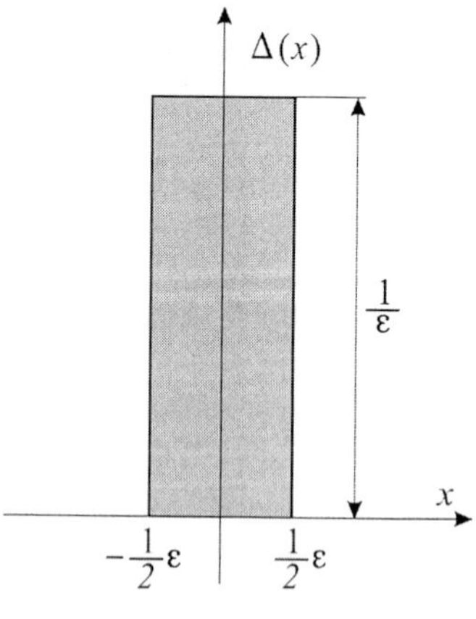

Figure A2.1 A square function.

$$\left.\begin{aligned}&\int_{-\infty}^{\infty}\delta(\mathbf{r}-\mathbf{a})\Phi(\mathbf{r})d\mathbf{r}=\Phi(\mathbf{a}),\\&\int_{-\infty}^{\infty}\delta(\mathbf{r})d\mathbf{r}=1,\end{aligned}\right\}\tag{A2.3}$$

in which $d\mathbf{r}=dxdydz$. The three-dimensional Dirac delta function $\delta(\mathbf{r})$ is equal to the product of each one-dimensional function, $\delta(x)\delta(y)\delta(z)$. The Dirac delta function can also be expressed in integral form using the Fourier integral theorem. We show an integral expression for the Dirac delta function in the following.

We first show the definition for the Fourier transform. The Fourier transform $F(k)$ of a certain function $f(x)$ can be defined by

$$F(k)=\frac{1}{(2\pi)^{1/2}}\int_{-\infty}^{\infty}f(x)e^{-ikx}dx\ .\tag{A2.4}$$

If $f(x)$ is single valued and periodic with at most a finite number of finite discontinuities, maxima, and minima, and the following quantity is finite:

$$\int_{-\infty}^{\infty}|f(x)|dx\,,\tag{A2.5}$$

then the following equation holds

$$f(x)=\frac{1}{(2\pi)^{1/2}}\int_{-\infty}^{\infty}F(x)e^{ikx}dk\ .\tag{A2.6}$$

This is called the Fourier integral of $f(x)$; a more detailed discussion concerning Fourier transforms and integrals should be referred to in general textbooks of mathematics. If f is a function of the positions $\mathbf{r}$ (=(x,y,z)), then Eqs. (A2.4) and (A2.6) must be replaced with the following equations:

$$F(\mathbf{k})=\frac{1}{(2\pi)^{3/2}}\int_{-\infty}^{\infty}f(\mathbf{r})e^{-i\mathbf{k}\cdot\mathbf{r}}d\mathbf{r}\,,\tag{A2.7}$$

$$f(\mathbf{r})=\frac{1}{(2\pi)^{3/2}}\int_{-\infty}^{\infty}F(\mathbf{k})e^{i\mathbf{k}\cdot\mathbf{r}}d\mathbf{k}\ .\tag{A2.8}$$

The Fourier transform $\hat{\delta}(\mathbf{k})$ of the Dirac delta function $\delta(\mathbf{r})$ can be straightforwardly obtained from Eq. (A2.7) as

$$\hat{\delta}(\mathbf{k}) = \frac{1}{(2\pi)^{3/2}} \int_{-\infty}^{\infty} \delta(\mathbf{r}) e^{-i\mathbf{k}\cdot\mathbf{r}} d\mathbf{r} = \frac{1}{(2\pi)^{3/2}}. \tag{A2.9}$$

Hence the definition of the Dirac delta function δ(**r**) can be obtained in integral form from Eqs. (A2.8) and (A2.9) as

$$\delta(\mathbf{r}) = \frac{1}{(2\pi)^{3/2}} \int_{-\infty}^{\infty} \frac{1}{(2\pi)^{3/2}} e^{i\mathbf{k}\cdot\mathbf{r}} d\mathbf{k} = \frac{1}{(2\pi)^{3}} \int_{-\infty}^{\infty} e^{i\mathbf{k}\cdot\mathbf{r}} d\mathbf{k}. \tag{A2.10}$$

Finally, we summarize useful expressions of Fourier transforms for flow quantities such as the pressure and flow velocity. If the pressure and flow velocity are denoted by P and **u**, respectively, the Fourier transforms of ∇P, $\nabla^2\mathbf{u}$, and $\nabla(\nabla\cdot\mathbf{u})$ are written as

$$\left.\begin{aligned}
&\frac{1}{(2\pi)^{3/2}} \int_{-\infty}^{\infty} \nabla P e^{-i\mathbf{k}\cdot\mathbf{r}} d\mathbf{r} = i\mathbf{k}\hat{P}(\mathbf{k},t), \\
&\frac{1}{(2\pi)^{3/2}} \int_{-\infty}^{\infty} \nabla^2 \mathbf{u} e^{-i\mathbf{k}\cdot\mathbf{r}} d\mathbf{r} = -k^2 \mathbf{J}(\mathbf{k},t), \\
&\frac{1}{(2\pi)^{3/2}} \int_{-\infty}^{\infty} \nabla(\nabla\cdot\mathbf{u}) e^{-i\mathbf{k}\cdot\mathbf{r}} d\mathbf{r} = -\mathbf{k}[\mathbf{k}\cdot\mathbf{J}(\mathbf{k},t)],
\end{aligned}\right\} \tag{A2.11}$$

in which $\hat{P}(\mathbf{k},t)$ and $\mathbf{J}(\mathbf{k},t)$ are the Fourier transforms of the pressure $P(\mathbf{r},t)$ and the fluid velocity $\mathbf{u}(\mathbf{r},t)$, respectively. We here derive only the first equation because the procedure for the other equations in Eq. (A2.11) is very similar. By setting **k** as $\mathbf{k}=(k_x,k_y,k_z)$, the x-component in the first equation of Eq. (A2.11) can be reformed as

$$\begin{aligned}
&\frac{1}{(2\pi)^{3/2}} \int_{-\infty}^{\infty}\int_{-\infty}^{\infty}\int_{-\infty}^{\infty} \frac{\partial P}{\partial x} e^{-i(k_x x+k_y y+k_z z)} dxdydz \\
&= \frac{1}{(2\pi)^{3/2}} \int_{-\infty}^{\infty}\int_{-\infty}^{\infty} \left[P e^{-i(k_x x+k_y y+k_z z)}\right]_{-\infty}^{\infty} dydz + \frac{ik_x}{(2\pi)^{3/2}} \int_{-\infty}^{\infty}\int_{-\infty}^{\infty}\int_{-\infty}^{\infty} P e^{-i\mathbf{k}\cdot\mathbf{r}} dxdydz \\
&\qquad = ik_x \frac{1}{(2\pi)^{3/2}} \int_{-\infty}^{\infty} P e^{-i\mathbf{k}\cdot\mathbf{r}} d\mathbf{r} = ik_x \hat{P}(\mathbf{k},t),
\end{aligned} \tag{A2.12}$$

in which we have taken into account the fact that the first term on the right-hand side vanishes because of $\sin(k_x x)\big|_{x=\infty} = \cos(k_x x)\big|_{x=\infty} = 0$ [1]. Similarly, expressions for y- and z-components are obtained, and finally the combination of such expressions into one vector form leads to the first equation in Eq. (A2.11).

References

1. R.P. Feynman, R.B. Leighton, and M.L. Sands, "The Feynman Lectures on Physics," Vol. 2, Addison Wesley Publishing Company, Reading, 1970

A3. THE LENNARD-JONES POTENTIAL

The Lennard-Jones molecule is frequently used as a model molecule for simulations of a molecular system. This model potential is also useful for modeling colloidal particles in some cases.

Various model potentials have been presented to date as an interaction energy between molecules. In particular, the Lennard-Jones 12-6 potential is well known as a model potential for spherical or nearly spherical molecules such as *Ar* (argon) molecules. If the separation between two molecules is denoted by r, the Lennard-Jones 12-6 potential $u(r)$ can be expressed as

$$u(r) = 4\varepsilon\left\{\left(\frac{\sigma}{r}\right)^{12} - \left(\frac{\sigma}{r}\right)^{6}\right\}, \tag{A3.1}$$

in which ε and σ are constants; ε is the depth of the potential well, and σ is the particle separation giving zero potential energy (that is, the effective particle diameter). Figure A3.1 shows the potential curve of this model potential. The Lennard-Jones potential has a steeply rising repulsive wall due to $(\sigma/r)^{12}$ and a long-range attractive tail due to $-(\sigma/r)^{6}$. It is well known that simulation results with the Lennard-Jones potential are in good agreement with

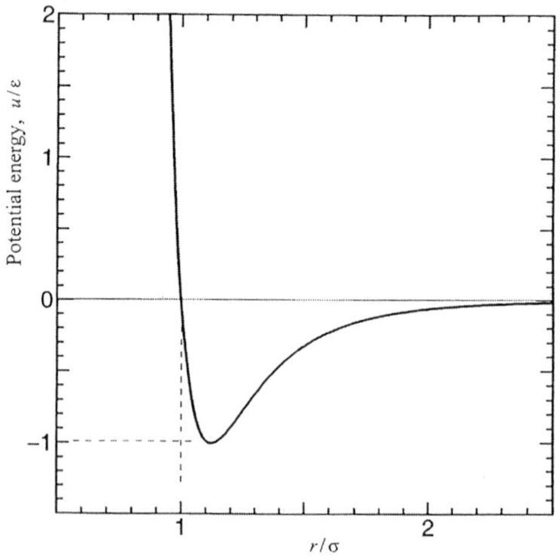

Figure A3.1 Lennard-Jones potential.

representative experimental results for *Ar* molecule systems with ε/k=119.8K and σ =3.405× 10^{-10} m, in which k is Boltzmann's constant. Using Eq. (A3.1), the force $\mathbf{f}(\mathbf{r}_{ij})$ exerted on particle i by particle j can be straightforwardly derived as

$$\mathbf{f}(\mathbf{r}_{ij}) = -\frac{\partial u(r_{ij})}{\partial \mathbf{r}_{ij}} = 24\varepsilon\left\{2\left(\frac{\sigma}{r_{ij}}\right)^{12} - \left(\frac{\sigma}{r_{ij}}\right)^{6}\right\}\frac{\mathbf{r}_{ij}}{r_{ij}^{2}}, \tag{A3.2}$$

in which $\mathbf{r}_{ij}=\mathbf{r}_i-\mathbf{r}_j$ and $r_{ij} = |\,\mathbf{r}_{ij}\,|$.

For Lennard-Jones systems, each quantity is usually nondimensionalized using σ, ε, m or combined values of these three basic quantities, and a nondimensional system is used in simulations. The following representative values are usually used: σ for distances, $1/\sigma^3$ for number densities, m/σ^3 for densities, ε/k for temperatures, ε for energies, ε/σ for forces, $(\varepsilon/m)^{1/2}$ for velocities, $\sigma(m/\varepsilon)^{1/2}$ for time, ε/σ^3 for pressure, k/m for specific heat,σ^3/ε for compressibility, $(m\varepsilon)^{1/2}/\sigma^2$ for viscosities, $(\varepsilon/m)^{1/2}k/\sigma^2$ for thermal conductivity coefficients, and $\sigma(\varepsilon/m)^{1/2}$ for diffusion coefficients.

The Lennard-Jones potential can be straightforwardly handled, so much simulation data have been accumulated to date for this model potential. Ree [1] analyzed such simulations data and presented the least-square expressions for the pressure P and the energy per unit mass, e:

$$P^* = n^*T^*\left\{1+\sum_{i=1}^{4} B_i x^i + B_{10}x^{10} - \sum_{i=1}^{5}\left(\frac{iC_i x^i}{T^{*1/2}} - \frac{C_i x^i}{T^*}\right)\right\}, \tag{A3.3}$$

$$e^* = T^*\left\{\frac{3}{2}+\sum_{i=1}^{4}\frac{B_i}{4}x^i + \frac{B_{10}}{4}x^{10} - \frac{1}{T^{*1/2}}\sum_{i=1}^{5}\left(\frac{i}{4}+\frac{1}{2}\right)C_i x^i + \frac{1}{T^*}\sum_{i=1}^{5}\left(\frac{1}{4}+\frac{1}{i}\right)D_i x^i\right\}, \tag{A3.4}$$

in which $x=n^*/T^{*1/4}$, and the quantities with the superscript * are dimensionless. The constants in Eqs. (A3.3) and (A3.4) are tabulated in Table A3.1. These approximate expressions were made using data for $0.05 \le n^* \le 0.96$ and $0.76 \le T^* \le 2.698$, and therefore one has to be careful in using Eqs. (A3.3) and (A3.4) for other ranges of n^* and T^*. Figure A3.2 shows the phase diagram for the Lennard-Jones system. The contours for constant pressures in Fig. A3.2 have been drawn using Eq. (A3.3).

References

1. F.H. Ree, Analytic Representation of Thermodynamic Data for the Lennard-Jones Fluid, J. Chem. Phys., 73 (1980) 5401

Table A3.1 Values of coefficients in Ree's expressions.

i	B_i	C_i	D_i
1	3.629	5.3692	-3.4921
2	7.2641	6.5797	18.6980
3	10.4924	6.1745	-35.5049
4	11.459	-4.2685	31.8151
5	—	1.6841	-11.1953
10	2.17619	—	—

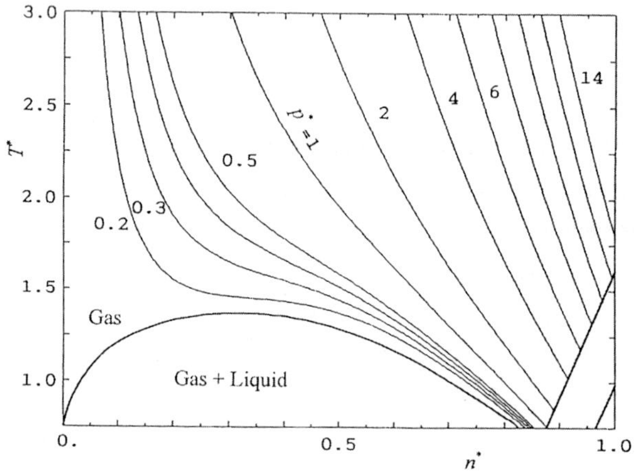

Figure A3.2 Phase diagram for Lennard-Jones system.

A4. EXPRESSIONS OF RESISTANCE AND MOBILITY FUNCTIONS FOR SPHERICAL PARTICLES

We have already shown the expressions for some resistance and mobility functions in Sec. 5.3.5. In this appendix, we show the expressions for the other resistance and mobility functions with the common variable $\xi=(r_{21}/a-2)=(s-2)$.

A4.1 Resistance Functions

A4.1.1 $Y_{\alpha\beta}^{B}$

For the nearly-touching case,

$$\left.\begin{aligned} Y_{11}^{B} &= 4\pi a^2\left\{-\frac{1}{4}\ln\xi^{-1}+0.2390-\frac{1}{8}\xi\ln\xi^{-1}\right\}, \\ Y_{12}^{B} &= -4\pi a^2\left\{-\frac{1}{4}\ln\xi^{-1}+0.0017-\frac{1}{8}\xi\ln\xi^{-1}\right\}. \end{aligned}\right\} \tag{A4.1}$$

For the far-separating case,

$$Y_{11}^{B} = 4\pi a^2\sum_{k=0}^{\infty}\left(\frac{1}{2s}\right)^{2k+1} f_{2k+1}^{Y}, \qquad Y_{12}^{B} = -4\pi a^2\sum_{k=0}^{\infty}\left(\frac{1}{2s}\right)^{2k} f_{2k}^{Y}, \tag{A4.2}$$

in which

$$\begin{gathered} f_0^Y = f_1^Y = 0,\ f_2^Y = -6,\ f_3^Y = -9,\ f_4^Y = -27/2,\ f_5^Y = -273/4, \\ f_6^Y = -1683/8,\ f_7^Y = -17625/16,\ f_8^Y = -129003/32, \\ f_9^Y = -1017825/64,\ f_{10}^Y = -7087107/128,\ f_{11}^Y = -47478057/256. \end{gathered} \tag{A4.3}$$

A4.1.2 $X_{\alpha\beta}^{C}$ and $Y_{\alpha\beta}^{C}$

For the nearly-touching case,

$$\left.\begin{aligned} X_{11}^{C} &= 8\pi a^3\left\{\frac{1}{8}\zeta(3,\frac{1}{2})-\frac{1}{8}\xi\ln\xi^{-1}\right\}, \\ X_{12}^{C} &= -8\pi a^3\left\{\frac{1}{8}\zeta(3,1)-\frac{1}{8}\xi\ln\xi^{-1}\right\}, \end{aligned}\right\} \tag{A4.4}$$

$$\left.\begin{aligned} Y_{11}^{C} &= 8\pi a^{3}\left\{\frac{1}{5}\ln\xi^{-1} + 0.7028 + \frac{47}{250}\xi\ln\xi^{-1}\right\}, \\ Y_{12}^{C} &= 8\pi a^{3}\left\{\frac{1}{20}\ln\xi^{-1} - 0.0274 + \frac{31}{250}\xi\ln\xi^{-1}\right\}, \end{aligned}\right\} \tag{A4.5}$$

in which the Riemann zeta function $\zeta(x,y)$ is defined by

$$\zeta(x,y) = \sum_{k=0}^{\infty}(k+y)^{-x}\,. \tag{A4.6}$$

For the far-separating case,

$$X_{11}^{C} = 8\pi a^{3}\sum_{k=0}^{\infty}\left(\frac{1}{2s}\right)^{2k} f_{2k}^{X}, \quad X_{12}^{C} = -8\pi a^{3}\sum_{k=0}^{\infty}\left(\frac{1}{2s}\right)^{2k+1} f_{2k+1}^{X}, \tag{A4.7}$$

$$Y_{11}^{C} = 8\pi a^{3}\sum_{k=0}^{\infty}\left(\frac{1}{2s}\right)^{2k} f_{2k}^{Y}, \quad Y_{12}^{C} = 8\pi a^{3}\sum_{k=0}^{\infty}\left(\frac{1}{2s}\right)^{2k+1} f_{2k+1}^{Y}, \tag{A4.8}$$

in which

$$\left.\begin{aligned} &f_{0}^{X} = 1,\ f_{1}^{X} = f_{2}^{X} = 0,\ f_{3}^{X} = 8,\ f_{4}^{X} = f_{5}^{X} = 0,\ f_{6}^{X} = 64,\ f_{7}^{X} = 0, \\ &\quad f_{8}^{X} = 768,\ f_{9}^{X} = 512,\ f_{10}^{X} = 6144,\ f_{11}^{X} = 12288, \end{aligned}\right\} \tag{A4.9}$$

$$\left.\begin{aligned} &f_{0}^{Y} = 1,\ f_{1}^{Y} = f_{2}^{Y} = 0,\ f_{3}^{Y} = 4,\ f_{4}^{Y} = 12,\ f_{5}^{Y} = 18,\ f_{6}^{Y} = 283,\ f_{7}^{Y} = 369/2, \\ &\quad f_{8}^{Y} = 11955/4,\ f_{9}^{Y} = 5945/8,\ f_{10}^{Y} = 511755/16,\ f_{11}^{Y} = 448833/32. \end{aligned}\right\} \tag{4.10}$$

A4.1.3 $X_{\alpha\beta}{}^{G}$ and $Y_{\alpha\beta}{}^{G}$

For the nearly-touching case,

$$\left.\begin{aligned} X_{11}^{G} &= 4\pi a^{2}\left\{\frac{3}{8}\xi^{-1} + \frac{27}{80}\ln\xi^{-1} - 0.469 + \frac{117}{560}\xi\ln\xi^{-1} + \mathrm{O}(\xi)\right\}, \\ X_{12}^{G} &= -4\pi a^{2}\left\{\frac{3}{8}\xi^{-1} + \frac{27}{80}\ln\xi^{-1} - 0.195 + \frac{117}{560}\xi\ln\xi^{-1} + \mathrm{O}(\xi)\right\}, \end{aligned}\right\} \tag{A4.11}$$

$$\left.\begin{aligned} Y_{11}^{G} &= 4\pi a^{2}\left\{\frac{1}{8}\ln\xi^{-1} - 0.142 + \frac{1}{16}\xi\ln\xi^{-1} + \mathrm{O}(\xi)\right\}, \\ Y_{12}^{G} &= -4\pi a^{2}\left\{\frac{1}{8}\ln\xi^{-1} - 0.103 + \frac{1}{16}\xi\ln\xi^{-1} + \mathrm{O}(\xi)\right\}. \end{aligned}\right\} \tag{A4.12}$$

For the far-separating case,

$$X_{11}^{G} = 4\pi a^2 \sum_{k=0}^{\infty} \left(\frac{1}{2s}\right)^{2k+1} f_{2k+1}^{X}, \quad X_{12}^{G} = -4\pi a^2 \sum_{k=0}^{\infty} \left(\frac{1}{2s}\right)^{2k} f_{2k}^{X}, \tag{A4.13}$$

$$Y_{11}^{G} = 4\pi a^2 \sum_{k=0}^{\infty} \left(\frac{1}{2s}\right)^{2k+1} f_{2k+1}^{Y}, \quad Y_{12}^{G} = -4\pi a^2 \sum_{k=0}^{\infty} \left(\frac{1}{2s}\right)^{2k} f_{2k}^{Y}, \tag{A4.14}$$

in which

$$\left.\begin{aligned} &f_0^X = f_1^X = 0,\ f_2^X = 15,\ f_3^X = 45,\ f_4^X = 39,\ f_5^X = 597,\ f_6^X = 2331, \\ &f_7^X = 6021,\ f_8^X = 34347,\ f_9^X = 101205,\ f_{10}^X = 458859,\ f_{11}^X = 1886037, \end{aligned}\right\} \tag{A4.15}$$

$$\left.\begin{aligned} &f_0^Y = f_1^Y = f_2^Y = f_3^Y = 0,\ f_4^Y = 32,\ f_5^Y = 108,\ f_6^Y = 162,\ f_7^Y = 531, \\ &f_8^Y = 1305/2,\ f_9^Y = 2475/4,\ f_{10}^Y = 81697/8,\ f_{11}^Y = 936379/16. \end{aligned}\right\} \tag{A4.16}$$

A4.1.4 $Y_{\alpha\beta}^{H}$

For the nearly-touching case,

$$\left.\begin{aligned} Y_{11}^{H} &= 8\pi a^3 \left\{\frac{1}{40}\ln\xi^{-1} - 0.074 + \frac{137}{2000}\xi\ln\xi^{-1} + O(\xi)\right\}, \\ Y_{12}^{H} &= 8\pi a^3 \left\{\frac{1}{10}\ln\xi^{-1} - 0.030 + \frac{113}{2000}\xi\ln\xi^{-1} + O(\xi)\right\}. \end{aligned}\right\} \tag{A4.17}$$

For the far-separating case,

$$Y_{11}^{H} = 8\pi a^3 \sum_{k=0}^{\infty} \left(\frac{1}{2s}\right)^{2k} f_{2k}^{Y}, \quad Y_{12}^{H} = 8\pi a^3 \sum_{k=0}^{\infty} \left(\frac{1}{2s}\right)^{2k+1} f_{2k+1}^{Y}, \tag{A4.18}$$

in which

$$\left.\begin{aligned} &f_0^Y = f_1^Y = f_2^Y = 0,\ f_3^Y = 10,\ f_4^Y = f_5^Y = 0,\ f_6^Y = -96,\ f_7^Y = 216, \\ &f_8^Y = 324,\ f_9^Y = 7078,\ f_{10}^Y = 14969,\ f_{11}^Y = 196299/2. \end{aligned}\right\} \tag{A4.19}$$

A4.1.5 $X_{\alpha\beta}^{K}$, $Y_{\alpha\beta}^{K}$, and $Z_{\alpha\beta}^{K}$

For the nearly-touching case,

$$\left.\begin{aligned} X_{11}^{K} &= \frac{20}{3}\pi a^3 \left\{\frac{3}{20}\xi^{-1} + \frac{27}{200}\ln\xi^{-1} + K_{11}^{X} + \frac{353}{2800}\xi\ln\xi^{-1} + O(\xi)\right\}, \\ X_{12}^{K} &= \frac{20}{3}\pi a^3 \left\{\frac{3}{20}\xi^{-1} + \frac{27}{200}\ln\xi^{-1} + K_{12}^{X} + \frac{493}{2800}\xi\ln\xi^{-1} + O(\xi)\right\}, \end{aligned}\right\} \tag{A4.20}$$

$$\left.\begin{aligned} Y_{11}^{K} &= \frac{20}{3}\pi a^{3}\left\{\frac{3}{25}\ln\xi^{-1} + K_{11}^{Y} + \frac{57}{2500}\xi\ln\xi^{-1} + \mathrm{O}(\xi)\right\}, \\ Y_{12}^{K} &= \frac{20}{3}\pi a^{3}\left\{\frac{3}{100}\ln\xi^{-1} + K_{12}^{Y} + \frac{159}{1250}\xi\ln\xi^{-1} + \mathrm{O}(\xi)\right\}, \end{aligned}\right\} \tag{A4.21}$$

in which

$$K_{11}^{X} + K_{12}^{X} = 0.5712, \quad K_{11}^{Y} + K_{12}^{Y} = 0.6760. \tag{A4.22}$$

For the far-separating case,

$$X_{11}^{K} = \frac{20}{3}\pi a^{3}\sum_{k=0}^{\infty}\left(\frac{1}{2s}\right)^{2k} f_{2k}^{X}, \quad X_{12}^{K} = \frac{20}{3}\pi a^{3}\sum_{k=0}^{\infty}\left(\frac{1}{2s}\right)^{2k+1} f_{2k+1}^{X}, \tag{A4.23}$$

$$Y_{11}^{K} = \frac{20}{3}\pi a^{3}\sum_{k=0}^{\infty}\left(\frac{1}{2s}\right)^{2k} f_{2k}^{Y}, \quad Y_{12}^{K} = \frac{20}{3}\pi a^{3}\sum_{k=0}^{\infty}\left(\frac{1}{2s}\right)^{2k+1} f_{2k+1}^{Y}, \tag{A4.24}$$

in which

$$\left.\begin{aligned} f_0^X = 1,\ f_1^X = f_2^X = 0,\ f_3^X = 40,\ f_4^X = 60,\ f_5^X = -204, \\ f_6^X = 1372,\ f_7^X = 3636,\ f_8^X = 9765.6, \\ f_9^X = 65600.8,\ f_{10}^X = 93746.4,\ f_{11}^X = 873626.4, \end{aligned}\right\} \tag{A4.25}$$

$$\left.\begin{aligned} f_0^Y = 1,\ f_1^Y = f_2^Y = 0,\ f_3^Y = -20,\ f_4^Y = 0,\ f_5^Y = 256, \\ f_6^Y = 640,\ f_7^Y = 0, f_8^Y = 2099.2, \\ f_9^Y = -12339.2,\ f_{10}^Y = -33676.8,\ f_{11}^Y = -126134.4. \end{aligned}\right\} \tag{A4.26}$$

From the relation in Eq. (5.81), $Z_{\alpha\beta}{}^{K}$ can approximately be evaluated from $z_{\alpha}{}^{k}$, which will be shown later

A4.2 Mobility Functions

A4.2.1 $y_{\alpha\beta}{}^{b}$

For the nearly-touching case,

$$\left.\begin{aligned} (4\pi a^2)y_{11}^{b} &= \frac{0.13368(\ln\xi^{-1})^2 + 0.19945\ln\xi^{-1} - 0.79238}{(\ln\xi^{-1})^2 + 6.04250\ln\xi^{-1} + 6.32549} + \mathrm{O}(\xi\ln\xi), \\ (4\pi a^2)y_{12}^{b} &= \frac{-0.13368(\ln\xi^{-1})^2 - 0.92720\ln\xi^{-1} - 0.18805}{(\ln\xi^{-1})^2 + 6.04250\ln\xi^{-1} + 6.32549} + \mathrm{O}(\xi\ln\xi). \end{aligned}\right\} \tag{A4.27}$$

For the far-separating case,

$$y_{11}^{b} = (4\pi a^2)^{-1}\sum_{k=0}^{\infty}\left(\frac{1}{2s}\right)^{2k+1} f_{2k+1}^{y}, \quad y_{12}^{b} = (4\pi a^2)^{-1}\sum_{k=0}^{\infty}\left(\frac{1}{2s}\right)^{2k} f_{2k}^{y}, \tag{A4.28}$$

in which

$$\left.\begin{aligned} &f_0^y = f_1^y = 0,\ f_2^y = -2,\ f_3^y = f_4^y = f_5^y = f_6^y = 0, f_7^y = 208, \\ &f_8^y = 0,\ f_9^y = 2432,\ f_{10}^y = -1280,\ f_{11}^y = 22272. \end{aligned}\right\} \tag{A4.29}$$

A4.2.2 $x_{\alpha\beta}^{\ c}$ and $y_{\alpha\beta}^{\ c}$

For the nearly-touching case,

$$\left.\begin{aligned} (8\pi a^3)y_{11}^c &= \frac{0.26736(\ln\xi^{-1})^2 + 5.60896\ln\xi^{-1} + 9.28111}{(\ln\xi^{-1})^2 + 6.04250\ln\xi^{-1} + 6.32549}, \\ (8\pi a^3)y_{12}^c &= \frac{0.26736(\ln\xi^{-1})^2 - 1.05770\ln\xi^{-1} + 0.29981}{(\ln\xi^{-1})^2 + 6.04250\ln\xi^{-1} + 6.32549}. \end{aligned}\right\} \tag{A4.30}$$

The $x_{\alpha\beta}^{\ c}$ can be calculated from Eq. (5.76) using the values of $X_{\alpha\beta}^{\ C}$.

For the far-separating case,

$$x_{11}^{c} = (8\pi a^3)^{-1}\sum_{k=0}^{\infty}\left(\frac{1}{s}\right)^{2k} f_{2k}^{x}, \quad x_{12}^{c} = (8\pi a^3)^{-1}\sum_{k=0}^{\infty}\left(\frac{1}{s}\right)^{2k+1} f_{2k+1}^{x}, \tag{A4.31}$$

$$y_{11}^{c} = (8\pi a^3)^{-1}\sum_{k=0}^{\infty}\left(\frac{1}{2s}\right)^{2k} f_{2k}^{y}, \quad y_{12}^{c} = (8\pi a^3)^{-1}\sum_{k=0}^{\infty}\left(\frac{1}{2s}\right)^{2k+1} f_{2k+1}^{y}, \tag{A4.32}$$

in which

$$\left.\begin{aligned} &f_0^x = 1,\ f_1^x = f_2^x = 0,\ f_3^x = 1,\ f_4^x = f_5^x = f_6^x = f_7^x = 0, \\ &f_8^x = -3,\ f_9^x = 0,\ f_{10}^x = -6,\ f_{11}^x = 0, \end{aligned}\right\} \tag{A4.33}$$

$$\left.\begin{aligned} &f_0^y = 1,\ f_1^y = f_2^y = 0,\ f_3^y = -4,\ f_4^y = f_5^y = 0,\ f_6^y = -240,\ f_7^y = 0, \\ &f_8^y = -2496,\ f_9^y = 4800,\ f_{10}^y = -18432,\ f_{11}^y = 61440. \end{aligned}\right\} \tag{A4.34}$$

A4.2.3 $x_{\alpha\beta}^{\ g}$ and $y_{\alpha\beta}^{\ g}$

For the nearly-touching case,

$$x_{11}^{g} = 2a(0.1792 - 0.8703\xi), \quad x_{12}^{g} = 2a(-0.3208 + 0.9184\xi), \tag{A4.35}$$

$$\left.\begin{aligned} y_{11}^{g} &= 2a\frac{0.0145(\ln\xi^{-1})^2+0.0786\ln\xi^{-1}-0.3193}{(\ln\xi^{-1})^2+6.04250\ln\xi^{-1}+6.32549}+\mathrm{O}(\xi\ln\xi),\\ y_{12}^{g} &= 2a\frac{-0.0869(\ln\xi^{-1})^2-0.2956\ln\xi^{-1}+0.1584}{(\ln\xi^{-1})^2+6.04250\ln\xi^{-1}+6.32549}+\mathrm{O}(\xi\ln\xi).\end{aligned}\right\} \tag{A4.36}$$

For the far-separating case,

$$x_{11}^{g} = 2a\sum_{k=0}^{\infty}\left(\frac{1}{2s}\right)^{2k+1} f_{2k+1}^{x}, \quad x_{12}^{g} = -2a\sum_{k=0}^{\infty}\left(\frac{1}{2s}\right)^{2k} f_{2k}^{x}, \tag{A4.37}$$

$$y_{11}^{g} = 2a\sum_{k=0}^{\infty}\left(\frac{1}{2s}\right)^{2k+1} f_{2k+1}^{y}, \quad y_{12}^{g} = -2a\sum_{k=0}^{\infty}\left(\frac{1}{2s}\right)^{2k} f_{2k}^{y}, \tag{A4.38}$$

in which

$$\left.\begin{aligned} &f_0^x = f_1^x = 0,\ f_2^x = 5,\ f_3^x = 0,\ f_4^x = -32,\ f_5^x = 200,\ f_6^x = 0\\ &f_7^x = -1120, f_8^x = 8000,\ f_9^x = -13056,\ f_{10}^x = 3200,\ f_{11}^x = 220160,\end{aligned}\right\} \tag{A4.39}$$

$$\left.\begin{aligned} &f_0^y = f_1^y = f_2^y = f_3^y = 0,\ f_4^y = 32/3,\ f_5^y = f_6^y = 0, f_7^y = 160/3,\\ &f_8^y = 0,\ f_9^y = -768,\ f_{10}^y = -3200/3,\ f_{11}^y = -3072.\end{aligned}\right\} \tag{A4.40}$$

A4.2.4 $y_{\alpha\beta}^{h}$

For the nearly-touching case,

$$\left.\begin{aligned} y_{11}^{h} &= \frac{-0.1014(\ln\xi^{-1})^2+0.0764\ln\xi^{-1}-0.7905}{(\ln\xi^{-1})^2+6.04250\ln\xi^{-1}+6.32549}+\mathrm{O}(\xi\ln\xi),\\ y_{12}^{h} &= \frac{0.3986(\ln\xi^{-1})^2+1.0762\ln\xi^{-1}-0.3510}{(\ln\xi^{-1})^2+6.04250\ln\xi^{-1}+6.32549}+\mathrm{O}(\xi\ln\xi).\end{aligned}\right\} \tag{A4.41}$$

For the far-separating case,

$$y_{11}^{h} = \sum_{k=0}^{\infty}\left(\frac{1}{2s}\right)^{2k} f_{2k}^{y}, \quad y_{12}^{h} = \sum_{k=0}^{\infty}\left(\frac{1}{2s}\right)^{2k+1} f_{2k+1}^{y}, \tag{A4.42}$$

in which

$$\left.\begin{aligned} &f_0^y = f_1^y = f_2^y = 0,\ f_3^y = 10,\ f_4^y = f_5^y = 0,\ f_6^y = -200,\\ &f_7^y = f_8^y = 0,\ f_9^y = 4000,\ f_{10}^y = 12800,\ f_{11}^y = 64000.\end{aligned}\right\} \tag{A4.43}$$

A4.2.5 x_α^k, y_α^k, and z_α^k

For the nearly-touching case,

$$\left.\begin{aligned} x_1^k &= -\frac{20}{3}\pi a^3(1.910-3.85\xi),\\ y_1^k &= -\frac{20}{3}\pi a^3\frac{1.1456(\ln\xi^{-1})^2+6.1694\ln\xi^{-1}+3.7112}{(\ln\xi^{-1})^2+6.04250\ln\xi^{-1}+6.32549}+\mathrm{O}(\xi(\ln\xi)^3),\\ z_1^k &= -\frac{20}{3}\pi a^3(0.9527+0.0914\xi-0.081\xi^2).\end{aligned}\right\} \quad \text{(A4.44)}$$

For the far-separating case,

$$\left.\begin{aligned} x_1^k &= -\frac{20}{3}\pi a^3\sum_{l=0}^{\infty}\left(\frac{1}{2s}\right)^l f_l^x, \qquad y_1^k = -\frac{20}{3}\pi a^3\sum_{l=0}^{\infty}\left(\frac{1}{2s}\right)^l f_l^y,\\ z_1^k &= -\frac{20}{3}\pi a^3\sum_{l=0}^{\infty}\left(\frac{1}{2s}\right)^l f_l^z,\end{aligned}\right\} \quad \text{(A4.45)}$$

in which

$$\left.\begin{aligned} &f_0^x=1,\ f_1^x=f_2^x=0,\ f_3^x=40,\ f_4^x=0,\ f_5^x=-384,\ f_6^x=1600,\\ &f_7^x=0, f_8^x=-5760,\ f_9^x=64000,\ f_{10}^x=-135168,\ f_{11}^x=153600,\end{aligned}\right\} \quad \text{(A4.46)}$$

$$\left.\begin{aligned} &f_0^y=1,\ f_1^y=f_2^y=0,\ f_3^y=-20,\ f_4^y=0,\ f_5^y=256,\ f_6^y=400, f_7^y=0,\\ &f_8^y=1280,\ f_9^y=-39998/5,\ f_{10}^y=-43008,\ f_{11}^y=-767988/5,\end{aligned}\right\} \quad \text{(A4.47)}$$

$$\left.\begin{aligned} &f_0^z=1,\ f_1^z=f_2^z=f_3^z=f_4^z=0,\ f_5^z=-64,\ f_6^z=f_7^z=0,\\ &f_8^z=800,\ f_9^z=0,\ f_{10}^z=3072,\ f_{11}^z=0.\end{aligned}\right\} \quad \text{(A4.48)}$$

References

1. S. Kim and S.J. Karrila, “Microhydrodynamics: Principles and Selected Applications,” Butterworth-Heinemann, Stoneham, 1991

A5. DIFFUSION COEFFICIENTS OF CIRCULAR CYLINDER PARTICLES AND THE RESISTANCE FUNCTIONS OF SPHEROCYLINDERS

In addition to spheres and spheroids, a cylinder is quite useful as a particle model. We here show approximate expressions for diffusion coefficients. It is noted that, even for cylindrical particles, the diffusion matrix **D** can be related to the mobility matrix **M** by Eq. (9.77).

When a single cylindrical particle moves in a fluid, its motion can be decomposed into the translational and rotational motion. Additionally, the translational motion can be decomposed into motion perpendicular and parallel to the rod axis, and the rotational motion can be decomposed into rotational motion around the rod axis and around the line perpendicular to the rod axis through the center of mass.

We use the following notations. The translational coefficients perpendicular and parallel to the particle axis are denoted by $D_\perp{}^T$ and $D_\parallel{}^T$, respectively. Also the rotational coefficient around the line, perpendicular to the particle axis through the center of mass, is denoted by D^R. Tirado et al. [1-3] presented the approximate expressions for such diffusion coefficients by numerical calculations for a cylindrical particle with a diameter d and length l:

$$D_\perp^T = \frac{kT}{\eta}\cdot\frac{1}{4\pi l}\left(\ln r_p + 0.839 + \frac{0.185}{r_p} + \frac{0.233}{r_p^2}\right), \tag{A5.1}$$

$$D_\parallel^T = \frac{kT}{\eta}\cdot\frac{1}{2\pi l}\left(\ln r_p - 0.207 + \frac{0.980}{r_p} - \frac{0.133}{r_p^2}\right), \tag{A5.2}$$

$$D^R = \frac{kT}{\eta}\cdot\frac{3}{\pi l^3}\left(\ln r_p - 0.662 + \frac{0.917}{r_p} - \frac{0.050}{r_p^2}\right), \tag{A5.3}$$

in which k is Boltzmann's constant, T is the temperature, η is the viscosity of the base liquid, and r_p is the aspect ratio expressed as $r_p=l/d$. Equations (A5.1) to (A5.3) are valid for $2\leq r_p\leq 20$. It is noted from Eqs. (5.13) and (5.14) that the resistance functions are related to the diffusion functions by $(X^A)^{-1}=(\eta/kT)D_\parallel{}^T$, $(Y^A)^{-1}=(\eta/\mathrm{kT})D_\perp{}^T$, and $(Y^C)^{-1}=(\eta/kT)D^R$.

Also, the following approximate expressions for the resistance functions were shown for r_p>>1 in Ref. [4]:

$$X^A = \frac{4\pi a}{\ln 2r_p + \ln 2 - 3/2}, \tag{A5.4}$$

$$Y^A = \frac{8\pi a}{\ln 2r_p + \ln 2 - 1/2}, \tag{A5.5}$$

$$X^C = \frac{8\pi a^3}{r_p^2}, \quad Y^C = \frac{8\pi a^3}{3}\left\{\frac{1}{\ln r_p}\left(1+\frac{\ln 2-1}{\ln r_p}\right)+\frac{3\times 5.45}{8\pi r_p^2}\right\}, \tag{A5.6}$$

$$Y^H = \frac{8\pi a^3}{3}\left\{\frac{1}{\ln r_p}\left(1+\frac{\ln 2-1}{\ln r_p}\right)-\frac{3\times 5.45}{8\pi r_p^2}\right\}. \tag{A5.7}$$

For a slender spheroidal particle, the approximate expressions for the resistance functions can be derived from Eqs. (5.19) to (5.21) for $r_p >> 1$:

$$X^A = 4\pi a\left[\frac{2}{2\ln 2r_p - 1} - \frac{4\ln 2r_p - 3}{\{4(\ln 2r_p)^2 - 4\ln 2r_p + 1\}r_p^2}\right], \tag{A5.8}$$

$$Y^A = 8\pi a\left[\frac{2}{2\ln 2r_p + 1} - \frac{1}{\{4(\ln 2r_p)^2 + 4\ln 2r_p + 1\}r_p^2}\right], \tag{A5.9}$$

$$X^C = \frac{16\pi a^3}{3r_p^2}, \quad Y^C = \frac{8\pi a^3}{3}\left[\frac{2}{2\ln 2r_p - 1} + \frac{1}{\{4(\ln 2r_p)^2 - 4\ln 2r_p + 1\}r_p^2}\right], \tag{A5.10}$$

$$Y^H = \frac{8\pi a^3}{3}\left[\frac{2}{2\ln 2r_p - 1} - \frac{8\ln 2r_p - 5}{\{4(\ln 2r_p)^2 - 4\ln 2r_p + 1\}r_p^2}\right], \tag{A5.11}$$

in which $r_p = a/b$, $2a$ is the length of the longer axis, and $2b$ is the length of the shorter axis. The neglect of higher-order terms in Eqs. (A5.8) to (A5.11) leads to the expressions in Ref. [4].

References

1. M.M. Tirado and J.G. de la Torre, Translational Friction Coefficients of Rigid, Symmetric Top Macromolecules: Application to Circular Cylinders, J. Chem. Phys., 71 (1979) 2581
2. M.M. Tirado and J.G. de la Torre, Rotational Dynamics of Rigid, Symmetric Top Macromolecules: Application to Circular Cylinders, J. Chem. Phys., 73 (1980) 1986
3. M.M. Tirado, C.L. Martinez, and J.G. de la Torre, Comparison of Theories for the Translational and Rotational Diffusion Coefficients of Rod-like Macromolecules: Application to Short DNA Fragments, J. Chem. Phys., 81 (1984) 2047
4. H. Brenner, Rheology of a Dilute Suspension of Axisymmetric Brownian Particles, Int. J. Multiphase Flow, 1 (1974) 195

A6. DERIVATION OF EXPRESSIONS FOR LONG-RANGE INTERACTIONS (THE EWALD SUM)

A6.1 Interactions between Charged Particles

We derive Eq. (11.21) in this appendix [1]. First we calculate the interaction energy $E_i^{(b)}$ between a point charge q_i and the canceling charge distributions, as shown in Fig. 11.9(b). If point charges in the central cell or in the simulation region are considered, the charge distribution ρ(**r**) at an arbitrary position **r** in the central cell can be written using the Dirac delta function as

$$\rho(\mathbf{r}) = \sum_{j=1}^{N} q_j \delta(\mathbf{r}-\mathbf{r}_j). \tag{A6.1}$$

If the central cell is cubic with the side length L (volume V) and also if periodic boundary conditions are used, the charge distribution (in the whole area including replicated cells outside the central cell) can be regarded as a periodic function of the charge distribution in (A6.1), with a period L in each direction. Hence the charge density ρ(**r**) at an arbitrary position **r** in the whole area can be expressed in a Fourier series as

$$\rho(\mathbf{r}) = \sum_{\mathbf{h}} \boldsymbol{\alpha}_{\mathbf{h}} e^{i2\pi\mathbf{h}\cdot\mathbf{r}}, \tag{A6.2}$$

in which

$$\begin{aligned}\boldsymbol{\alpha}_{\mathbf{h}} &= \frac{1}{L^3}\int_V \rho(\mathbf{r}) e^{-i2\pi\mathbf{h}\cdot\mathbf{r}} d\mathbf{r} = \frac{1}{L^3}\int_V \sum_{j=1}^{N} q_j \delta(\mathbf{r}-\mathbf{r}_j) e^{-i2\pi\mathbf{h}\cdot\mathbf{r}} d\mathbf{r} \\ &= \frac{1}{L^3}\sum_{j=1}^{N} q_j e^{-i2\pi\mathbf{h}\cdot\mathbf{r}_j}.\end{aligned} \tag{A6.3}$$

In these equations, $\mathbf{h}=\mathbf{n}/L$ and $\mathbf{n}=(n_x,n_y,n_z)$ with $n_x,n_y,n_z=0,\pm1,\pm2,\ldots$

Similarly, we show the Fourier series of the canceling distribution $\rho^{(b)}(\mathbf{r})$ at the position **r** in the central cell. The canceling distribution $\rho^{(b)}(\mathbf{r})$ can be written using Eq. (11.20) as

$$\rho^{(b)}(\mathbf{r}) = \sum_{j=1}^{N} q_j \sigma(\mathbf{r}-\mathbf{r}_j) = \int \sum_{j=1}^{N} q_j \delta(\mathbf{r}-\mathbf{r}_j-\mathbf{r}')\sigma(\mathbf{r}')d\mathbf{r}'. \tag{A6.4}$$

As in the previous case, by regarding this distribution as expanding the whole area beyond the central cell, the distribution at an arbitrary position **r** in the whole area can be derived in a Fourier series as

$$\rho^{(b)}(\mathbf{r}) = \sum_{\mathbf{h}(\neq 0)} \gamma_{\mathbf{h}} e^{i2\pi\mathbf{h}\cdot\mathbf{r}}, \tag{A6.5}$$

in which

$$\gamma_{\mathbf{h}} = \frac{1}{L^3}\int_V \left(\int \sum_{j=1}^{N} q_j \delta(\mathbf{r}-\mathbf{r}_j-\mathbf{r}')\sigma(\mathbf{r}')d\mathbf{r}' \right) e^{-i2\pi\mathbf{h}\cdot\mathbf{r}} d\mathbf{r}$$
$$= \frac{1}{L^3}\int \sum_{j=1}^{N} q_j \sigma(\mathbf{r}') e^{-i2\pi\mathbf{h}\cdot(\mathbf{r}_j+\mathbf{r}')} d\mathbf{r}' = \alpha_{\mathbf{h}} \int \sigma(\mathbf{r}') e^{-i2\pi\mathbf{h}\cdot\mathbf{r}'} d\mathbf{r}' = \alpha_{\mathbf{h}}\beta_{\mathbf{h}}. \tag{A6.6}$$

From the formulae concerning the Laplace transformation of a function including exponential and sine functions, $\beta_{\mathbf{h}}$ in Eq. (A6.6) can straightforwardly be evaluated by taking the z'-axis along the vector $\mathbf{h}$:

$$\beta_{\mathbf{h}} = \exp(-\pi^2 h^2 / \kappa^2). \tag{A6.7}$$

Removing the case of $\mathbf{h}$=0 results from the condition of neutrality.

Hence, the interaction energy $E_i^{(b)\prime}$ between the point charge q_i at $\mathbf{r}_i$ and the canceling charge distributions (including the canceling distribution of q_i itself) can be written as

$$E_i^{(b)\prime} = q_i \int \frac{\rho^{(b)}(\mathbf{r}_i+\mathbf{r}')}{r'} d\mathbf{r}' = q_i \sum_{\mathbf{h}(\neq 0)} \alpha_{\mathbf{h}}\beta_{\mathbf{h}} \int \frac{e^{i2\pi\mathbf{h}\cdot(\mathbf{r}_i+\mathbf{r}')}}{r'} d\mathbf{r}'$$
$$= q_i \sum_{\mathbf{h}(\neq 0)} \alpha_{\mathbf{h}}\beta_{\mathbf{h}} e^{i2\pi\mathbf{h}\cdot\mathbf{r}_i} \int \frac{e^{i2\pi\mathbf{h}\cdot\mathbf{r}'}}{r'} d\mathbf{r}'. \tag{A6.8}$$

The integral term in this equation can be straightforwardly calculated using the polar coordinate system with the zenithal angle θ taken from the direction of $\mathbf{h}$. That is,

$$\int \frac{e^{i2\pi\mathbf{h}\cdot\mathbf{r}'}}{r'} d\mathbf{r}' = 2\pi \int_0^{\infty}\int_0^{\pi} \frac{e^{i2\pi h r'\cos\theta}}{r'} r'^2 \sin\theta d\theta dr' = 1/\pi h^2 \quad , \tag{A6.9}$$

in which the relation of $\cos(2\pi h r')|_{r'=\infty}$=0 has been used [2]. The expression for $E_i^{(b)\prime}$ is, therefore, written as

$$E_i^{(b)\prime} = q_i \sum_{\mathbf{h}(\neq 0)} \alpha_{\mathbf{h}}\beta_{\mathbf{h}} e^{i2\pi\mathbf{h}\cdot\mathbf{r}_i} \frac{1}{\pi h^2}$$
$$= \frac{1}{\pi L^3} \sum_{\mathbf{h}(\neq 0)} \sum_{j=1}^{N} q_i q_j \frac{1}{h^2} \exp(-\pi^2 h^2/\kappa^2) \cos(2\pi\mathbf{h}\cdot\mathbf{r}_{ij}) \tag{A6.10}$$
$$= \frac{1}{\pi L^3} \sum_{\mathbf{k}(\neq 0)} \sum_{j=1}^{N} q_i q_j \frac{4\pi^2}{k^2} \exp(-k^2/4\kappa^2) \cos(\mathbf{k}\cdot\mathbf{r}_{ij}),$$

in which $\mathbf{k}$=2$\pi\mathbf{h}$. The interaction energy between the point charge of particle i and its canceling distribution is included in $E_i^{(b)\prime}$ in Eq. (A6.10) and this interaction energy $E_i^{(b)\prime\prime}$ can be written as

$$E_i^{(b)}{}'' = q_i \int q_i \frac{\sigma(\mathbf{r}')}{r'} d\mathbf{r}' = 2\kappa q_i^2 / \pi^{1/2}. \tag{A6.11}$$

Hence, the interaction energy $E_i^{(b)}$, between the point charge q_i and the canceling charge distribution without the contribution from its own canceling distribution, is finally written as

$$\begin{aligned} E_i^{(b)} &= E_i^{(b)}{}' - E_i^{(b)}{}'' \\ &= \frac{1}{\pi L^3} \sum_{\mathbf{k}(\neq 0)} \sum_{j=1}^{N} q_i q_j \frac{4\pi^2}{k^2} \exp(-k^2/4\kappa^2) \cos(\mathbf{k}\cdot\mathbf{r}_{ij}) - \frac{2\kappa q_i^2}{\pi^{1/2}}. \end{aligned} \tag{A6.12}$$

Next, we consider the interaction energy $E_i^{(a)}$ between the point charge q_i and the other point charges screened by the screening distribution as shown in Fig. 11.9(a). This can be written as

$$\begin{aligned} E_i^{(a)} &= q_i \sum_{j=1}^{N} \sum_{\mathbf{n}}{}' q_j \left\{ \frac{1}{|\mathbf{r}_j + L\mathbf{n} - \mathbf{r}_i|} - \int \frac{\sigma(\mathbf{r} - \mathbf{r}_j - L\mathbf{n})}{|\mathbf{r} - \mathbf{r}_i|} d\mathbf{r} \right\} \\ &= \sum_{j=1}^{N} \sum_{\mathbf{n}}{}' q_i q_j \left\{ \frac{1}{|\mathbf{r}_{ji} + L\mathbf{n}|} - \int \frac{\sigma(\mathbf{r}')}{|\mathbf{r}_{ji} + L\mathbf{n} + \mathbf{r}'|} d\mathbf{r}' \right\}, \end{aligned} \tag{A6.13}$$

in which the superscript prime means that the case of $j=i$ is excluded for $\mathbf{n}=0$. By means of the polar coordinate system with the z'-axis along the vector $-(\mathbf{r}_{ji}+L\mathbf{n})$ and the zenithal angle from the z'-axis, the integral term on the right-hand side in Eq. (A6.13) can be reformed as

$$\begin{aligned} \int \frac{\sigma(\mathbf{r}')}{|\mathbf{r}_{ji} + L\mathbf{n} + \mathbf{r}'|} d\mathbf{r}' &= 2\pi \int_0^{\infty} \int_0^{\pi} \frac{\sigma(\mathbf{r}')}{\sqrt{(r'\sin\theta)^2 + \left(|\mathbf{r}_{ji} + L\mathbf{n}| - r'\cos\theta\right)^2}} r'^2 \sin\theta d\theta dr' \\ &= 2\pi \frac{\kappa^3}{\pi^{3/2}} \int_0^{\infty} r'^2 \exp(-\kappa^2 r'^2) dr' \int_0^{\pi} \frac{\sin\theta}{\sqrt{r'^2 - 2|\mathbf{r}_{ji} + L\mathbf{n}| r' \cos\theta + |\mathbf{r}_{ji} + L\mathbf{n}|^2}} d\theta, \end{aligned} \tag{A6.14}$$

in which the integral term with respect to θ is calculated as

$$\begin{aligned} \int_0^{\pi} \frac{\sin\theta}{\sqrt{r'^2 - 2lr'\cos\theta + l^2}} d\theta &= \int_{-1}^{1} \frac{1}{\sqrt{r'^2 + l^2 - 2lr't}} dt \\ &= \frac{1}{lr'} \left\{ \sqrt{(r'+l)^2} - \sqrt{(r'-l)^2} \right\} \\ &= \begin{cases} 2/l & (\text{for } r' \leq l), \\ 2/r' & (\text{for } r' > l). \end{cases} \end{aligned} \tag{A6.15}$$

For simplicity, the notation l has been used for $|\mathbf{r}_{ji}+L\mathbf{n}|$ in this equation. Substitution of this equation into Eq. (A6.14) leads to the following equation:

$$2\pi\frac{\kappa^3}{\pi^{3/2}}\left\{\frac{2}{l}\int_0^l r'^2\exp(-\kappa^2 r'^2)dr' + 2\int_l^\infty r'\exp(-\kappa^2 r'^2)dr'\right\}$$

$$= 2\pi\frac{\kappa^3}{\pi^{3/2}}\left\{\left(-\frac{1}{\kappa^2}e^{-\kappa^2 l^2} + \frac{1}{l\kappa^2}\int_0^l e^{-\kappa^2 r'^2}dr'\right) + \left(\frac{1}{\kappa^2}e^{-\kappa^2 l^2}\right)\right\} \qquad \text{(A6.16)}$$

$$= \mathrm{erf}(\kappa l)/l = \mathrm{erf}\left(\kappa\left|\mathbf{r}_{ji} + L\mathbf{n}\right|\right)/\left|\mathbf{r}_{ji} + L\mathbf{n}\right|,$$

in which the error function erf(x) is defined as erf(x)=1-erfc(x), and erfc(x) is the complementary error function expressed in Eq. (11.22). With these expressions, Eq. (A6.13) reduces to the following equation:

$$E_i^{(a)} = \sum_{j=1}^{N}\sum_{\mathbf{n}}{}' q_i q_j \left\{\frac{1}{\left|\mathbf{r}_{ji} + L\mathbf{n}\right|} - \frac{\mathrm{erf}\left(\kappa\left|\mathbf{r}_{ji} + L\mathbf{n}\right|\right)}{\left|\mathbf{r}_{ji} + L\mathbf{n}\right|}\right\}$$

$$= \sum_{j=1}^{N}\sum_{\mathbf{n}}{}' q_i q_j \frac{\mathrm{erfc}\left(\kappa\left|\mathbf{r}_{ji} + L\mathbf{n}\right|\right)}{\left|\mathbf{r}_{ji} + L\mathbf{n}\right|}. \qquad \text{(A6.17)}$$

Finally, the total interaction energy E_i between the point charge of particle i and the other point charges including virtual particles in the replicated area can be obtained as $E_i = E_i^{(a)} + E_i^{(b)}$ with Eqs. (A6.17) and (A6.12) for $E_i^{(a)}$ and $E_i^{(b)}$. Then, Eq. (11.21) for E can be obtained by the following equation:

$$E = \frac{1}{2}\sum_{i=1}^{N} E_i \,. \qquad \text{(A6.18)}$$

A6.2 Interactions between Electric Dipoles

An electric dipole (similarly, a magnetic dipole) can be regarded as a pair of positive and negative charges with the same magnitude and with an infinitesimally small separation. Hence, a system composed of N electric dipoles can be obtained by considering a system composed of N positive and N negative point charges and by taking the limit of the separations between each pair of positive and negative charges to zero. Now we define a positive point charge Q_i and its partner negative charge Q_{i+N} in the following:

$$\left.\begin{aligned} Q_i &= \frac{|\boldsymbol{\mu}_i|}{2\varepsilon} \quad \text{at } \mathbf{R}_i = \mathbf{r}_i + \varepsilon\hat{\boldsymbol{\mu}}_i \qquad (i = 1,2,\cdots,N), \\ Q_{i+N} &= -\frac{|\boldsymbol{\mu}_i|}{2\varepsilon} \quad \text{at } \mathbf{R}_{i+N} = \mathbf{r}_i - \varepsilon\hat{\boldsymbol{\mu}}_i \qquad (i = 1,2,\cdots,N), \end{aligned}\right\} \qquad \text{(A6.19)}$$

in which $\mathbf{R}_i$ and $\mathbf{R}_{i+N}$ are the position vectors of the positive and negative point charges which make an electric dipole in the limit of $\varepsilon \to 0$, and 2ε is the separation between a pair of point charges. Also $\hat{\boldsymbol{\mu}}_i$ is the unit vector denoting the direction of the electric dipole moment $\boldsymbol{\mu}_i$.

The dipole moment $\boldsymbol{\mu}_i$ is then defined by

$$\boldsymbol{\mu}_i = \lim_{\varepsilon\to 0} 2\varepsilon Q_i \hat{\boldsymbol{\mu}}_i \qquad (i = 1,2,\cdots,N). \tag{A6.20}$$

With the definition of point charges in Eq. (A6.19), the interaction energies $\widetilde{E}$ of point charges in the central cell with the other charges in the whole area including the replicated cells can be expressed, using the Ewald sum method in Eq. (11.21), as

$$\begin{aligned}\widetilde{E} = \frac{1}{2}\sum_{i=1}^{2N}\sum_{j=1}^{2N}\Bigg\{&\sum_{\mathbf{n}}{}'' Q_i Q_j \frac{\mathrm{erfc}\left(\kappa\left|\mathbf{R}_{ji}+L\mathbf{n}\right|\right)}{\left|\mathbf{R}_{ji}+L\mathbf{n}\right|} \\ &+\frac{1}{\pi L^3}\sum_{\mathbf{k}(\neq 0)} Q_i Q_j \frac{4\pi^2}{k^2}\exp(-k^2/4\kappa^2)\cos(\mathbf{k}\cdot\mathbf{R}_{ji})\Bigg\} \\ &-\frac{\kappa}{\pi^{1/2}}\sum_{i=1}^{2N} Q_i^2 - \sum_{i=1}^{N} Q_i \int \frac{Q_{i+N}\sigma(\mathbf{r})}{\left|\mathbf{R}_{i+N}+\mathbf{r}-\mathbf{R}_i\right|}d\mathbf{r},\end{aligned} \tag{A6.21}$$

in which the superscript double primes attached to the notation Σ means that the cases of $j=i$ and $j=i\pm N$ are excluded for $\mathbf{n}=0$. The second term on the right-handle side in Eq. (A6.21) includes the interactions of the point charge i with the canceling distribution of the partner charge $i+N$, so the fourth term is a correction term for excluding these unnecessary interactions. In the following, we reform Eq. (A6.21) such that the expression in the limit of $\varepsilon\to 0$ can be straightforwardly obtained.

For simplicity of mathematical manipulation, we introduce the following notation α_{ij}:

$$\alpha_{ij} = \mathrm{erfc}\left(\kappa\left|\mathbf{R}_{ji}+L\mathbf{n}\right|\right)\Big/\left|\mathbf{R}_{ji}+L\mathbf{n}\right|. \tag{A6.22}$$

With this notation, the first term in Eq. (A6.21), denoted by $E^{(1)}$, is written as

$$\begin{aligned}E^{(1)} &= \frac{1}{2}\sum_{i=1}^{N}\sum_{j=1}^{N}\sum_{\mathbf{n}}{}'\left\{Q_iQ_j\alpha_{ij} + Q_iQ_{j+N}\alpha_{i,j+N} + Q_{i+N}Q_j\alpha_{i+N,j} + Q_{i+N}Q_{j+N}\alpha_{i+N,j+N}\right\} \\ &= \frac{1}{2}\sum_{i=1}^{N}\sum_{j=1}^{N}\sum_{\mathbf{n}}{}' Q_iQ_j\left(\alpha_{ij} - \alpha_{i,j+N} - \alpha_{i+N,j} + \alpha_{i+N,j+N}\right).\end{aligned} \tag{A6.23}$$

Using Eq. (A6.19), the following approximate expression can be obtained:

$$\begin{aligned}1\Big/\left|\mathbf{R}_{ji}+L\mathbf{n}\right| &= 1\Big/\left|\mathbf{r}_{ji}+\varepsilon(\hat{\boldsymbol{\mu}}_j-\hat{\boldsymbol{\mu}}_i)+L\mathbf{n}\right| \\ &= 1\Big/\sqrt{a^2+2\varepsilon\,\mathbf{a}\cdot(\hat{\boldsymbol{\mu}}_j-\hat{\boldsymbol{\mu}}_i)+\varepsilon^2(\hat{\boldsymbol{\mu}}_j-\hat{\boldsymbol{\mu}}_i)^2} \\ &= \frac{1}{a}\left\{1+2\varepsilon\,\mathbf{a}\cdot(\hat{\boldsymbol{\mu}}_j-\hat{\boldsymbol{\mu}}_i)/a^2+\varepsilon^2(\hat{\boldsymbol{\mu}}_j-\hat{\boldsymbol{\mu}}_i)^2/a^2\right\}^{-1/2} \\ &= \frac{1}{a}\left\{1-\varepsilon\,\mathbf{a}\cdot(\hat{\boldsymbol{\mu}}_j-\hat{\boldsymbol{\mu}}_i)/a^2-\varepsilon^2(\hat{\boldsymbol{\mu}}_j-\hat{\boldsymbol{\mu}}_i)^2/2a^2+3\varepsilon^2\left\{\mathbf{a}\cdot(\hat{\boldsymbol{\mu}}_j-\hat{\boldsymbol{\mu}}_i)\right\}^2/2a^4\right\},\end{aligned} \tag{A6.24}$$

in which $\mathbf{a}$ is $(\mathbf{r}_{ji}+L\mathbf{n})$ with $a=|\mathbf{a}|$, and terms higher-order than ε^2 have been neglected: similar neglect of the higher-order terms will be conducted in the following procedures. The complementary error function can be written in an approximate expression:

$$\begin{aligned}&\mathrm{erfc}\left(\kappa\left|\mathbf{R}_{ji}+L\mathbf{n}\right|\right)=\mathrm{erfc}\left(\kappa\left|\mathbf{a}+\varepsilon(\hat{\boldsymbol{\mu}}_j-\hat{\boldsymbol{\mu}}_i)\right|\right)\\&=\mathrm{erfc}\left[\kappa a\left\{1+\varepsilon\ \mathbf{a}\cdot(\hat{\boldsymbol{\mu}}_j-\hat{\boldsymbol{\mu}}_i)/a^2+\varepsilon^2(\hat{\boldsymbol{\mu}}_j-\hat{\boldsymbol{\mu}}_i)^2/2a^2-\varepsilon^2\left\{\mathbf{a}\cdot(\hat{\boldsymbol{\mu}}_j-\hat{\boldsymbol{\mu}}_i)\right\}^2/2a^4\right\}\right].\end{aligned} \tag{A6.25}$$

From mathematical formulae, the error function erf(x) can be expanded in the following series:

$$\mathrm{erf}(x)=\frac{2}{\pi^{1/2}}\sum_{\mathrm{p=0}}^{\infty}\frac{(-1)^p x^{2p+1}}{p!(2p+1)}. \tag{A6.26}$$

With this equation, erf($x+\delta x$) can be approximated as

$$\begin{aligned}\mathrm{erf}(x+\delta x)&=\frac{2}{\pi^{1/2}}\sum_{\mathrm{p=0}}^{\infty}\frac{(-1)^p(x+\delta x)^{2p+1}}{p!(2p+1)}\\&=\frac{2}{\pi^{1/2}}\sum_{\mathrm{p=0}}^{\infty}\frac{(-1)^p}{p!(2p+1)}x^{2p+1}\left\{1+(2p+1)\frac{\delta x}{x}+\frac{(2p+1)2p}{2!}\left(\frac{\delta x}{x}\right)^2\right\}\\&=\mathrm{erf}(x)+\frac{2}{\pi^{1/2}}e^{-x^2}\delta x-\frac{2}{\pi^{1/2}}xe^{-x^2}(\delta x)^2.\end{aligned} \tag{A6.27}$$

Hence the consideration of Eqs. (A6.25) to (A6.27) leads to the following approximate expression for erfc($\kappa\,|\,\mathbf{R}_{ji}+L\mathbf{n}\,|$):

$$\begin{aligned}\mathrm{erfc}\left(\kappa\left|\mathbf{R}_{ji}+L\mathbf{n}\right|\right)=\mathrm{erfc}(\kappa a)-\frac{2}{\pi^{1/2}}e^{-\kappa^2a^2}\Bigg[&\kappa\varepsilon\frac{\mathbf{a}\cdot(\hat{\boldsymbol{\mu}}_j-\hat{\boldsymbol{\mu}}_i)}{a}+\frac{\kappa\varepsilon^2}{2}\frac{(\hat{\boldsymbol{\mu}}_j-\hat{\boldsymbol{\mu}}_i)^2}{a}\\&-\frac{\kappa\varepsilon^2}{2}\frac{\left\{\mathbf{a}\cdot(\hat{\boldsymbol{\mu}}_j-\hat{\boldsymbol{\mu}}_i)\right\}^2}{a^3}\Bigg]+\frac{2\kappa^3\varepsilon^2}{\pi^{1/2}}e^{-\kappa^2a^2}\frac{\left\{\mathbf{a}\cdot(\hat{\boldsymbol{\mu}}_j-\hat{\boldsymbol{\mu}}_i)\right\}^2}{a}.\end{aligned} \tag{A6.28}$$

The expression for α_{ij} can, therefore, be obtained using Eqs. (A6.24) and (A6.28) as

$$\begin{aligned}\alpha_{ij}=\frac{\mathrm{erfc}\left(\kappa\left|\mathbf{R}_{ji}+L\mathbf{n}\right|\right)}{\left|\mathbf{R}_{ji}+L\mathbf{n}\right|}&=\frac{1}{a}\mathrm{erfc}(\kappa a)-\varepsilon\frac{2\kappa}{\pi^{1/2}}e^{-\kappa^2a^2}\frac{\mathbf{a}\cdot(\hat{\boldsymbol{\mu}}_j-\hat{\boldsymbol{\mu}}_i)}{a^2}\\&-\varepsilon\ \mathrm{erfc}(\kappa a)\frac{\mathbf{a}\cdot(\hat{\boldsymbol{\mu}}_j-\hat{\boldsymbol{\mu}}_i)}{a^3}-\varepsilon^2\frac{\kappa}{\pi^{1/2}}e^{-\kappa^2a^2}\frac{(\hat{\boldsymbol{\mu}}_j-\hat{\boldsymbol{\mu}}_i)^2}{a^2}\\&-\varepsilon^2\frac{1}{2}\mathrm{erfc}(\kappa a)\frac{(\hat{\boldsymbol{\mu}}_j-\hat{\boldsymbol{\mu}}_i)^2}{a^3}+\varepsilon^2\frac{3\kappa}{\pi^{1/2}}e^{-\kappa^2a^2}\frac{\left\{\mathbf{a}\cdot(\hat{\boldsymbol{\mu}}_j-\hat{\boldsymbol{\mu}}_i)\right\}^2}{a^4}\\&+\varepsilon^2\frac{2\kappa^3}{\pi^{1/2}}e^{-\kappa^2a^2}\frac{\left\{\mathbf{a}\cdot(\hat{\boldsymbol{\mu}}_j-\hat{\boldsymbol{\mu}}_i)\right\}^2}{a^2}+\varepsilon^2\frac{3}{2}\mathrm{erfc}(\kappa a)\frac{\left\{\mathbf{a}\cdot(\hat{\boldsymbol{\mu}}_j-\hat{\boldsymbol{\mu}}_i)\right\}^2}{a^5}.\end{aligned} \tag{A6.29}$$

If ($\hat{\boldsymbol{\mu}}_j,\hat{\boldsymbol{\mu}}_i$) is replaced with ($-\hat{\boldsymbol{\mu}}_j,\hat{\boldsymbol{\mu}}_i$), ($\hat{\boldsymbol{\mu}}_j,-\hat{\boldsymbol{\mu}}_i$), or ($-\hat{\boldsymbol{\mu}}_j,-\hat{\boldsymbol{\mu}}_i$) in Eq. (A6.29), then the expression for $\alpha_{i,j+N}$, $\alpha_{i+N,j}$, or $\alpha_{i+N,j+N}$ is obtained, respectively. If these equations are substituted into Eq. (A6.23), the following final expression for $E^{(1)}$ can be obtained:

$$E^{(1)} = \frac{1}{2}\sum_{i=1}^{N}\sum_{j=1}^{N}\sum_{\mathbf{n}}{}' 4\varepsilon^2 Q_i Q_j \Big[A\big(|\mathbf{r}_{ji}+L\mathbf{n}|\big)(\hat{\boldsymbol{\mu}}_i\cdot\hat{\boldsymbol{\mu}}_j) \\ - B\big(|\mathbf{r}_{ji}+L\mathbf{n}|\big)\{\hat{\boldsymbol{\mu}}_i\cdot(\mathbf{r}_{ji}+L\mathbf{n})\}\{\hat{\boldsymbol{\mu}}_j\cdot(\mathbf{r}_{ji}+L\mathbf{n})\}\Big], \qquad \text{(A6.30)}$$

in which

$$\left.\begin{aligned} A(r) &= \mathrm{erfc}(\kappa r)/r^3 + (2\kappa/\pi^{1/2})\exp(-\kappa^2 r^2)/r^2, \\ B(r) &= 3\mathrm{erfc}(\kappa r)/r^5 + (2\kappa/\pi^{1/2})(2\kappa^2+3/r^2)\exp(-\kappa^2 r^2)/r^2. \end{aligned}\right\} \qquad \text{(A6.31)}$$

Now we derive the approximate expression of the second term in Eq. (A6.21), denoted by $E^{(2)}$. If we introduce the following notation:

$$\beta_{ij} = \cos(\mathbf{k}\cdot\mathbf{R}_{ji}), \qquad \text{(A6.32)}$$

$E^{(2)}$ can be rewritten as

$$E^{(2)} = \frac{1}{2}\sum_{i=1}^{N}\sum_{j=1}^{N}\frac{1}{\pi L^3}\sum_{\mathbf{k}(\neq 0)}\frac{4\pi^2}{k^2}\exp(-k^2/4\kappa^2)\big(Q_iQ_j\beta_{ij} \\ + Q_iQ_{j+N}\beta_{i,j+N} + Q_{i+N}Q_j\beta_{i+N,j} + Q_{i+N}Q_{j+N}\beta_{i+N,j+N}\big) \\ = \frac{1}{2}\sum_{i=1}^{N}\sum_{j=1}^{N}\frac{1}{\pi L^3}\sum_{\mathbf{k}(\neq 0)}\frac{4\pi^2}{k^2}\exp(-k^2/4\kappa^2)Q_iQ_j\big(\beta_{ij}-\beta_{i,j+N}-\beta_{i+N,j}+\beta_{i+N,j+N}\big). \qquad \text{(A6.33)}$$

With Eq. (A6.19), Eq. (A6.32) can be reformed as

$$\beta_{ij} = \cos\{\mathbf{k}\cdot(\mathbf{r}_j+\varepsilon\,\hat{\boldsymbol{\mu}}_j-\mathbf{r}_i-\varepsilon\,\hat{\boldsymbol{\mu}}_i)\} = \cos\{\mathbf{k}\cdot\mathbf{r}_{ji}+\varepsilon\,\mathbf{k}\cdot(\hat{\boldsymbol{\mu}}_j-\hat{\boldsymbol{\mu}}_i)\} \\ = \cos(\mathbf{k}\cdot\mathbf{r}_{ji})\left[1-\frac{\varepsilon^2}{2}\{\mathbf{k}\cdot(\hat{\boldsymbol{\mu}}_j-\hat{\boldsymbol{\mu}}_i)\}^2\right] - \sin(\mathbf{k}\cdot\mathbf{r}_{ji})\{\varepsilon\,\mathbf{k}\cdot(\hat{\boldsymbol{\mu}}_j-\hat{\boldsymbol{\mu}}_i)\}. \qquad \text{(A6.34)}$$

If $(\hat{\boldsymbol{\mu}}_j, \hat{\boldsymbol{\mu}}_i)$ is replaced with $(-\hat{\boldsymbol{\mu}}_j, \hat{\boldsymbol{\mu}}_i)$, $(\hat{\boldsymbol{\mu}}_j, -\hat{\boldsymbol{\mu}}_i)$, or $(-\hat{\boldsymbol{\mu}}_j, -\hat{\boldsymbol{\mu}}_i)$, then the expression for $\beta_{i,j+N}$, $\beta_{i+N,j}$, or $\beta_{i+N,j+N}$ is obtained, respectively. Hence, by substituting these equations into Eq. (A6.33), we can obtain the following final form for $E^{(2)}$:

$$E^{(2)} = \frac{1}{2}\sum_{i=1}^{N}\sum_{j=1}^{N}\frac{1}{\pi L^3}\sum_{\mathbf{k}(\neq 0)}\frac{4\pi^2}{k^2}\exp(-k^2/4\kappa^2)4\varepsilon^2 Q_iQ_j(\mathbf{k}\cdot\hat{\boldsymbol{\mu}}_i)(\mathbf{k}\cdot\hat{\boldsymbol{\mu}}_j)\cos(\mathbf{k}\cdot\mathbf{r}_{ji}). \qquad \text{(A6.35)}$$

The third term in Eq. (A6.21), denoted by $E^{(3)}$, can be reformed as

$$E^{(3)} = -\frac{\kappa}{\pi^{1/2}}\sum_{i=1}^{N}(Q_i^2+Q_{i+N}^2) = -\frac{2\kappa}{\pi^{1/2}}\sum_{i=1}^{N}Q_i^2. \qquad \text{(A6.36)}$$

Finally, we derive the expression for the fourth term in Eq. (A6.21), denoted by $E^{(4)}$. With Eqs. (A6.19) and (11.20), this term can be written as

$$E^{(4)} = -\sum_{i=1}^{N} Q_i \int \frac{-Q_i\sigma(\mathbf{r})}{\left|\mathbf{r}_i - \varepsilon\hat{\boldsymbol{\mu}}_i + \mathbf{r} - \mathbf{r}_i - \varepsilon\hat{\boldsymbol{\mu}}_i\right|} d\mathbf{r} = \frac{\kappa^3}{\pi^{3/2}} \sum_{i=1}^{N} Q_i^2 \int \frac{e^{-\kappa^2 r^2}}{\left|\mathbf{r} - 2\varepsilon\hat{\boldsymbol{\mu}}_i\right|} d\mathbf{r} . \tag{A6.37}$$

The following approximation, similar to Eq. (A6.24), is valid:

$$1/\left|\mathbf{r} - 2\varepsilon\hat{\boldsymbol{\mu}}_i\right| = 1/r + 2\varepsilon(\mathbf{r}\cdot\hat{\boldsymbol{\mu}}_i)/r^3 - 2\varepsilon^2/r^3 + 6\varepsilon^2(\mathbf{r}\cdot\hat{\boldsymbol{\mu}}_i)^2/r^5 . \tag{A6.38}$$

By substituting this equation into Eq. (A6.37) and conducting the integration, the following equation can be obtained:

$$E^{(4)} = \frac{\kappa^3}{\pi^{3/2}} \sum_{i=1}^{N} Q_i^2 \frac{2\pi}{\kappa^2} = \frac{2\kappa}{\pi^{1/2}} \sum_{i=1}^{N} Q_i^2 . \tag{A6.39}$$

It should be noted that the third and fourth terms on the right-hand side in Eq. (A6.38) cancel with each other after the integration, so that these terms make no contribution to $E^{(4)}$. Finally, the expression for $\widetilde{E}$ can be written as

$$\begin{aligned}
\widetilde{E} &= E^{(1)} + E^{(2)} + E^{(3)} + E^{(4)} = E^{(1)} + E^{(2)} \\
&= \frac{1}{2}\sum_{i=1}^{N}\sum_{j=1}^{N}\Bigg[\sum_{\mathbf{n}}{}' 4\varepsilon^2 Q_i Q_j \Big\{A\big(\left|\mathbf{r}_{ji} + L\mathbf{n}\right|\big)(\hat{\boldsymbol{\mu}}_j\cdot\hat{\boldsymbol{\mu}}_i) \\
&\quad - B\big(\left|\mathbf{r}_{ji} + L\mathbf{n}\right|\big)\big((\mathbf{r}_{ji} + L\mathbf{n})\cdot\hat{\boldsymbol{\mu}}_j\big)\big((\mathbf{r}_{ji} + L\mathbf{n})\cdot\hat{\boldsymbol{\mu}}_i\big)\Big\} \\
&\quad + \frac{1}{\pi L^3}\sum_{\mathbf{k}(\neq 0)} \frac{4\pi^2}{k^2}\exp(-k^2/4\kappa^2) 4\varepsilon^2 Q_i Q_j (\mathbf{k}\cdot\hat{\boldsymbol{\mu}}_j)(\mathbf{k}\cdot\hat{\boldsymbol{\mu}}_i)\cos(\mathbf{k}\cdot\mathbf{r}_{ji})\Bigg].
\end{aligned} \tag{A6.40}$$

Hence, in the limit of $\varepsilon\to 0$ with Eq. (A6.20), Eq. (A6.40) reduces to the expression for the interaction energies E of the central cell which is composed of N electric dipoles:

$$\begin{aligned}
E &= \lim_{\varepsilon\to 0}\widetilde{E} = \frac{1}{2}\sum_{i=1}^{N}\sum_{j=1}^{N}\Bigg[\sum_{\mathbf{n}}{}' \Big\{A\big(\left|\mathbf{r}_{ji} + L\mathbf{n}\right|\big)(\boldsymbol{\mu}_j\cdot\boldsymbol{\mu}_i) \\
&\quad - B\big(\left|\mathbf{r}_{ji} + L\mathbf{n}\right|\big)\big((\mathbf{r}_{ji} + L\mathbf{n})\cdot\boldsymbol{\mu}_j\big)\big((\mathbf{r}_{ji} + L\mathbf{n})\cdot\boldsymbol{\mu}_i\big)\Big\} \\
&\quad + \frac{1}{\pi L^3}\sum_{\mathbf{k}(\neq 0)} \frac{4\pi^2}{k^2}\exp(-k^2/4\kappa^2)(\mathbf{k}\cdot\boldsymbol{\mu}_j)(\mathbf{k}\cdot\boldsymbol{\mu}_i)\cos(\mathbf{k}\cdot\mathbf{r}_{ji})\Bigg].
\end{aligned} \tag{A6.41}$$

According to the similarity between electric and magnetic dipoles, the interaction energy for a system of magnetic dipoles can be obtained by replacing the electric dipole moment $\boldsymbol{\mu}_i$ with the magnetic dipole moment $\mathbf{m}_i$ and multiplying it by $(\mu_0/4\pi)$, in which μ_0 is the permeability of free space.

References

1. S.W. de Leeuw, J.W. Perran, and E.R. Smith, Simulation of Electrostatic Systems in Periodic Boundary Conditions. I. Lattice Sums and Dielectric Constants, Proc. Roy. Soc. London A, 373 (1980) 27
2. R.P. Feynman, R.B. Leighton, and M.L. Sands, "The Feynman Lectures on Physics," Vol. 2, Addison Wesley Publishing Company, Reading, 1970

A7. UNIT SYSTEMS USED IN MAGNETIC MATERIALS

The CGS unit system and the SI unit system which was developed from the MKSA unit system are generally used in the field of magnetic materials. Although the CGS unit system is commonly used in the business world, the SI unit system is almost always used in textbooks on magnetic materials. Using quantities expressed in different unit systems at the same time will lead to wrong expressions for physical quantities, so one must adhere to the same unit system for handling equations or physical values of magnetic materials. Many textbooks on magnetic materials have transformation tables from one unit system to another. We here summarize the two unit systems based on the MKSA system. In the first unit system, the magnetization **M** corresponds to the magnetic field **H** in units, and, in the second unit system, **M** corresponds to the magnetic flux density **B**. Some of the typical quantities, used in magnetic materials, are tabulated below.

	$\mathbf{B}=\mu_0(\mathbf{H}+\mathbf{M})$	$\mathbf{B}=\mu_0\mathbf{H}+\mathbf{M}$
Magnetic field strength **H**	[A/m]	[A/m]
Magnetization strength **M**	[A/m]	[Wb/m²]
Magnetic flux density **B**	[T] (=[Wb/m²])	[T] (=[Wb/m²])
Permeability of free space μ_0	$\mu_0=4\pi\times10^{-7}$ [H/m] (=[Wb/(A·m)]	$\mu_0=4\pi\times10^{-7}$ [H/m] (=[Wb/(A·m)]
Magnetic charge q	[A·m]	[Wb] (=[N·m/A])
Magnetic moment **m**	[A·m²]	[Wb·m] (=[N·m²/A])
Potential energy U	$U=-\mu_0\mathbf{m}\cdot\mathbf{H}$ [J] (=[Wb·A])	$U=-\mathbf{m}\cdot\mathbf{H}$ [J] (=[Wb·A])
Torque **T**	$\mathbf{T}=\mu_0\mathbf{m}\times\mathbf{H}$ [N·m] (=[Wb·A])	$\mathbf{T}=\mathbf{m}\times\mathbf{H}$ [N·m] (=[Wb·A])
Magnetic field induced by magnetic charge, **H**	$\mathbf{H}=\frac{q}{4\pi r^2}\cdot\frac{\mathbf{r}}{r}$ [A/m]	$\mathbf{H}=\frac{q}{4\pi\mu_0 r^2}\cdot\frac{\mathbf{r}}{r}$ [A/m]
Magnetic force acting between two magnetic charges, **F**	$\mathbf{F}=\frac{\mu_0 qq'}{4\pi r^2}\cdot\frac{\mathbf{r}}{r}$ [N] (=[Wb·A/m])	$\mathbf{F}=\frac{qq'}{4\pi\mu_0 r^2}\cdot\frac{\mathbf{r}}{r}$ [N] (=[Wb·A/m])
Magnetic interaction between two magnetic moments, U	$U=\frac{\mu_0}{4\pi r^3}\{\mathbf{m}_1\cdot\mathbf{m}_2-\frac{3}{r^2}\times(\mathbf{m}_1\cdot\mathbf{r})(\mathbf{m}_2\cdot\mathbf{r})\}$ [J] (=[Wb·A])	$U=\frac{1}{4\pi\mu_0 r^3}\{\mathbf{m}_1\cdot\mathbf{m}_2-\frac{3}{r^2}\times(\mathbf{m}_1\cdot\mathbf{r})(\mathbf{m}_2\cdot\mathbf{r})\}$ [J] (=[Wb·A])
Combined units: [H]=[Wb/A], [T]=[Wb/m²], [J]=[N·m] Equivalent units: [N]=[Wb·A/m]		

In this book we are using the first unit system of **M** corresponding to **H** in units.

A8. THE VIRIAL EQUATION OF STATE

The derivation process of the virial equation of state, which is an expression for the pressure in a molecular system, is very useful for understanding the expression for the osmotic pressure in colloidal systems. We, therefore, show the derivation of the virial equation of state in detail in this appendix.

The virial equation of state can generally be derived by considering the interactions between particles, and also, between particles and the wall of a system. However, there is no such system wall in molecular simulations, since the periodic boundary condition is used as shown in Sec. 11.2. Hence we derive the virial equation from a different approach [1].

We consider an equilibrium system with a volume V (L the side length) composed of N particles. As shown in Fig. A8.1, the pressure is defined as the force per unit area acting normally on an imaginary plane which is taken perpendicular to the x-axis. In other words, the pressure may be regarded as the force per unit area compressing the surface of a solid system, by the particles on the left-hand side, as if the right-hand side were solid. Hence, in the case of Fig. A8.1, the pressure is defined as positive when it acts in the x-direction. The pressure is due to the momentum transfer and the interactions between particles. First we derive the pressure P_K due to the former contribution and then P_U due to the latter contribution.

We focus our attention on particle i in the left-hand side area and assume that this particle crosses the imaginary plane during an infinitesimal time, collides with a particle in the right-hand side area, and then returns to the left-hand side area. In this situation, the force exerted by particle i on the imaginary plane can be evaluated by deriving the momentum change during the time interval. However, if the system is in equilibrium, we can evaluate this force statistically without considering the detailed particle motion. That is, the momenta of the incoming and outgoing motion of particles across the imaginary plane can be treated separately to evaluate the force exerted on the imaginary plane.

As shown in Fig. A8.1, we consider a particle crossing the infinitesimal small area dS with velocity $\mathbf{v}=(v_x,v_y,v_z)$ during the time interval dt. The probability that the particle has such a velocity is given by the Maxwellian distribution in Eq. (2.60) as $f(\mathbf{v})d\mathbf{v}$. The number of particles having such a velocity and crossing the element of the area dS is, on average, the ratio of the value $v_x dtdS$ of the inclined cylinder in Fig. A8.1 to the volume V/N occupied by one particle. Hence, the x-component p_{in}^x of the incoming momentum, normal to dS, carried by the particles which cross dS with the velocity $\mathbf{v}$ from the left side per unit area in unit time, can be expressed as

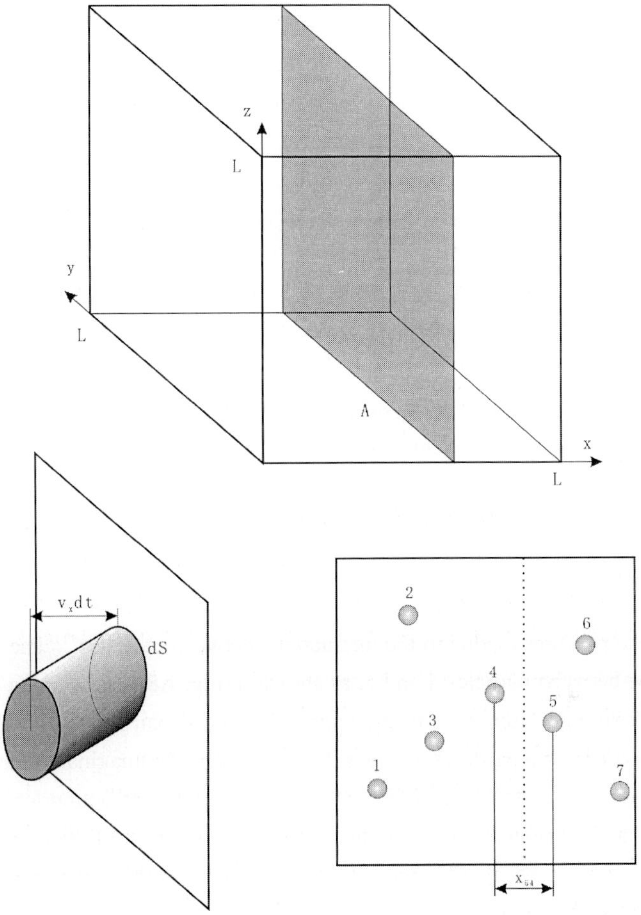

Figure A8.1 Imaginary plane inserted normal to the *x*-axis for deriving the virial equation of state.

$$p_{in}^{x} = \left\{ \frac{v_x dtdS}{V/N} \cdot f(\mathbf{v})d\mathbf{v} \cdot mv_x \right\} \Big/ (dtdS) = m\frac{N}{V} v_x{}^2 f(\mathbf{v})d\mathbf{v} . \tag{A8.1}$$

The integration of this equation with respect to the velocity components leads to the following equation:

$$\sigma_{in}^{x} = \int_{-\infty}^{\infty}\int_{-\infty}^{\infty}\int_{0}^{\infty} p_{in}^{x} dv_x dv_y dv_z = m\frac{N}{V} \int_{-\infty}^{\infty}\int_{-\infty}^{\infty}\int_{0}^{\infty} v_x{}^2 f(\mathbf{v})d\mathbf{v} , \tag{A8.2}$$

in which σ_{in}^{x} is the x-component of the incoming momentum carried by the particles from the left side per unit area in unit time. Similarly, the x-component of the outgoing momentum from the right-hand side per unit area in unit time, σ_{out}^{x}, is expressed as

$$\sigma_{out}^{x} = m\frac{N}{V}\int_{-\infty}^{\infty}\int_{-\infty}^{\infty}\int_{-\infty}^{0} v_x^{\,2} f(\mathbf{v})d\mathbf{v} . \tag{A8.3}$$

Thus, by taking into account that both σ_{in}^{x} and σ_{out}^{x} are taken as positive, the pressure P_K^x is obtained as

$$P_K^x = \sigma_{in}^x + \sigma_{out}^x = m\frac{N}{V}\int_{-\infty}^{\infty}\int_{-\infty}^{\infty}\int_{-\infty}^{\infty} v_x^{\,2} f(\mathbf{v})d\mathbf{v} = m\frac{N}{V}\left\langle v_x^{\,2}\right\rangle . \tag{A8.4}$$

Since the pressure is generally defined as the average of P_K^x, P_K^y, and P_K^z, the pressure P_K due to the kinetic energy of particles is finally obtained as

$$P_K = \frac{1}{3}(P_K^x + P_K^y + P_K^z) = \frac{2N}{3V}\left\langle \frac{m(v_x^{\,2} + v_y^{\,2} + v_z^{\,2})}{2} \right\rangle = \frac{N}{V}kT . \tag{A8.5}$$

The last expression has been obtained by conducting the averaging procedure with Eq. (2.60).

Next we consider the pressure P_U due to the interactions between particles. The imaginary plane is assumed to be between particles 4 and 5, as shown in Fig. A8.1, in which a particle at a lefter position has a younger number. In the approximation of pair interactions, a pair of particles alone, such as these two particles on the different sides of the imaginary plane, make a contribution to the pressure. In the case of Fig. A8.1, therefore, the interaction force between particles 1 and 4 does not contribute to the pressure. If the x-component of the force acting on particle i by particle j is denoted by f_{ij}^x, then the x-component of the force exerted on the imaginary plane in Fig. A8.1 is written as

$$-\sum_{i=1}^{4}\sum_{j=5}^{N} f_{ij}^x . \tag{A8.6}$$

This force is constant even if the imaginary plane is at any position between x_4 and x_5. By setting the imaginary plane at all positions between the left and right walls to evaluate the above-mentioned force at each position, and by averaging such forces, the x-component τ^x of the force acting on the imaginary plane per unit area can be written as

$$\tau^x = -\left\{\left(\frac{x_{21}}{L}\sum_{j=2}^{N} f_{1j}^x\right) + \left(\frac{x_{32}}{L}\sum_{i=1}^{2}\sum_{j=3}^{N} f_{ij}^x\right) + \cdots + \left(\frac{x_{N,N-1}}{L}\sum_{i=1}^{N-1} f_{i,N}^x\right)\right\}\Big/ L^2 , \tag{A8.7}$$

in which $x_{ij}=x_i-x_j$. The reformation of the right-hand side in Eq. (A8.7) leads to

$$\begin{aligned}
\tau^x V &= -\big[x_{21}(f_{12}^x + f_{13}^x + \cdots + f_{1N}^x) + x_{32}(f_{13}^x + f_{14}^x + \cdots + f_{1N}^x + f_{23}^x + f_{24}^x + \cdots + f_{2N}^x) \\
&\qquad + \cdots + x_{N,N-1}(f_{1N}^x + f_{2N}^x + \cdots + f_{N-1,N}^x)\big] \\
&= \left\{x_{12} f_{12}^x + (x_{12} + x_{23}) f_{13}^x + \cdots + (x_{12} + x_{23} + \cdots + x_{N-1,N}) f_{1N}^x\right\} \\
&\qquad + \left\{x_{23} f_{23}^x + (x_{23} + x_{34}) f_{24}^x + \cdots + (x_{23} + x_{34} + \cdots + x_{N-1,N}) f_{2N}^x\right\} + \cdots \\
&= (x_{12} f_{12}^x + x_{13} f_{13}^x + \cdots + x_{1N} f_{1N}^x) + (x_{23} f_{23}^x + x_{24} f_{24}^x + \cdots + x_{2N} f_{2N}^x) + \cdots \\
&= \sum_{i=1}^{N-1} \sum_{j=i+1}^{N} x_{ij} f_{ij}^x = \sum_{i=1}^{N} \sum_{\substack{j=1 \\ (i<j)}}^{N} x_{ij} f_{ij}^x .
\end{aligned} \tag{A8.8}$$

Thus the arithmetic average of τ^x, τ^y, and τ^z gives the expression for P_U:

$$P_U = \frac{1}{3}\left\langle \tau^x + \tau^y + \tau^z \right\rangle = \frac{1}{3V}\left\langle \sum_i \sum_{\substack{j \\ (i<j)}} (x_{ij} f_{ij}^x + y_{ij} f_{ij}^y + z_{ij} f_{ij}^z) \right\rangle = \frac{1}{3V}\left\langle \sum_i \sum_{\substack{j \\ (i<j)}} \mathbf{r}_{ij} \cdot \mathbf{f}_{ij} \right\rangle . \tag{A8.9}$$

After all, the expression for the pressure P can finally be obtained as the sum of Eqs. (A8.5) and (A8.9):

$$P = \frac{N}{V} kT + \frac{1}{3V}\left\langle \sum_i \sum_{\substack{j \\ (i<j)}} \mathbf{r}_{ij} \cdot \mathbf{f}_{ij} \right\rangle . \tag{A8.10}$$

This is called the virial equation of state.

If we use the notation W denoted by

$$W = \frac{1}{3} \sum_i \sum_{\substack{j \\ (i<j)}} \mathbf{r}_{ij} \cdot \mathbf{f}_{ij} , \tag{A8.11}$$

then W is called the internal virial. By taking account of the following relation:

$$\begin{aligned}
\sum_i \mathbf{r}_i \cdot \mathbf{f}_i &= \sum_i \sum_{j(\neq i)} \mathbf{r}_i \cdot \mathbf{f}_{ij} = \frac{1}{2} \sum_i \sum_{j(\neq i)} (\mathbf{r}_i \cdot \mathbf{f}_{ij} + \mathbf{r}_j \cdot \mathbf{f}_{ji}) \\
&= \frac{1}{2} \sum_i \sum_{j(\neq i)} (\mathbf{r}_i \cdot \mathbf{f}_{ij} - \mathbf{r}_j \cdot \mathbf{f}_{ij}) = \frac{1}{2} \sum_i \sum_{j(\neq i)} \mathbf{r}_{ij} \cdot \mathbf{f}_{ij} = \sum_i \sum_{\substack{j \\ (i<j)}} \mathbf{r}_{ij} \cdot \mathbf{f}_{ij} ,
\end{aligned} \tag{A8.12}$$

W can be rewritten as

$$W = \frac{1}{3} \sum_i \mathbf{r}_i \cdot \mathbf{f}_i . \tag{A8.13}$$

References

1. J.M. Haile, "Molecular Dynamics Simulation: Elementary Methods," 332-339, John Wiley & Sons, New York, 1992

A9. RANDOM NUMBERS

A9.1 Uniform Random Numbers

In molecular simulations such as Monte Carlo and molecular dynamics methods, a random number sequence distributed uniformly from zero to unity is used for setting the initial velocities of molecules and dealing with the collision dynamics between molecules and material surfaces. Also, in Brownian dynamics methods, a uniform random number sequence is indispensable in generating random displacements of colloidal particles. In most cases, the random number generator with which a computer is equipped is generally used, but arithmetic methods for generating random numbers are still useful even today since random numbers generated by such methods are reproducible. We outline a representative arithmetic method of generating uniform random sequences below [1].

A uniform random number sequence is defined as a sequence of random numbers which have perfectly no correlation with each other. An arithmetic method generates a uniform random number sequence so that random numbers generated by the method satisfy this characteristic as much as possible. Random numbers generated by an arithmetic method are formally called pseudo-random numbers, but simply called just random numbers in most cases. As pointed out, the generation by an arithmetic method is reproducible, so we can generate perfectly the same random number sequence at any required time. We here explain a multiplicative congruential method, which is a representative method for generating random numbers arithmetically. However, the detailed discussion concerning the uniformity and randomness of the random number sequences is beyond the present objective, and therefore should be found elsewhere [1].

The multiplicative congruential method uses the following arithmetic equation to generate a sequence of random numbers which are uniformly distributed in the range (0,1):

$$x_n = \lambda x_{n-1} (\text{mod} \quad P), \tag{A9.1}$$

in which x_n is the remainder when λx_{n-1} is divided by P, and λ and P are both large positive integers. With an initial integer value x_0, Eq. (A9.1) generates a random number sequence $(x_1, x_2, \ldots, x_n, \ldots)$. Since x_n lies in the range $(0, P\text{-}1)$, the values x_n/P are used as a pseudo-random number sequence distributed uniformly between 0 and 1.

Next we make some statements concerning appropriate values of λ and P, which give good characteristics of the uniformity, randomness, and periodicity of the random numbers. It is quite clear that such a random number sequence repeats itself after at most P steps and is therefore periodic. Hence a large value of P should be chosen to maximize the period of the sequence. For example, if P is taken as 2^m (m is a certain integer), then λ is chosen so that λ(mod 8) =5 or 3. The values of x_0 and P should be taken in such a way that x_0 is prime to P, that is, x_0 and P have no common divisonar without unity. With an appropriate choice of λ, x_0, and P satisfying the above-mentioned characteristics, we can obtain a random number sequence with good statistical properties and a long period.

Appropriate sets of the values of λ and P may be found in a textbook [1], so we here consider one concrete example to explain how to generate actually a uniform random number sequence on a computer; the FORTRAN program for the example, with $\lambda=5^{11}$, $P=2^{31}$ and $x_0=584287$, is shown in Appendix A11.2. There is ordinarily a maximum integer number which can be expressed on a computer. For example, many 32-bit word computers have adopted the expression of two's complement, in which 32 bits are used for expressing integers ranging from (-2^{31}) to (2^{31}-1); one bit is reserved for plus or minus sign. Hence, in primitive methods, values of P and λ are chosen so that λx_{n-1} in Eq. (A9.1) have to be smaller than the largest integer (2^{31}-1) on a computer. Such restriction can, however, be removed by taking account of the internal treatments on a computer when a calculated value becomes more than the maximum integer. That is, when a computer has adopted the expression of two's complement, it answers (-2^{31}) for the arithmetic operation ((2^{31}-1)+1). Similarly, it answers (-2^{31}+1) and (-2^{31}+2) for the arithmetic operations ((2^{31}-1)+2) and ((2^{31}-1)+3), respectively. Furthermore, we consider the case where a positive integer a is multiplied by another positive integer b. If $a\times b$ is over (2^{31}-1), the value c which a computer gives as an answer is different from a true value. In this case c has the following relation with its theoretical value ab(mod P):

$$ab(\text{mod } P)=\begin{cases} c & (\text{for } c\geq 0), \\ c+P & (\text{for } c<0), \end{cases} \tag{A9.2}$$

in which $P=2^{31}$ in this case. The FORTRAN program shown in Appendix A11.2 is based on this method.

A9.2 Non-Uniform Random Numbers

Besides uniform random number sequences, a non-uniform random number sequence which obeys a certain probability density is important in molecular-microsimulations. For example, when initial velocities of particles are assigned according to the Maxwellian distribution, a non-uniform random number sequence is indispensable. Such a non-uniform sequence can be generated using the above-mentioned uniform random number sequences. We here explain the two representative methods for generating non-uniform random number sequences, that is, the inverse transformation method and the rejection method.

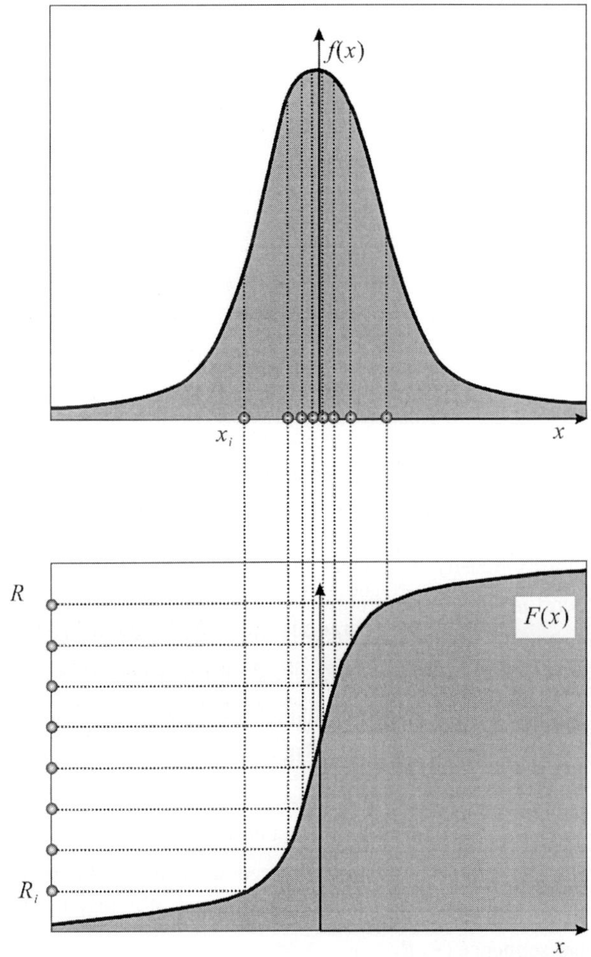

Figure A9.1 Inverse transformation method of generating random variables obeying an arbitrary distribution.

The inverse transformation method is applicable when the inverse function of a non-uniform distribution of random numbers can be expressed analytically. If a stochastic variable x obeys a probability density function $f(x)$, the cumulative distribution can be written as

$$F(x) = \int_{-\infty}^{x} f(x')dx' . \tag{A9.3}$$

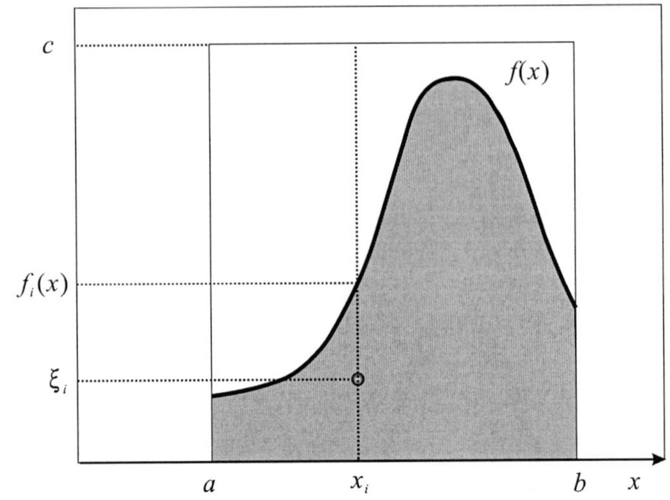

Figure A9.2 Rejection method of generating random variables obeying an arbitrary distribution.

Since $f(x)$ is the probability density, it is clear that $F(x)$ lies in the range of $0<F(x)<1$. As shown in Fig. A9.1, therefore, if a random number R_i sampled from a uniform random number sequence ranging from 0 to 1 is made equivalent to $F(x)$, the corresponding value of x_i can be obtained from the inverse transformation procedure of Eq. (A9.3). In a random number sequence $(x_1,x_2,\ldots)$ generated by this method, x with a large value of $f(x)$ will be sampled more frequently, that is, x is sampled according to the probability density $f(x)$. This is quite understandable from Fig. A9.1, but the detailed discussion concerning the mathematical aspect should be referred to in Ref. [1]. Generally, if the cumulative distribution shown in Eq. A9.3 with R instead of $F(x)$ can analytically be expressed as $x=g(R)$, the non-uniform random number sequence $(x_1,x_2,\ldots)$ obeying the probability density $f(x)$ can be generated using a uniform random number sequence $(R_1,R_2,\ldots)$.

We next outline the rejection method. This method is applicable to more general cases in which the inverse transformation cannot be obtained because of a complicated expression of the probability density function. As shown in Fig. A9.2, the probability density $f(x)$ is assumed to be defined for the range of $a\leq x\leq b$, and the constant c is also assumed to be equal or greater than the maximum value of $f(x)$. First, x is sampled using a uniform random number sequence ranging from 0 to 1, that is, $x_i=(b-a)R_i+a$, where R_i is a uniform random number. Next, another random number R_i' is generated to make $\xi_i=cR'$. If $f(x_i)>\xi_i$, x_i is accepted, and otherwise, x_i is rejected as a random number. This is an essential algorithm for the rejection method. It is seen from Fig. A9.2 that, in the above-mentioned rejection method, an arbitrary random number x_i is generated with higher probability for a larger value of $f(x_i)$. That is, random numbers are sampled according to the probability density.

Finally, we show the concrete methods of generating random number sequences for some representative probability densities which frequently appear in molecular-microsimulations of molecular or colloidal systems. In the case of the following normal distribution:

$$f(x)=\frac{1}{\sigma(2\pi)^{1/2}}\exp\left\{\frac{-(x-\langle x\rangle)^2}{2\sigma^2}\right\} \qquad (-\infty<x<\infty), \tag{A9.4}$$

a random number x can be generated as, with two uniform random numbers R_1 and R_2 ranging from 0 to 1,

$$x=\langle x\rangle+(-2\sigma^2\ln R_1)^{1/2}\cos 2\pi R_2 \qquad \text{or} \qquad =\langle x\rangle+(-2\sigma^2\ln R_1)^{1/2}\sin 2\pi R_2\,. \tag{A9.5}$$

This is called the Box-Müler method [2]. It is noted that $\langle y\rangle$ is the average value and σ is the square root of the variance, or the standard deviation.

In many cases of Brownian dynamics simulations, we have to deal with the following probability density of the bivariate normal distribution:

$$\rho(x,y)=\frac{1}{2\pi\sigma_x\sigma_y(1-c_{xy}^2)^{1/2}}\exp\left[-\frac{1}{2(1-c_{xy}^2)}\left\{\frac{(x-\langle x\rangle)^2}{\sigma_x^2}-2c_{xy}\frac{(x-\langle x\rangle)}{\sigma_x}\cdot\frac{(y-\langle y\rangle)}{\sigma_y}+\frac{(y-\langle y\rangle)^2}{\sigma_y^2}\right\}\right], \tag{A9.6}$$

in which $-\infty<x,y<\infty$, and c_{xy} is a constant satisfying $-1<c_{xy}<1$.

If Eq. (A9.6) is integrated with respect to x and the result is denoted by $\hat{\rho}(y)$, then $\hat{\rho}(y)$ can be straightforwardly obtained as

$$\hat{\rho}(y)=\int_{-\infty}^{\infty}\rho(x,y)dx=\frac{1}{(2\pi)^{1/2}\sigma_y}\exp\left\{\frac{-(y-\langle y\rangle)^2}{2\sigma_y^2}\right\}. \tag{A9.7}$$

This is a normal distribution with an average $\langle y\rangle$ and variance σ_y^2. The expression for a random variable y can, therefore, be determined by the Box-Müller method in Eq. (A9.5) using Eq. (A9.7):

$$y=\langle y\rangle+(-2\sigma_y^2\ln R_1)^{1/2}\cos 2\pi R_2\,. \tag{A9.8}$$

For a given value of y, x can be determined from the conditional probability distribution $\tilde{\rho}(x\,|\,y)$. From Eqs. (A9.6) and (A9.7), the following equation can be obtained:

$$\tilde{\rho}(x\mid y)=\rho(x,y)/\hat{\rho}(y)=\frac{1}{(2\pi)^{1/2}\sigma_x(1-c_{xy}^2)^{1/2}}\exp\left[-\frac{1}{2\sigma_x^2(1-c_{xy}^2)}\right.$$
$$\left.\times\left\{(x-\langle x\rangle)-c_{xy}\frac{\sigma_x}{\sigma_y}(y-\langle y\rangle)\right\}^2\right]. \tag{A9.9}$$

Finally, from the Box-Müller method with Eq. (A9.9), the random variable x can be obtained as

$$x=\langle x\rangle+c_{xy}\frac{\sigma_x}{\sigma_y}(y-\langle y\rangle)+(1-c_{xy}^2)^{1/2}(-2\sigma_x^2\ln R_3)^{1/2}\cos 2\pi R_4. \tag{A9.10}$$

Although first y and then x are determined in the above procedure, the determination in the inverse order is certainly possible from the symmetry of Eq. (A9.6) with respect to x and y.

In Brownian dynamics simulations based on the generalized Langevin equation, the random variables $x_1,x_2,\ldots,x_n$ have to satisfy the following multivariate normal distribution:

$$\rho(\mathbf{x})=\frac{1}{\{(2\pi)^n|\mathbf{D}|\}^{1/2}}\exp\left(-\frac{1}{2}\mathbf{x}\cdot\mathbf{C}\cdot\mathbf{x}\right), \tag{A9.11}$$

in which $\mathbf{x}$ is a vector expressed as $\mathbf{x}=[x_1,x_2,\ldots,x_n]$, $\mathbf{D}$ is a matrix with the ij-component D_{ij} $(=\langle x_i x_j\rangle)$, $\mathbf{C}$ is $\mathbf{D}^{-1}$, the inverse matrix of $\mathbf{D}$, and $|\mathbf{D}|$ is the determinant of $\mathbf{D}$. $\mathbf{D}$ is clearly symmetric from the definition, that is, $\mathbf{D}^t=\mathbf{D}$, so $\mathbf{C}$ is also symmetric and expressed as $\mathbf{C}^t=\mathbf{C}$. It is here assumed that $\langle x_i\rangle$ $(i=1,2,\ldots,n)=0$.

In the generalized Langevin equation, the random variables $\mathbf{x}'=[x_1,x_2,\ldots,x_{n-1}]$ before the present time are known, so that all we have to do is to derive the method of generating x_n satisfying the probability density function expressed in Eq. (A9.11) [3]. Since $\mathbf{x}'$ is known, x_n has to satisfy the following conditional probability density function $\tilde{\rho}(x_n\mid\mathbf{x}')$:

$$\tilde{\rho}(x_n\mid\mathbf{x}')=\rho(\mathbf{x})/\hat{\rho}(\mathbf{x}'), \tag{A9.12}$$

in which

$$\hat{\rho}(\mathbf{x}')=\int_{-\infty}^{\infty}\rho(\mathbf{x})dx_n. \tag{A9.13}$$

If $\mathbf{C}'$ is a $(n-1)\times(n-1)$ matrix made by removing the n-th row and column of $\mathbf{C}$, $\mathbf{c}$ is the vector made from the n-th row of $\mathbf{C}$, expressed as $\mathbf{c}=[C_{n1}, C_{n2}, \ldots, C_{nn}]$, and similarly $\mathbf{c}'=[C_{n1}, C_{n2}, \ldots, C_{n,n-1}]$, then Eq. (A9.11) can be reformed as

$$\rho(\mathbf{x})=\frac{1}{\{(2\pi)^n|\mathbf{D}|\}^{1/2}}\exp\left(-\frac{1}{2}\mathbf{x}'\cdot\mathbf{C}'\cdot\mathbf{x}'\right)\exp\left(-\frac{1}{2}C_{nn}x_n^2-\mathbf{x}'\cdot\mathbf{c}'x_n\right). \tag{A9.14}$$

Now, the integration in Eq. (A9.13) can be straightforwardly conducted to give the following equation:

$$\hat{\rho}(\mathbf{x}') = \frac{1}{\{(2\pi)^n|\mathbf{D}|\}^{1/2}} \exp\left(-\frac{1}{2}\mathbf{x}'\cdot\mathbf{C}'\cdot\mathbf{x}'\right)\cdot\left(\frac{2\pi}{C_{nn}}\right)^{1/2} \exp\left\{\frac{1}{2C_{nn}}(\mathbf{x}'\cdot\mathbf{c}')^2\right\}. \tag{A9.15}$$

By substituting this equation and Eq. (A9.14) into Eq. (A9.12), the following expression is obtained:

$$\widetilde{\rho}(x_n \mid \mathbf{x}') = \left(\frac{C_{nn}}{2\pi}\right)^{1/2} \exp\left\{-\frac{1}{2}C_{nn}\left(x_n + \frac{\mathbf{x}'\cdot\mathbf{c}'}{C_{nn}}\right)^2\right\}. \tag{A9.16}$$

Hence, with the random variable ζ_n which is generated by the Box-Müller method using the normal distribution with zero average and unit variance, the desired random variable x_n can finally be obtained as

$$x_n = \frac{1}{C_{nn}^{\ 1/2}}\zeta_n - \frac{1}{C_{nn}}\mathbf{x}'\cdot\mathbf{c}'. \tag{A9.17}$$

It is clear from these derivation procedures that this equation for a two-dimensional case (n=2) reduces to Eq. (A9.10), if we neglect the average terms in Eq. (A9.10); this can readily be verified.

We now return to the method of generating random variables $x_1, x_2, \ldots, x_n$ which obey the multivariate normal distribution expressed in Eq. (A9.11) [4,5]. The notations of **D**, **C**, etc., appearing below, are the same as in Eq. (A9.11). We only outline the final result without the detailed mathematical derivation. If the independent random variables generated by the Box-Müller method according to the normal distribution with zero average and unit variance are denoted by ($\zeta_1, \zeta_2, \ldots, \zeta_n$), then the random variables ($x_1, x_2, \ldots, x_n$) satisfying the multivariate normal distribution in Eq. (A9.11) can be obtained from ($\zeta_1, \zeta_2, \ldots, \zeta_n$) as

$$x_i = \sum_{j=1}^{i} L_{ij}\zeta_j \qquad (i = 1,2,\cdots,n), \tag{A9.18}$$

in which

$$\left.\begin{aligned} &L_{11} = D_{11}^{1/2}, \\ &L_{i1} = D_{i1}/L_{11} && (i>1), \\ &L_{ii} = (D_{ii} - \sum_{k=1}^{i-1} L_{ik}^2)^{1/2} && (i>1), \\ &L_{ij} = (D_{ij} - \sum_{k=1}^{j-1} L_{ik}L_{jk})/L_{jj} && (i>j>1). \end{aligned}\right\} \tag{A9.19}$$

It is straightforwardly verified that Eq. (A9.18) for the case of n=2 reduces to Eq. (A9.10) if we neglect the average terms in Eq. (A9.10).

The above method for a multivariate normal distribution is applicable under the condition that the matrix **D** is symmetric and positive definite. The condition for the matrix **D** being positive definite is equivalent to the following, all conditions being satisfied:

$$D_{11} > 0, \qquad \begin{vmatrix} D_{11} & D_{12} \\ D_{21} & D_{22} \end{vmatrix} > 0, \qquad \begin{vmatrix} D_{11} & D_{12} & D_{13} \\ D_{21} & D_{22} & D_{23} \\ D_{31} & D_{32} & D_{33} \end{vmatrix} > 0, \qquad \cdots. \tag{A9.20}$$

These conditions ensure that the square roots in Eqs. (A9.19) are not imaginary, but real.

References

1. J.M. Hammersley and D.C. Handscomb, "Monte Carlo Methods," Methuen & Co, London, 1964
2. G.E.P Box and M.E. Müller, A Note on the Generation of Random Normal Deviates, Ann. Math. Stat., 29 (1958) 610
3. L.G. Nilsson and J.A. Padro, A Time-Saving Algorithm for Generalized Langevin-Dynamics Simulations with Arbitrary Memory Kernels, Molec. Phys., 71 (1990) 355
4. D.L. Ermak and J.A. McCammon, Brownian Dynamics with Hydrodynamic Interactions, J. Chem. Phys., 69 (1978) 1352
5. M.P. Allen and D.J. Tildesley, "Computer Simulation of Liquids," 347-349, Clarendon Press, Oxford, 1987

A10. THE NUMERICAL CALCULATION OF RESISTANCE AND MOBILITY FUNCTIONS

To conduct Stokesian dynamics simulations of non-dilute colloidal dispersions, we have to take into account hydrodynamic interactions among colloidal particles. This means that the values of resistance functions or mobility functions for arbitrary particle-particle separations are indispensable in advancing the time step in simulations. Their tabulation as a function of particle-particle separations is quite useful from the viewpoint of computation time. Appropriate values of the resistance and mobility functions can be obtained from an interpolation procedure with the tabulated data and are used in actual Stokesian dynamics simulations. The data of the resistance and mobility functions for a system composed of two equal spherical particles, which were numerically obtained, have already been shown in Sec. 5.3. In this appendix, we outline the calculation method of the resistance and mobility functions for a two rigid spherical particle system. We concentrate our attention mainly on the side of the methodology of the calculation method and not discuss in detail the mathematical aspect such as the derivation of each equation, since the mathematical manipulation is not straightforward and is beyond the objective of the present book.

We use the coordinate system shown in Fig. A10.1. The disturbance velocity at a point **x** in Fig. A10.1 is expressed using Lamb's general solution as [1-3]

$$\begin{aligned}\mathbf{v}(\mathbf{x})-\mathbf{v}^{\infty}(\mathbf{x}) &= \sum_{n=1}^{\infty}\left\{\nabla\Phi_{-n-1}^{(1)}+\nabla\times(\mathbf{r}_1\chi_{-n-1}^{(1)})+\frac{(n+1)}{n(2n-1)}\mathbf{r}_1 p_{-n-1}^{(1)}-\frac{(n-2)}{2n(2n-1)}r_1^{\,2}\nabla p_{-n-1}^{(1)}\right\}\\ &+\sum_{n=1}^{\infty}\left\{\nabla\Phi_{-n-1}^{(2)}+\nabla\times(\mathbf{r}_2\chi_{-n-1}^{(2)})+\frac{(n+1)}{n(2n-1)}\mathbf{r}_2 p_{-n-1}^{(2)}-\frac{(n-2)}{2n(2n-1)}r_2^{\,2}\nabla p_{-n-1}^{(2)}\right\},\end{aligned} \tag{A10.1}$$

in which $\mathbf{v}^{\infty}$ is the ambient velocity field, $\mathbf{r}_\alpha=\mathbf{x}-\mathbf{x}_\alpha$, $r_\alpha=|\mathbf{r}_\alpha|$ for $\alpha=1,2$, and the spherical harmonics $p_{-n-1}^{(\alpha)}$, $\Phi_{-n-1}^{(\alpha)}$, and $\chi_{-n-1}^{(\alpha)}$ are expanded as follows:

$$\left.\begin{aligned} p_{-n-1}^{(\alpha)} &= \sum_{m=0}^{n} r_\alpha^{-n-1} p_n^m(\cos\theta_\alpha)\left\{a_{0n}^{(\alpha)}\delta_{0m}+a_{mn}^{(\alpha)}\sin m\varphi\right\},\\ \Phi_{-n-1}^{(\alpha)} &= \sum_{m=0}^{n} r_\alpha^{-n-1} p_n^m(\cos\theta_\alpha)\left\{b_{0n}^{(\alpha)}\delta_{0m}+b_{mn}^{(\alpha)}\sin m\varphi\right\},\\ \chi_{-n-1}^{(\alpha)} &= \sum_{m=0}^{n} r_\alpha^{-n-1} p_n^m(\cos\theta_\alpha) c_{mn}^{(\alpha)}\cos m\varphi .\end{aligned}\right\} \tag{A10.2}$$

Equation (A10.1) with Eq. (A10.2) satisfies the Stokes equation (4.20) and the equation of continuity (4.6). If the unknown constants of $a_{mn}^{(\alpha)}$, $b_{mn}^{(\alpha)}$, and $c_{mn}^{(\alpha)}$ are determined properly, the solution (A10.1) can satisfy the boundary condition on the particle surfaces. For each

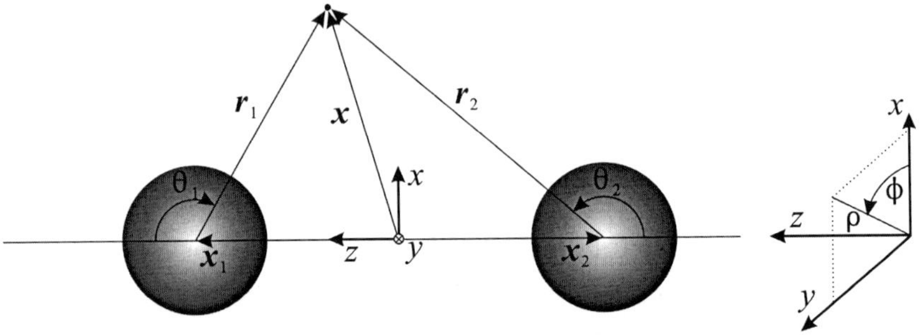

Figure A10.1 A coordinate system for two spherical particle interaction.

resistance function, only one particular value of m is actually required. The final goal of the present discussion is, therefore, to obtain algebraic equations for the unknown coefficients, whereby the resistance and mobility functions can be calculated numerically.

The disturbance velocity at the particle surface, $\mathbf{v}_s$, is equal to the difference between the imposed ambient velocity field and the velocity of the particle motion, and is expressed for the boundary condition at the surface of particle 1 as

$$\mathbf{v}_s = \sum_{l=1}^{2}\sum_{m=0}^{l}\Big[\nabla\left\{r_1^{\,l} P_l^m(\cos\theta_1)(A_{l0}\delta_{0m} + A_{lm}\sin m\varphi_1)\right\} + \nabla\times\left\{\mathbf{r}_1 r_1^{\,l} P_l^m(\cos\theta_1) B_{lm}\cos m\varphi_1\right\}\Big], \tag{A10.3}$$

in which $P_l^m(\cos\theta_1)$ is the associated Legendre function, (θ_1, φ_1) are the zenithal and azimuthal angles. The appropriate choice of the coefficients A_{lm} and B_{lm} gives all surface velocities associated with disturbance fields; for example, setting all coefficients zero except A_{10}=1 gives a translational velocity along the particle-particle axis [1,2].

The use of the cylindrical coordinate system (z,ρ,φ), shown in Fig A10.1, makes the mathematical manipulation more straightforward. The solution for the disturbance velocity, Eq. (A10.1), has to satisfy the boundary condition (A10.3). This requirement concerning the boundary condition leads to the following algebraic equations which must be satisfied by the unknown coefficients, $a_{mn}^{(\alpha)}$, $b_{mn}^{(\alpha)}$, and $c_{mn}^{(\alpha)}$:

$$\begin{aligned}\sum_{n=l}^{\infty}\Bigg[& a_{mn}^{(1)}\sum_{\alpha=1}^{2}(-S)^{\alpha-1}\left\{\frac{(n+1)}{2(2n-1)} r_\alpha^{-n}\xi_\alpha P_n^m(\xi_\alpha) - \frac{(n-2)}{2n(2n-1)} r_\alpha^{-n}(1-\xi_\alpha^2)(P_n^m(\xi_\alpha))'\right\} \\ & b_{mn}^{(1)}\sum_{\alpha=1}^{2}(-S)^{\alpha-1}\left\{-(n+1)r_\alpha^{-n-2}\xi_\alpha P_n^m(\xi_\alpha) + r_\alpha^{-n-2}(1-\xi_\alpha^2)(P_n^m(\xi_\alpha))'\right\} + mc_{mn}^{(1)} \\ & \times\sum_{\alpha=1}^{2}(-S)^{\alpha-1}\left\{r_\alpha^{-n-1}P_n^m(\xi_\alpha)\right\}\Bigg] = A_{lm}\left\{l\xi_1 P_l^m(\xi_1) + (1-\xi_1^2)(P_l^m(\xi_1))'\right\} + B_{lm} m P_l^m(\xi_1),\end{aligned} \tag{A10.4}$$

$$\sum_{n=l}^{\infty}\left[a_{mn}^{(1)}\sum_{\alpha=1}^{2}S^{\alpha-1}\left\{\frac{(n+1)}{2(2n-1)}r_{\alpha}^{-n}\sin\theta_{\alpha}P_{n}^{m}(\xi_{\alpha})+\frac{(n-2)}{2n(2n-1)}r_{\alpha}^{-n}\right.\right.$$
$$\left.\times\left\{\xi_{\alpha}P_{n}^{m+1}(\xi_{\alpha})+m\sin\theta_{\alpha}P_{n}^{m}(\xi_{\alpha})\right\}\right\}+b_{mn}^{(1)}\sum_{\alpha=1}^{2}S^{\alpha-1}r_{\alpha}^{-n-2}$$
$$\left.\times\left\{-(n+1)\sin\theta_{\alpha}P_{n}^{m}(\xi_{\alpha})-\xi_{\alpha}P_{n}^{m+1}(\xi_{\alpha})-m\sin\theta_{\alpha}P_{n}^{m}(\xi_{\alpha})\right\}-c_{mn}^{(1)}\sum_{\alpha=1}^{2}S^{\alpha-1}r_{\alpha}^{-n-1}P_{n}^{m+1}(\xi_{\alpha})\right]$$
$$=A_{lm}\left\{l\sin\theta_{1}P_{l}^{m}(\xi_{1})-\left(\xi_{1}P_{l}^{m+1}(\xi_{1})+m\sin\theta_{1}P_{l}^{m}(\xi_{1})\right)\right\}-B_{lm}P_{l}^{m+1}(\xi_{1}),$$

(A10.5)

$$\sum_{n=l}^{\infty}\left[-ma_{mn}^{(1)}\sum_{\alpha=1}^{2}S^{\alpha-1}\left\{\frac{(n-2)}{2n(2n-1)}r_{\alpha}^{-n}P_{n}^{m}(\xi_{\alpha})(\sin\theta_{\alpha})^{-1}\right\}\right.$$
$$\left.+mb_{mn}^{(1)}\sum_{\alpha=1}^{2}S^{\alpha-1}\left\{r_{\alpha}^{-n-2}P_{n}^{m}(\xi_{\alpha})(\sin\theta_{\alpha})^{-1}\right\}+c_{mn}^{(1)}\sum_{\alpha=1}^{2}S^{\alpha-1}\left\{r_{\alpha}^{-n-1}\sin\theta_{\alpha}(P_{n}^{m}(\xi_{\alpha}))'\right\}\right]$$
$$=A_{l}mP_{l}^{m}(\xi_{1})/\sin\theta_{1}+B_{lm}\sin\theta_{1}(P_{l}^{m}(\xi_{1}))',$$

(A10.6)

in which $\xi_\alpha=\cos\theta_\alpha$ and S is the symmetry parameter; S=1 for problems with mirror symmetry and S= -1 for problems with mirror anti-symmetry [1,2]. Equation (A10.4) comes directly from the z-component boundary condition, and Eq. (A10.6) from the φ-component boundary condition. Equation (A10.5) is obtained by subtracting the φ-component equation from the ρ-component equation. If the constants $a_{mn}^{(\alpha)}$, $b_{mn}^{(\alpha)}$, and $c_{mn}^{(\alpha)}$ can be determined by the algebraic equations (A10.4) to (A10.6), then the solution for the velocity field for the two equal-particle system may be obtained.

It is seen from Eq. (13.7) that the velocity field can be expanded as a function of forces, torques, stresslets, etc. The resistance functions can, therefore, be correlated to $a_{mn}^{(\alpha)}$, $b_{mn}^{(\alpha)}$, and $c_{mn}^{(\alpha)}$ by comparing Eq. (A10.1) with the equation expressed by forces, torques , stresslets, etc., such as Eq. (13.7). The final results for the resistance functions are as follows:

$$\left.\begin{aligned}
&X_{11}^{A*}=\frac{1}{3}\{a_{01}(1,-1)+a_{01}(1,1)\},\quad X_{12}^{A*}=\frac{1}{3}\{a_{01}(1,-1)-a_{01}(1,1)\},\\
&Y_{11}^{A*}=\frac{1}{3}\{a_{11}(1,-1)+a_{11}(1,1)\},\quad Y_{12}^{A*}=-\frac{1}{3}\{a_{11}(1,-1)-a_{11}(1,1)\},\\
&Y_{11}^{B*}=-\{c_{11}(1,-1)+c_{11}(1,1)\},\quad Y_{12}^{B*}=c_{11}(1,-1)-c_{11}(1,1),\\
&X_{11}^{G*}=-\frac{1}{4}\{a_{02}(1,-1)+a_{02}(1,1)\},\quad X_{12}^{G*}=-\frac{1}{4}\{a_{02}(1,-1)-a_{02}(1,1)\},\\
&Y_{11}^{G*}=-\frac{1}{4}\{a_{12}(1,-1)+a_{12}(1,1)\},\quad Y_{12}^{G*}=\frac{1}{4}\{a_{12}(1,-1)-a_{12}(1,1)\},\\
&X_{11}^{M*}+X_{12}^{M*}=\frac{1}{10}a_{02}(2,1),\quad Y_{11}^{M*}+Y_{12}^{M*}=\frac{1}{10}a_{12}(2,-1),\\
&Z_{11}^{M*}+Z_{12}^{M*}=\frac{1}{10}a_{22}(2,1),
\end{aligned}\right\}\tag{A10.7}$$

in which the superscript (1) of $a_{mn}^{(1)}$, $b_{mn}^{(1)}$, and $c_{mn}^{(1)}$ has been dropped for simplicity, and they are a function of l and S, expressed as $a_{mn}(l,S)$, $b_{mn}(l,S)$, and $c_{mn}(l,S)$. The results shown in Eq. (A10.7) have been obtained for the set of A_{lm}=1 and B_{lm}=0. From the other set of B_{lm}=1 and A_{lm}=0, the remaining resistance functions are derived as

$$\left.\begin{aligned}
&X_{11}^{C*}=\frac{1}{2}\{c_{01}(1,-1)+c_{01}(1,1)\},\quad X_{12}^{C*}=-\frac{1}{2}\{c_{01}(1,-1)-c_{01}(1,1)\},\\
&Y_{11}^{C*}=\frac{1}{2}\{c_{11}(1,-1)+c_{11}(1,1)\},\quad Y_{12}^{C*}=\frac{1}{2}\{c_{11}(1,-1)-c_{11}(1,1)\},\\
&Y_{11}^{H*}=-\frac{1}{8}\{a_{12}(1,-1)+a_{12}(1,1)\},\quad Y_{12}^{H*}=-\frac{1}{8}\{a_{12}(1,-1)-a_{12}(1,1)\}.
\end{aligned}\right\}\tag{A10.8}$$

These equations of Eq. (A10.7) and (A10.8) complete the calculation of the resistance functions.

To solve the algebraic equations of Eqs. (A10.4) to (A10.6), the summation of n has to be truncated at N terms. We consider a special case of l=m=S=1, A_{lm}=1, and B_{lm}=0, to understand the way of solving these equations. In this case, the 3N unknown coefficients, a_{11}, a_{12}, ..., a_{1N}, b_{11}, b_{12}, ..., b_{1N}, c_{11}, c_{12}, ..., c_{1N}, have to be determined by the truncated version of Eqs. (A10.4) to (A10.6). Thus N points at the particle surface, at which the solution of the velocity field, Eq. (A10.1), satisfies the boundary condition, are required. If we take such points as $\theta_1=k\pi/(N\text{-}1)$ (k=0,1,...,N-1), we can get the 3N algebraic equations necessary for calculating the 3N unknown coefficients.

Finally, it should be noted that more boundary points (that is, a larger value of N) are necessary as the two particles approach to a nearly touching situation; in other words, this means that the solutions do not converge rapidly as N increases, when the particles are almost touching. Once the resistance functions from these numerical procedures are obtained, the mobility functions are calculated from the resistance functions using the relationships between the resistance and mobility functions which have already been given in Sec. 5.3. A sample FORTRAN program for calculating the resistance functions is shown in Appendix A11.7.

References

1. S. Kim and R. T. Mifflin, The Resistance and Mobility Functions of Two Equal Spheres in Low-Reynolds-Number Flow, Phys. Fluids, 28 (1985) 2033
2. S. Kim and S. J. Karrila, "Microhydrodynamics," Butterworth-Heinemann, Stoneham, 1991
3. P. Ganatos, R. Pfeffer, and S. Weinbaum, A Numerical-Solution Technique for Three-Dimensional Stokes Flows with Application to the Motion of Strongly Interacting Spheres in a Plane, J. Fluid Mech., 84 (1978) 79

A11. SEVERAL FORTRAN SUBROUTINES FOR SIMULATIONS

In this appendix, we show useful sample FORTRAN programs for molecular-microsimulations. First we show several subroutine programs for setting initial configurations and computing forces and interaction energies between particles. Then we show a complete program based on the canonical Monte Carlo algorithm for evaluating the radial distribution function for a molecular system. Lastly a calculation program for resistance and mobility functions is shown. In these programs, the following variables are commonly used.

Common variables:

RX(I), *RY(I)*, *RZ(I)* : components of the position vector $\mathbf{r}_i^*$ of particle *i*
FX(I), *FY(I)*, *FZ(I)* : components of the force $\mathbf{f}_i^*$ acting on particle *i*
N : number of particles in a system
NDENS : number density of particles
TEMP : system temperature
RCOFF : cutoff radius
H : time interval
XL, *YL*, *ZL* : side length of a simulation box of cuboid
L : side length of a cubic simulation box
RAN(J) : uniform random number sequence ranging from 0 to 1 (*J*=1,2,…,*NRANMX*)

Lennard-Jones particles and the non-dimensional quantities expressed in Appendix A3 are used, unless otherwise specified

A11.1 Initial Configurations

A11.1.1 Two-dimensional system (SUBROUTINE INIPOSIT)

Particles are initially placed at hexagonal lattice points with the lattice constant a^*, as shown in Fig. A11.1 for a two-dimensional system. In this case, the lattice with the lengths $\sqrt{3}\,a^*$ in the *x*-axis and $2a^*$ in the *y*-axis is used as a unit cell. The simulation region is made by replicating this unit cell a certain number of times in each direction. Hence, the number of particles in a system, *N*, is restricted such as *N*=16, 36, 64, 100, 144, …, and so forth. Since the number density n^* is related to the lattice constant a^* as $n^*=4/(\sqrt{3}\,a^*\cdot 2a^*)$, a^* is expressed as $a^*=(2/(\sqrt{3}\,n^*))^{1/2}$ for a given value of n^*. If the simulation region is made by replicating the unit cell *P* times in each direction, then the relation of $4P^2=N$ is satisfied and then *P* is

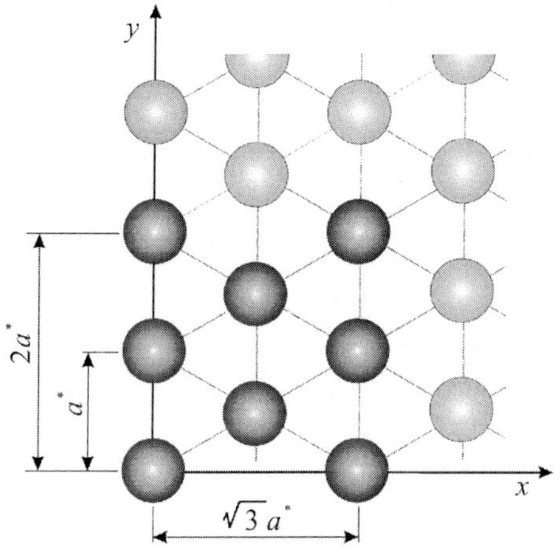

Figure A11.1 Close-packed configuration for two-dimensional system.

expressed as $P=(N/4)^{1/2}$ for a given value of N. The dimensions of the simulation region, therefore, become ($\sqrt{3}\,a^{*}P, 2a^{*}P$) in the x- and y-directions, respectively.

```
00010 C************************************************************************
00020 C   THIS SUBROUTINE IS FOR SETTING INITIAL POSITIONS OF PARTICLES
00030 C   AT CLOSE-PACKED LATTICE POINTS FOR TWO-DIMENSIONAL SYSTEM.
00040 C      0<RX(I)<XL  ,  0<RY(I)<YL
00050 C************************************************************************
00060 C**** SUB INIPOSIT *****
00070       SUBROUTINE INIPOSIT( N, NDENS, RCOFF, NP )
00080 C
00090       IMPLICIT REAL*8( A-H , O-Z )
00100 C
00110       COMMON /BLOCK1/ RX , RY
00120       COMMON /BLOCK6/ XL , YL
00130 C
00140       PARAMETER( NN=1000 )
00150 C
00160       REAL*8 RX(NN), RY(NN), XL, YL, NDENS, RCOFF
00170       INTEGER N     , NP
00180 C
00190       REAL*8 RXI, RYI, RX0, RY0 , A , AX , AY , C1
00200       INTEGER KX , KY , K  , P
00210 C
00220       A  = ( (2.D0/3.D0**0.5)/NDENS )**0.5
00230       P  = IDNINT( DSQRT(DBLE(N/4)) )
00240       XL = 3.D0**0.5*A*DBLE(P)
00250       YL = 2.D0*A*DBLE(P)
00260       IF( (XL/2.D0.LE.RCOFF) .OR. (YL/2.D0.LE.RCOFF) )THEN
00270         WRITE(NP,*)
00280         WRITE(NP,*) '++++ SIMULATION BOX IS TOO SMALL ++++'
00290         WRITE(NP,*)
00300         RETURN
00310       END IF
00320 C
00330       AX  = 3.D0**0.5*A
00340       AY  = 2.D0*A
00350       KX  = P
00360       KY  = P
```

```
00370      C1   = 0.01D0
00380 C                                          --- SET INITIAL POSITIONS ---
00390      K    = 0
00400 C
00410      DO 100 IFACE=1,4
00420       IF( IFACE.EQ.1 ) THEN
00430         RX0 = C1
00440         RY0 = C1
00450       ELSE IF( IFACE.EQ.2 ) THEN
00460         RX0 = C1
00470         RY0 = A + C1
00480       ELSE IF( IFACE.EQ.3 ) THEN
00490         RX0 = AX/2.D0 + C1
00500         RY0 = A/2.D0  + C1
00510       ELSE IF( IFACE.EQ.4 ) THEN
00520         RX0 = AX/2.D0 + C1
00530         RY0 = A*3.D0/2.D0 + C1
00540       END IF
00550       DO 40 J=0,KY-1
00560         RYI = DBLE(J)*AY + RY0
00570         IF( RYI .GE. YL )      GOTO 40
00580         DO 30 I=0,KX-1
00590          RXI = DBLE(I)*AX + RX0
00600          IF( RXI .GE. XL ) GOTO 30
00610 C
00620          K=K+1
00630          RX(K)  = RXI
00640          RY(K)  = RYI
00650    30    CONTINUE
00660    40   CONTINUE
00670   100 CONTINUE
00680       RETURN
00690       END
```

A11.1.2 Three-dimensional system (SUBROUTINE INIPOSIT)

Particles are initially placed at face-centered-cubic lattice points as shown in Fig. 11.1(a) for a three-dimensional system. The number density n^* is related to a lattice constant a^* as $n^*=4/a^{*3}$, and then a^* is expressed as $a^*=(4/n^*)^{1/3}$ for a given value of n^*. If the simulation region is made by replicating the unit cell Q times in each direction, then the relation of $4Q^3=N$ is satisfied and therefore Q is expressed as $Q=(N/4)^{1/3}$ for a given value of N.

```
00010 C**********************************************************************
00020 C   THIS SUBROUTINE IS FOR SETTING INITIAL POSITIONS OF PARTICLES
00030 C   AT CLOSE-PACKED LATTICE POINTS FOR THREE-DIMENSIONAL SYSTEM.
00040 C      0<RX(I)<L  ,   0<RY(I)<L  ,   0<RZ(I)<L
00050 C**********************************************************************
00060 C**** SUB INIPOSIT *****
00070      SUBROUTINE INIPOSIT( N , NDENS , L )
00080 C
00090      IMPLICIT REAL*8( A-H , O-Z )
00100 C
00110      COMMON /BLOCK1/ RX , RY , RZ
00120 C
00130      PARAMETER( NN=2050 , PI=3.141592653589793D0 )
00140 C
00150      REAL*8  RX(NN), RY(NN), RZ(NN), NDENS , L
00160      INTEGER N
00170 C
00180      REAL*8  RXI, RYI, RZI, RX0, RY0, RZ0 , C0
00190      INTEGER Q , K , IX , IY , IZ
00200 C
00210      C0 = ( 4.D0/NDENS )**(1./3.)
00220      Q  = IDNINT( DBLE(N/4)**(1./3.) )
```

```
00230        L  =  C0*DBLE(Q)
00240 C                                    --- SET INITIAL POSITIONS ---
00250        K  =  0
00260 C
00270        DO 100 IFACE=1,4
00280 C
00290         IF( IFACE.EQ.1 ) THEN
00300           RX0 = 0.0001D0
00310           RY0 = 0.0001D0
00320           RZ0 = 0.0001D0
00330         ELSE IF( IFACE.EQ.2 ) THEN
00340           RX0 = C0/2.D0
00350           RY0 = C0/2.D0
00360           RZ0 = 0.0001D0
00370         ELSE IF( IFACE.EQ.3 ) THEN
00380           RX0 = C0/2.D0
00390           RY0 = 0.0001D0
00400           RZ0 = C0/2.D0
00410         ELSE IF( IFACE.EQ.4 ) THEN
00420           RX0 = 0.0001D0
00430           RY0 = C0/2.D0
00440           RZ0 = C0/2.D0
00450         END IF
00460 C
00470         DO 50 IZ=0,Q-1
00480           RZI = DBLE(IZ)*C0 + RZ0
00490           IF( RZI .GE. L )          GOTO 50
00500           DO 40 IY=0,Q-1
00510             RYI = DBLE(IY)*C0 + RY0
00520             IF( RYI .GE. L )        GOTO 40
00530             DO 30 IX=0,Q-1
00540               RXI = DBLE(IX)*C0 + RX0
00550               IF( RXI .GE. L )    GOTO 30
00560 C
00570               K=K+1
00580               RX(K)  =  RXI
00590               RY(K)  =  RYI
00600               RZ(K)  =  RZI
00610    30        CONTINUE
00620    40      CONTINUE
00630    50    CONTINUE
00640   100 CONTINUE
00650 C
00660        N = K
00670        RETURN
00680        END
```

A11.2 Random Numbers (SUBROUTINE RANCAL)

A uniform random number sequence is generated in terms of the method shown in Appendix A9. This subroutine is for a computer in which the expression of two's complement is adopted, so Eq. (A9.2) is used.

```
00010 C**************************************************************************
00020 C    THIS SUBROUTINE IS FOR GENERATING UNIFORM RANDOM NUMBERS
00030 C    (SINGLE PRECISION).
00040 C      N     : NUMBER OF RANDOM NUMBERS TO GENERATE
00050 C      IX    : INITIAL VALUE OF RANDOM NUMBERS (POSITIVE INTEGER)
00060 C            : LAST GENERATED VALUE IS KEPT
00070 C      X(N)  : GENERATED RANDOM NUMBERS (0<X(N)<1)
00080 C**************************************************************************
00090 C**** SUB RANCAL ****
00100       SUBROUTINE RANCAL( N, IX, X )
00110 C
00120       DIMENSION  X(N)
```

```
00130       DATA INTEGMX/2147483647/
00140       DATA INTEGST,INTEG/584287,48828125/
00150 C
00160       AINTEGMX = REAL( INTEGMX )
00170 C
00180       IF ( IX.LT.0 ) PAUSE
00190       IF ( IX.EQ.0 ) IX = INTEGST
00200       DO 30 I=1,N
00210          IX = IX*INTEG
00220          IF (IX) 10, 20, 20
00230    10    IX   = (IX+INTEGMX)+1
00240    20    X(I) = REAL(IX)/AINTEGMX
00250    30 CONTINUE
00260       RETURN
00270       END
```

A11.3 The Cell Index Method for a Two-dimensional System

A square is used as a simulation region, as shown in Fig. 11.6. This simulation region is divided into $P{\times}P$ subcells (or cutoff cells). Each cutoff cell has the name list of particles belonging to itself in $TABLE(*,GRP)$ and the number of such particles is saved in $TMX(GRP)$. The GRP is the name of a cutoff cell and, for example, for the case of $P=6$, the cutoff cells are named as in Fig. 11.6. Particle i has the name of the cutoff cell to which it belongs in $GRPX(I)$ and $GRPY(I)$. If particle i belongs to GRP, then GRP is expressed as $GRP=GRPX(I)+(GRPY(I)-1)*P$. The x-value of the length L of the cutoff cells is used to judge which cutoff cell a particle belongs to. These values are saved in $GRPLX(GRP)$ $(GRP=1,2,\ldots,P)$. Since the cutoff cells are square, $GRPLX(*)$ is also used for the judgment concerning the y-axis. The subroutine INIPOSIT for setting an initial configuration which is called in the program is different from that shown in Appendix A11.1.1. The subroutine INIPOSIT is for setting particles at square lattice points and can straightforwardly be developed from the program shown in Appendix A11.1.2.

```
00010 C************************************************************
00020 C*  THIS PROGRAM IS PART OF THE MAIN PROGRAM WHICH IS FOR
00030 C*  INTRODUCING THE CELL INDEX METHOD FOR TWO-DIMENSIONAL SYSTEM.
00040 C************************************************************
00050 C    GRPY(I),GRPX(I)   : GROUP TO WHICH PARTICLE I BELONGS
00060 C    P             : NUMBER OF CUT-OFF CELLS IN EACH DIRECTION
00070 C    TMX(GRP)     : TOTAL NUMBER OF PARTICLES BELONGING TO GROUP(GRP)
00080 C    TABLE(*,GRP): NAME OF PARTICLE BELONGING TO GROUP(GRP)
00090 C    GRPLX(P)     : IS USED TO DETERMINE THE CELL TO WHICH
00100 C                    PARTICLES BELONG
00110 C    0<RX(I)<L  ,   0<RY(I)<L
00120 C-----------------------------------------------------------
00130 C
00140       IMPLICIT REAL*8 (A-H,O-Z)
00150 C
00160       COMMON /BLOCK1/ RX   , RY
00170       COMMON /BLOCK3/ GRPX , GRPY
00180       COMMON /BLOCK5/ TMX  , TABLE
00190       COMMON /BLOCK6/ P    , GRPLX
00200 C                     --- NN : NUM. OF PARTICLES      ---
00210 C                     --- PP : NUM. OF CUT-OFF CELLS  ---
00220       INTEGER NN , PP , PP2 , TT
00230       PARAMETER( NN=1000 , PP=15 , PP2=225 , TT=500 )
```

```
00240 C
00250       REAL*8   RX(NN) , RY(NN) , GRPLX(PP) , NDENS  , L
00260       INTEGER  GRPX(NN) , GRPY(NN) , TMX(PP2) , TABLE(TT,PP2) , N , P
00270                .
00280                .
00290                .
00300 C    ----------------------------------------------------------------
00310 C    ------------------ SET CELL INDEX METHOD  ------------------
00320 C    ----------------------------------------------------------------
00330 C                                        --- MAKE P*P CELLS ---
00340       CALL CELLSET( N , NDENS , L , RCOFF )
00350 C                                  --- SET INITIAL POSITIONS ---
00360       CALL INIPOSIT( N , NDENS , RCOFF , NP )
00370 C                                  --- CELL NAME OF PARTICLES ---
00380       CALL GROUP( N )
00390 C                          --- PARTICLE NAMES OF EACH CELL ---
00400       CALL TABLECAL( N , P )
00410                .
00420                .
00430                .
00440       STOP
00450       END
00460 C**********************************************************************
00470 C   THIS SUBROUTINE IS FOR DISTRIBUTING A MAIN CELL INTO
00480 C   MANY SUB-CELLS.
00490 C**********************************************************************
00500 C*** SUB CELLSET ****
00510       SUBROUTINE CELLSET( N ,NDENS , L , RCOFF )
00520 C
00530       IMPLICIT REAL*8 (A-H,O-Z)
00540 C
00550       COMMON /BLOCK6/  P   , GRPLX
00560 C
00570       INTEGER  PP
00580       PARAMETER( PP=15 )
00590 C
00600       INTEGER  P
00610       REAL*8   GRPLX(PP) , NDENS , L , C0
00620 C
00630       L = DSQRT( DBLE(N)/NDENS )
00640       P = INT(L/RCOFF)
00650       IF(P .LE. 2) PAUSE
00660 C
00670       C0 = L/DBLE(P)
00680       DO 10 I=1,P
00690         GRPLX(I) = C0*DBLE(I)
00700    10 CONTINUE
00710       RETURN
00720       END
00730 C**********************************************************************
00740 C   THIS SUBROUTINE IS FOR CHECKING THE SUB-CELL TO WHICH EACH
00750 C   PARTICLE BELONGS.
00760 C**********************************************************************
00770 C**** SUB GROUP *****
00780       SUBROUTINE GROUP( N )
00790 C
00800       IMPLICIT REAL*8 (A-H,O-Z)
00810 C
00820       COMMON /BLOCK1/  RX , RY
00830       COMMON /BLOCK3/  GRPX, GRPY
00840       COMMON /BLOCK6/  P   , GRPLX
00850 C
00860       INTEGER  NN , PP
00870       PARAMETER( NN=1000 , PP=15 )
00880 C
00890       INTEGER  GRPX(NN) , GRPY(NN) , N      , P
00900       REAL*8   RX(NN)   , RY(NN)   , GRPLX(PP)
00910 C
00920       DO 30 I=1,N
00930 C                                        ----- X AXIS -----
00940         DO 10 J=1,P
00950           IF( GRPLX(J) .GT. RX(I) ) THEN
```

```
00960              GRPX(I) = J
00970              GOTO 15
00980            END IF
00990    10    CONTINUE
01000          GRPX(I) = P
01010  C                                    ----- Y AXIS -----
01020    15    DO 20 J=1,P
01030              IF( GRPLX(J) .GT. RY(I) ) THEN
01040                GRPY(I) = J
01050                GOTO 30
01060              END IF
01070    20    CONTINUE
01080          GRPY(I) = P
01090  C
01100    30 CONTINUE
01110          RETURN
01120          END
01130  C*********************************************************************
01140  C    THIS SUBROUTINE IS FOR CHECKING THE PARTICLE NAMES WHICH EACH
01150  C    SUB-CELL HAS.
01160  C*********************************************************************
01170  C**** SUB TABLECAL *****
01180          SUBROUTINE TABLECAL( N , P )
01190  C
01200          IMPLICIT REAL*8 (A-H,O-Z)
01210  C
01220          COMMON /BLOCK3/  GRPX, GRPY
01230          COMMON /BLOCK5/  TMX , TABLE
01240  C
01250          INTEGER  NN , PP2 , TT
01260          PARAMETER( NN=1000 , PP2=225 , TT=500  )
01270  C
01280          INTEGER  GRPX(NN) , GRPY(NN) , TMX(PP2) , TABLE(TT,PP2)
01290          INTEGER  N   , P , GX  , GY , GP
01300  C
01310          DO 10 GY=1,P
01320          DO 10 GX=1,P
01330            GP = GX + (GY-1)*P
01340            TMX(GP)    = 0
01350            TABLE(1,GP) = 0
01360    10 CONTINUE
01370  C
01380          DO 20 I=1,N
01390            GX = GRPX(I)
01400            GY = GRPY(I)
01410            GP = GX + (GY-1)*P
01420            TMX(GP) = TMX(GP) + 1
01430            TABLE( TMX(GP),GP ) = I
01440    20 CONTINUE
01450          RETURN
01460          END
```

A11.4 Calculation of Forces and Interaction Energies

A11.4.1 Calculation of interaction energies (SUBROUTINE ENECAL)

The interaction energy of particle i with a partner particle is saved in $E(*,I)$ and the name of the partners is saved in $ETABLE(*,I)$. The total interaction energy of particle i is stored in $ETOT(I)$. In this subroutine, DO loop has the structure of (DO 100 I=1,N; DO 40 J=1,N), but a more efficient structure of (DO 100 I=1,N-1; DO 40 J=I+1,N) can straightforwardly be developed from the subroutine of forces shown in the next section. Forces and energies are usually computed in the same subroutine. In Line number 450, 460, etc., the treatment of the periodic boundary condition is carried out.

```
00010 C*********************************************************************
00020 C    THIS SUBROUTINE IS FOR CALCULATING ENERGIES BETWEEN PARTICLES
00030 C    FOR THREE-DIMENSIONAL LENNARD-JONES SYSTEM.
00040 C      ETOT(I)     : ENERGY OF PARTICLE I
00050 C      E(*,I)      : INTERACTION ENERGY OF PARTICLE I WITH THE OTHERS
00060 C      ETMX(I)     : TOTAL NUMBER OF PARTICLES INTERACTING WITH I
00070 C      ETABLE(*,I) : NAME OF PARTICLES INTERACTING WITH I
00080 C      RCOFF2 = RCOFF**2
00090 C*********************************************************************
00100 C**** SUB ENECAL *****
00110       SUBROUTINE ENECAL( N , RCOFF2 )
00120 C
00130       IMPLICIT REAL*8 (A-H,O-Z)
00140 C
00150       COMMON /BLOCK1/  RX , RY , RZ
00160       COMMON /BLOCK4/  E    , ETOT , ETMX , ETABLE
00170       COMMON /BLOCK8/  NDENS, TEMP , RCOFF , L
00180 C
00190       INTEGER NN , TT
00200       PARAMETER( NN=2050 , TT=600 )
00210 C
00220       INTEGER  N  , ETMX(NN) , ETABLE(TT,NN)
00230       REAL*8   RX(NN), RY(NN), RZ(NN), E(TT,NN), ETOT(NN), NDENS, L
00240 C
00250       REAL*8   RXIJ , RYIJ , RZIJ , RIJSQ , L2
00260 C
00270       L2 = L/2.D0
00280       DO 5 I=1,N
00290        ETMX(I) = 0
00300        ETABLE(1,I) = 0
00310        DO 3 J=1,TT
00320         E(J,I) = 0.D0
00330      3  CONTINUE
00340      5 CONTINUE
00350 C
00360       DO 100 I=1,N
00370 C
00380        ETOT(I)= 0.D0
00390 C
00400 C                                     ---- CALCULATE ENERGY ---
00410         DO 40 J=1,N
00420          IF( J .EQ. I )           GOTO 40
00430 C
00440           RXIJ = RX(J) - RX(I)
00450           IF( RXIJ .GE.  L2 ) RXIJ = RXIJ - L
00460           IF( RXIJ .LT. -L2 ) RXIJ = RXIJ + L
00470           IF( DABS(RXIJ) .GE. RCOFF ) GOTO 40
00480           RYIJ = RY(J) - RY(I)
00490           IF( RYIJ .GE.  L2 ) RYIJ = RYIJ - L
00500           IF( RYIJ .LT. -L2 ) RYIJ = RYIJ + L
00510           IF( DABS(RYIJ) .GE. RCOFF ) GOTO 40
00520           RZIJ = RZ(J) - RZ(I)
00530           IF( RZIJ .GE.  L2 ) RZIJ = RZIJ - L
00540           IF( RZIJ .LT. -L2 ) RZIJ = RZIJ + L
00550           IF( DABS(RZIJ) .GE. RCOFF ) GOTO 40
00560 C
00570           RIJSQ= RXIJ**2 + RYIJ**2 + RZIJ**2
00580           IF( RIJSQ .GE. RCOFF2 )     GOTO 40
00590 C
00600           SR2  = 1.D0/RIJSQ
00610           SR6  = SR2*SR2*SR2
00620           EIJ  = SR6*( SR6-1.D0 )
00630 C
00640           ETOT(I) = ETOT(I) + EIJ*4.D0
00650 C
00660           ETMX(I) = ETMX(I) + 1
00670           ETABLE(ETMX(I),I) = J
00680           E(ETMX(I),I) = EIJ*4.D0
```

```
   40       CONTINUE
C
  100 CONTINUE
      RETURN
      END
```

A11.4.2 Calculation of forces (SUBROUTINE FORCECAL)

From the action-reaction law, the force exerted on particle j by particle i is equal to the force with the negative sign exerted on particle i by particle j. Hence, when the force acting on particle i is computed, the counterpart force is also saved at the partner's variable. This means that $N(N\text{-}1)/2$ computations are sufficient for evaluating forces acting on particles. This is why the DO loop has the structure of (DO 100 I=1,N-1; DO 50 J=I+1,N). In Line numbers 400, etc., the periodic boundary condition is dealt with. By considering the program for interaction energies shown above, it is straightforward to add a program for computations of interaction energies to the force subroutine.

```
00010 C*********************************************************************
00020 C   THIS SUBROUTINE IS FOR CALCULATING FORCES BETWEEN PARTICLES
00030 C   FOR THREE-DIMENSIONAL LENNARD-JONES SYSTEM.
00040 C      RCOFF2 = RCOFF**2
00050 C*********************************************************************
00060 C**** SUB FORCECAL *****
00070       SUBROUTINE FORCECAL( N, RCOFF, RCOFF2, RX, RY, RZ, FX, FY, FZ )
00080 C
00090       IMPLICIT REAL*8 ( A-H , O-Z )
00100 C
00110       COMMON /BLOCK8/ XL    , YL    , ZL
00120 C
00130       PARAMETER( NN=2050 )
00140 C
00150       INTEGER  N
00160       REAL*8   RX(NN)  , RY(NN)  , RZ(NN)  , FX(NN)  , FY(NN)  , FZ(NN)
00170 C
00180       REAL*8   RXI  , RYI  , RZI  , RXIJ , RYIJ , RZIJ , RIJSQ
00190       REAL*8   FXI  , FYI  , FZI  , FXIJ , FYIJ , FZIJ , FIJ
00200       REAL*8   SR2  , SR6  , SR12
00210 C
00220       DO 10 I=1,N
00230         FX(I) = 0.D0
00240         FY(I) = 0.D0
00250         FZ(I) = 0.D0
00260    10 CONTINUE
00270 C
00280       DO 100 I=1,N-1
00290 C
00300         RXI = RX(I)
00310         RYI = RY(I)
00320         RZI = RZ(I)
00330         FXI = FX(I)
00340         FYI = FY(I)
00350         FZI = FZ(I)
00360 C
00370         DO 50 J=I+1,N
00380 C
00390           RXIJ = RXI  - RX(J)
00400           RXIJ = RXIJ - DNINT(RXIJ/XL)*XL
00410           IF( DABS(RXIJ) .GE. RCOFF )     GOTO 50
00420           RYIJ = RYI  - RY(J)
```

```
00430              RYIJ = RYIJ - DNINT(RYIJ/YL)*YL
00440              IF( DABS(RYIJ) .GE. RCOFF )        GOTO 50
00450              RZIJ = RZI  - RZ(J)
00460              RZIJ = RZIJ - DNINT(RZIJ/ZL)*ZL
00470              IF( DABS(RZIJ) .GE. RCOFF )        GOTO 50
00480  C
00490              RIJSQ= RXIJ*RXIJ + RYIJ*RYIJ + RZIJ*RZIJ
00500              IF( RIJSQ .GE. RCOFF2 )            GOTO 50
00510  C
00520              SR2  = 1.D0/RIJSQ
00530              SR6  = SR2*SR2*SR2
00540              SR12 = SR6*SR6
00550              FIJ  = ( 2.D0*SR12 - SR6 )/RIJSQ
00560              FXIJ = FIJ*RXIJ
00570              FYIJ = FIJ*RYIJ
00580              FZIJ = FIJ*RZIJ
00590              FXI  = FXI  + FXIJ
00600              FYI  = FYI  + FYIJ
00610              FZI  = FZI  + FZIJ
00620  C
00630              FX(J) = FX(J) - FXIJ
00640              FY(J) = FY(J) - FYIJ
00650              FZ(J) = FZ(J) - FZIJ
00660  C
00670    50    CONTINUE
00680  C
00690              FX(I) = FXI
00700              FY(I) = FYI
00710              FZ(I) = FZI
00720  C
00730    100 CONTINUE
00740  C
00750         DO 120 I=1,N
00760              FX(I) = FX(I)*24.D0
00770              FY(I) = FY(I)*24.D0
00780              FZ(I) = FZ(I)*24.D0
00790    120 CONTINUE
00800         RETURN
00810         END
```

A11.5 Calculation of Forces, Torques, and Viscosity for Ferromagnetic Colloidal Dispersions (SUBROUTINE FORCE)

We show the subroutine for computing forces, torques, and viscosity for a model dispersion shown in Sec. 12.1.1. In the subroutine, the nondimensional forces and torques in Eqs. (12.12) to (12.15) and the following nondimensional viscosity are computed:

$$\eta_{yx}^{m*} = \frac{\eta_{yx}^m}{\eta} = -\frac{6\pi}{V^*}\sum_{i=1}^{N}\sum_{\substack{j=1\\(j>i)}}^{N} y_{ij}^* F_{ijx}^* + \frac{4\pi}{V^*}\sum_{i=1}^{N} T_{iz}^* . \tag{A11.1}$$

When the subroutine FORCE is called with *ITREE*=1, the forces and torques alone are computed, and when with *ITREE*=2, the viscosity is also computed in addition. DO loop has the structure of (DO 100 *I*=1, *N*-1; DO 50 *J*=*I*+1,*N*) because of the action-reaction law. The treatment of the Lees-Edwards boundary condition is conducted in Lines 1040 to 1140. In Lines 1470 to 1520, repulsive forces are computed when the surfactant layers of two particles overlap. In Lines 1990 to 2140, the remainders of the torques due to the magnetic field is computed to obtain the total torques.

This subroutine program is written for beginners, so some improvements may be necessary. For example, the calculation of (3.D0*C2) in Lines 1790-1810 is carried out three times, but once may be sufficient. This kind of optimization may be accomplished automatically by many compilers.

```
00010 C*******************************************************************
00020 C*    THIS SUBROUTINE IS FOR CALCULATING FORCES AND TORQUES
00030 C*    FOR A PARTICLE MODEL WITH A POINT DIPOLE MOMENT.
00040 C*------------------------------------------------------------------
00050 C*    N     : NUMBER OF PARTICLES
00060 C*    D     : DIAMETER OF PARTICLE INCLUDING SURFACTANT LAYER
00070 C*            ( =2 FOR THIS CASE )
00080 C*    NDENS : NUMBER DENSITY
00090 C*    RM    : NONDIMENSIONAL PARAMETER OF PARTICLE-PARTICLE INTERACT
00100 C*            (MAGNETIC PARTICLE-PARTICLE FORCE) / (VISCOUS FORCE)
00110 C*    RH    : NONDIMENSIONAL PARAMETER OF FIELD-PARTICLE INTERACT
00120 C*            (MAGNETIC FIELD-PARTICLE FORCE) / (VISCOUS FORCE)
00130 C*    RV    : NONDIMENSIONAL PARAMETER OF PARTICLE-PARTICLE INTERACT
00140 C*            (STERIC PARTICLE-PARTICLE FORCE) / (VISCOUS FORCE)
00150 C*    RCOFF : CUTOFF RADIUS FOR CALCULATION OF INTERACTION
00160 C*    RCOFF2: 2*RCOFF
00170 C*    RCOFFMOB : CUTOFF RADIUS FOR CALCULATION OF MOBILITY FUNCTIONS
00180 C*    L     : DIMENSIONS OF SIMULATION REGION
00190 C*    H     : TIME INTERVAL FOR SIMULATIONS
00200 C*    TM    : THICKNESS OF SURFACTANT LAYER
00210 C*    (HX,HY,HZ)        : APPLIED MAGNETIC FIELD (UNIT VECTOR)
00220 C*                                                          *
00230 C*    RX(N),RY(N),RZ(N) : PARTICLE POSITION (-L/2<RX,RY,RZ<L/2)
00240 C*    NX(N),NY(N),NZ(N) : DIRECTION OF MAGNETIC MOMENT
00250 C*    FX(N),FY(N),FZ(N) : PARTICLE FORCE
00260 C*    TORQX(N),TORQY(N),TORQZ(N) : PARTICLE TORQUE
00270 C*    VISYXMM,VISYXMH,VISXYMM,VISXYMH : VISCOSITIES
00280 C*    DX,CORY : LEES-EDWARDS BOUNDARY CONDITION
00290 C*******************************************************************
00300 C
00310 C          --- CALCULATE FORCES AND TORQUES, WITHOUT VISCOSITIES ---
00320 CCC   ITREE = 1
00330 CCC   CALL FORCE( RCOFF2,ITREE,VISYXMM,VISYXMH,VISXYMM,VISXYMH )
00340 C                  --- CALCULATE FORCES, TORQUES AND VISCOSITIES ---
00350 CCC   ITREE = 2
00360 CCC   CALL FORCE( RCOFF2,ITREE,VISYXMM,VISYXMH,VISXYMM,VISXYMH )
00370 C
00380 C**** SUB FORCE *****
00390       SUBROUTINE FORCE( RCOFF2,ITREE,VISYXMM,VISYXMH,VISXYMM,VISXYMH )
00400 C
00410       IMPLICIT REAL*8 (A-H,O-Z)
00420 C
00430       COMMON /BLOCK1/  RX , RY , RZ
00440       COMMON /BLOCK2/  NX , NY , NZ
00450       COMMON /BLOCK5/  FX , FY , FZ
00460       COMMON /BLOCK6/  TORQX , TORQY , TORQZ
00470       COMMON /BLOCK7/  N , NDENS , RCOFF ,  L , H , TM , D , RCOFFMOB
00480       COMMON /BLOCK8/  RM , RH , RV , HX , HY , HZ
00490       COMMON /BLOCK19/ DX , CORY
00500 C
00510       INTEGER  NN
00520       PARAMETER( NN=1000 , PI=3.141592653589793D0 )
00530 C
00540 C
00550       REAL*8   RX(NN) , RY(NN) , RZ(NN) , NX(NN) , NY(NN) , NZ(NN)
00560       REAL*8   FX(NN) , FY(NN) , FZ(NN)
00570       REAL*8   TORQX(NN) , TORQY(NN) , TORQZ(NN)
00580       REAL*8   NDENS  , L
00590 C
00600       REAL*8   RXI , RYI , RZI , RXIJ , RYIJ , RZIJ
```

```
00610      REAL*8   NXI , NYI , NZI , NXJ  , NYJ  , NZJ
00620      REAL*8   FXI , FYI , FZI , FXIJ , FYIJ , FZIJ
00630      REAL*8   TORQXI , TORQYI , TORQZI , TORQXIJ , TORQYIJ , TORQZIJ
00640      REAL*8   TXIJ , TYIJ , TZIJ , RIJ , RIJ2 , RIJ3 , RIJ4
00650      REAL*8   RM8, RMN, RMN2
00660      REAL*8   C0 , C1 , C2 , C3 , C1X , C1Y , C1Z , C2X , C2Y, C2Z
00670 C
00680      RM8 = 8.D0*RM
00690      RMN = 2.D0/(1.D0+TM)
00700      RMN2 = RMN**2
00710      DO 10 I=1,N
00720        FX(I) = 0.D0
00730        FY(I) = 0.D0
00740        FZ(I) = 0.D0
00750        TORQX(I) = 0.D0
00760        TORQY(I) = 0.D0
00770        TORQZ(I) = 0.D0
00780   10 CONTINUE
00790      IF( ITREE .EQ. 2 ) THEN
00800        VISYXMM = 0.D0
00810        VISYXMH = 0.D0
00820        VISXYMM = 0.D0
00830        VISXYMH = 0.D0
00840      END IF
00850 C
00860 C
00870      DO 100 I=1,N-1
00880 C
00890        RXI = RX(I)
00900        RYI = RY(I)
00910        RZI = RZ(I)
00920        NXI = NX(I)
00930        NYI = NY(I)
00940        NZI = NZ(I)
00950        FXI = FX(I)
00960        FYI = FY(I)
00970        FZI = FZ(I)
00980        TORQXI = TORQX(I)
00990        TORQYI = TORQY(I)
01000        TORQZI = TORQZ(I)
01010 C
01020        DO 50 J=I+1,N
01030 C
01040          RZIJ = RZI  - RZ(J)
01050          RZIJ = RZIJ - DNINT(RZIJ/L)*L
01060          IF( DABS(RZIJ) .GE. RCOFF )      GOTO 50
01070          RXIJ = RXI  - RX(J)
01080          RYIJ = RYI  - RY(J)
01090          CORY = - DNINT( RYIJ/L )
01100          RYIJ = RYIJ + CORY*L
01110          IF( DABS(RYIJ) .GE. RCOFF )      GOTO 50
01120          RXIJ = RXIJ + CORY*DX
01130          RXIJ = RXIJ - DNINT( RXIJ/L )*L
01140          IF( DABS(RXIJ) .GE. RCOFF )      GOTO 50
01150 C
01160          RIJ2 = RXIJ*RXIJ + RYIJ*RYIJ + RZIJ*RZIJ
01170          IF( RIJ2 .GE. RCOFF2 )       GOTO 50
01180 C
01190          IF( RIJ2 .LT. RMN2 ) THEN
01200            RIJ = DSQRT(RIJ2)
01210            RXIJ = RMN*RXIJ/RIJ
01220            RYIJ = RMN*RYIJ/RIJ
01230            RZIJ = RMN*RZIJ/RIJ
01240            RIJ2 = RMN2
01250          END IF
01260          RIJ = DSQRT(RIJ2)
01270          RIJ3 = RIJ*RIJ2
01280          RIJ4 = RIJ2**2
01290          TXIJ = RXIJ/RIJ
01300          TYIJ = RYIJ/RIJ
01310          TZIJ = RZIJ/RIJ
01320          NXJ = NX(J)
```

```
01330           NYJ = NY(J)
01340           NZJ = NZ(J)
01350 C
01360           C1    = NXI*NXJ  + NYI*NYJ   + NZI*NZJ
01370           C2    = NXI*TXIJ + NYI*TYIJ  + NZI*TZIJ
01380           C3    = NXJ*TXIJ + NYJ*TYIJ  + NZJ*TZIJ
01390 C                                    --- MAGNETIC FORCE ---
01400           FXIJ = - ( RM8/RIJ4) * (  ( - C1 + 5.D0*C2*C3 )*TXIJ
01410      &                             - ( C3*NXI + C2*NXJ )  )
01420           FYIJ = - ( RM8/RIJ4) * (  ( - C1 + 5.D0*C2*C3 )*TYIJ
01430      &                             - ( C3*NYI + C2*NYJ )  )
01440           FZIJ = - ( RM8/RIJ4) * (  ( - C1 + 5.D0*C2*C3 )*TZIJ
01450      &                             - ( C3*NZI + C2*NZJ )  )
01460 C                                    --- STERIC REPULSION ---
01470           IF( RIJ .LT. 2.D0 ) THEN
01480             C0 = DLOG( 2.D0 / RIJ )
01490             FXIJ = FXIJ + RV*TXIJ*C0
01500             FYIJ = FYIJ + RV*TYIJ*C0
01510             FZIJ = FZIJ + RV*TZIJ*C0
01520           END IF
01530 C
01540           FXI  = FXI  + FXIJ
01550           FYI  = FYI  + FYIJ
01560           FZI  = FZI  + FZIJ
01570 C
01580           FX(J) = FX(J) - FXIJ
01590           FY(J) = FY(J) - FYIJ
01600           FZ(J) = FZ(J) - FZIJ
01610 C                                    --- PART OF TORQUES ---
01620           C1X  = NYI*NZJ  - NZI*NYJ
01630           C1Y  = NZI*NXJ  - NXI*NZJ
01640           C1Z  = NXI*NYJ  - NYI*NXJ
01650           C2X  = NYI*TZIJ - NZI*TYIJ
01660           C2Y  = NZI*TXIJ - NXI*TZIJ
01670           C2Z  = NXI*TYIJ - NYI*TXIJ
01680 C
01690           TORQXIJ = - ( C1X - 3.D0*C3*C2X )/RIJ3
01700           TORQYIJ = - ( C1Y - 3.D0*C3*C2Y )/RIJ3
01710           TORQZIJ = - ( C1Z - 3.D0*C3*C2Z )/RIJ3
01720           TORQXI  = TORQXI + TORQXIJ
01730           TORQYI  = TORQYI + TORQYIJ
01740           TORQZI  = TORQZI + TORQZIJ
01750 C
01760           C2X  = NYJ*TZIJ - NZJ*TYIJ
01770           C2Y  = NZJ*TXIJ - NXJ*TZIJ
01780           C2Z  = NXJ*TYIJ - NYJ*TXIJ
01790           TORQX(J) = TORQX(J) - ( -C1X - 3.D0*C2*C2X )/RIJ3
01800           TORQY(J) = TORQY(J) - ( -C1Y - 3.D0*C2*C2Y )/RIJ3
01810           TORQZ(J) = TORQZ(J) - ( -C1Z - 3.D0*C2*C2Z )/RIJ3
01820 C                                    --- CAL. VISCOSITIES ---
01830           IF( ITREE .EQ. 2 ) THEN
01840             VISYXMM = VISYXMM + RYIJ*FXIJ
01850             VISXYMM = VISXYMM + RXIJ*FYIJ
01860           END IF
01870 C
01880    50  CONTINUE
01890 C
01900           FX(I) = FXI
01910           FY(I) = FYI
01920           FZ(I) = FZI
01930           TORQX(I) = TORQXI
01940           TORQY(I) = TORQYI
01950           TORQZ(I) = TORQZI
01960 C
01970   100 CONTINUE
01980 C                                    --- FINAL FORM OF TORQUES ---
01990         DO 120 I=1,N
02000           NXI = NX(I)
02010           NYI = NY(I)
02020           NZI = NZ(I)
02030           C1X  = NYI*HZ  - NZI*HY
02040           C1Y  = NZI*HX  - NXI*HZ
```

```
      C1Z   = NXI*HY  - NYI*HX
C
      TORQX(I) = RM*TORQX(I)*2.D0 + RH*C1X
      TORQY(I) = RM*TORQY(I)*2.D0 + RH*C1Y
      TORQZ(I) = RM*TORQZ(I)*2.D0 + RH*C1Z
C                                     --- CAL. VISCOSITIES ---
      IF( ITREE .EQ. 2 ) THEN
        VISYXMH = VISYXMH + TORQZ(I)
      END IF
  120 CONTINUE
    IF( ITREE .EQ. 2 ) THEN
      VISYXMM = - VISYXMM*(6.D0*PI)/L**3
      VISXYMM = - VISXYMM*(6.D0*PI)/L**3
      VISYXMH =   VISYXMH*(4.D0*PI)/L**3
      VISXYMH = - VISYXMH
    END IF
                                                RETURN
                                                END
```

A11.6 Calculation of the Radial Distribution Function by Means of the Canonical Ensemble Algorithm (MCRADIA1.FORT)

We here show a full program for computing the radial distribution function in a three-dimensional Lennard-Jones system. The subroutines called, such as ENECAL, are the same as the ones already shown before. The variables such as *ETOT*(*) have already been explained in the preceding section for ENECAL. The main loop starts after an initial configuration is given. The maximum displacement of particles, per one step, is *DLTA*. A new candidate position (*RXCAN*, *RYCAN*, *RZCAN*) of the particle of interest is determined by random numbers. Then the interaction of the particle at the new position with the ambient particles is examined, and the name of interacting particles, the interaction energies, and the total number of such interacting particles are saved in *PNUM*(*), *ECAN*(*), and *PMX*, respectively. Whether or not such a movement is accepted is determined by the Metropolis transition probability. If the movement is accepted, then both the energy data of the particle and the interaction energies of the old and new partner particles are replaced with new values. This renewing procedure of energies is straightforwardly applicable to renewing the data of *TMX*(*) and *TABLE*(**,*) for the cell index method. When the counter of MC steps, *MCSMPL*, becomes greater than *NRADIAL*, the sampling procedure starts to compute the radial distribution function. The sampling data is accumulated in *SUMRAD*(*) and, after the main loop finishes, the average valuc is computed. The average for each radial is saved in *RADIAL*(*) and lastly printed out in a file. The evaluation method of the radial distribution function may be found in Sec. 11.5.

```
C***********************************************************************
C*
C*           MCRADIA1.FORT
C*
C*    -----------------------------------------------------------
```

```
00060 C*    - NVT MONTE CARLO SIMULATION OF RADIAL DISTRIBUTION   -
00070 C*    - FUNCTIONS FOR THREE-DIMENSIONAL LENNARD-JONES SYSTEM.-
00080 C*    ---------------------------------------------------------
00090 C*             1. CALCULATION OF RADIAL DISTRIBUTION FUNCTIONS
00100 C*             2. NOT USING THE CELL INDEX METHOD
00110 C*
00120 C*    COMMAND PROC. (FOR HITAC-VOS3)
00130 C*     10 ALLOC DD(FT09F001) DS(@AA1.DATA) REN REU : PROCESS
00140 C*     20 ALLOC DD(FT06F001) DS(*)
00150 C*     30 ALLOC DD(FT10F001) DS(AA11.DATA) REN REU : PARA.,MEAN ENE
00160 C*     40 ALLOC DD(FT11F001) DS(AA21.DATA) REN REU : RADIAL FUNC.
00170 C*     50 ALLOC DD(FT19F001) DS(AA091.DATA) SHR REU: POSITION(OLD)
00180 C*     60 ALLOC DD(FT21F001) DS(AA001.DATA) REN REU: POSITION
00190 C*     70 ALLOC DD(FT22F001) DS(AA011.DATA) REN REU
00200 C*     80 RUN MCRADIA1.FORT
00210 C*     90 FREE ALL
00220 C*     95 END
00230 C*
00240 C*                                    VER.1  BY A.SATOH , '95 3/2
00250 C*****************************************************************
00260 C    ETOT(I)    : ENERGY OF PARTICLE I
00270 C    E(*,I)     : INTERACTION ENERGY OF PARTICLE I WITH THE OTHERS
00280 C    ETMX(I)    : TOTAL NUMBER OF PARTICLES INTERACTING WITH I
00290 C    ETABLE(*,I) : NAME OF PARTICLES INTERACTING WITH I
00300 C    MEANENE(**) : MEAN ENERGY OF SYSTEM AT EACH MC STEP
00310 C    DLTA    : MAXIMUM MOVEMENT DISTANCE FOR USUAL MC METHOD
00320 C    DR      : RADIAL INTERVAL FOR CALCULATING RAD.DIST.FUNC.
00330 C    RMX     : MAXIMUM RADIAL
00340 C    R0(*)   : IS USED TO DETERMINE THE UNIT TO WHICH EACH PARTICLE
00350 C              BELONGS ( 1=< * =< NR0MX )
00360 C    R(*)    : REPRESENTATIVE POINT FOR EACH UNIT FOR RAD.DIST.FUNC.
00370 C              ( 1=< * =< NRMX )
00380 C    RADIAL  : RAD.DIST.FUNC.
00390 C    SUMRAD  : SUMMATION OF RAD.DIST.FUNC.
00400 C    0<RX(I)<L ,  0<RY(I)<L ,  0<RZ(I)<L
00410 C---------------------------------------------------------------
00420       IMPLICIT REAL*8 (A-H,O-Z)
00430 C
00440       COMMON /BLOCK1/ RX  , RY  , RZ
00450       COMMON /BLOCK4/ E   , ETOT , ETMX , ETABLE
00460       COMMON /BLOCK7/ NRAN , RAN  , IX
00470       COMMON /BLOCK8/ NDENS, TEMP , RCOFF,  L
00480       COMMON /BLOCK9/ DLTA
00490       COMMON /BLOCK10/ MEANENE
00500       COMMON /BLOCK12/ DR , RMX , CRAD, NRMX , NR0MX
00510       COMMON /BLOCK13/ R  , R0
00520       COMMON /BLOCK14/ RADIAL , SUMRAD
00530 C
00540       INTEGER NN , TT , SS , NNS
00550       PARAMETER( NN=2050 , TT=600 , SS=200 )
00560       PARAMETER( NNS=100000 )
00570       PARAMETER( NRANMX=500000   , PI=3.141592653589793D0 )
00580 C
00590       REAL*8  RX(NN) , RY(NN) , RZ(NN) , E(TT,NN) , ETOT(NN)
00600       REAL*8  NDENS   , L
00610       REAL    MEANENE(NNS)
00620 C
00630       INTEGER ETMX(NN) , ETABLE(TT,NN) , N
00640 C
00650       REAL    RAN(NRANMX)
00660       INTEGER NRAN , IX , NRANCHK
00670 C
00680       REAL*8  RXCAN , RYCAN , RZCAN , RXIJ , RYIJ , RZIJ , RIJSQ
00690       REAL*8  EIJ  , ETOTCAN, ECAN(NN), RCOFF2 , L2 , C1  , C3
00700 C
00710       INTEGER PNUM(NN), PMX
00720       INTEGER MCSMPL  , MCSMPL1 , MCSMPL2
00730       INTEGER MCSMPLST, MCSMPLMX, SMPLSTEP
00740       INTEGER NGRAPH1 , NOPT    , NP  , IC1  , ETMXMX
00750 C
00760       INTEGER NRMX  , NR0MX , NRADIAL, RADCOUNT
00770       REAL*8  R0(SS) , R(SS)  , RADIAL(SS) , SUMRAD(SS) , R0CHKSQ
```

```
00780 C
00790                                                      NP=9
00800 C
00810              OPEN(9, FILE='@AA1.DATA',STATUS='UNKNOWN',TYPE='TEXT')
00820              OPEN(10,FILE='AA11.DATA',STATUS='UNKNOWN',TYPE='TEXT')
00830              OPEN(11,FILE='AA21.DATA',STATUS='UNKNOWN',TYPE='TEXT')
00840              OPEN(21,FILE='AA001.DATA',STATUS='UNKNOWN',TYPE='TEXT')
00850              OPEN(22,FILE='AA011.DATA',STATUS='UNKNOWN',TYPE='TEXT')
00860 C
00870 C                                       --- PARAMETER (1) ---
00880 C                           +  N=32,108,256,500,864,1372,2048,... +
00890 C                           +  NDENS=0.1 , 0.65 , 1.2            +
00900     N       = 256
00910     NDENS   = 0.65D0
00920     TEMP    = 1.2D0
00930     RCOFF   = 2.5D0
00940     DLTA    = 0.11D0
00950 C                                       --- PARAMETER (3) ---
00960 C                               +++ MCSMPLMX.GE.200 ! +++
00970     MCSMPLST = 1
00980     MCSMPLMX = 100000
00990     NRADIAL  = 10000
01000 C                                       --- PARAMETER (4) ---
01010     NGRAPH1  = 50000
01020     SMPLSTEP = 1
01030     NOPT     = 20
01040 C                                       --- PARAMETER (5) ---
01050     NCHK     = 50
01060     NACCEPT  = 0
01070 C                                       --- PARAMETER (6) ---
01080     C0 = ( 4.D0/NDENS )**(1./3.)
01090     IQ = IDNINT( DBLE(N/4)**(1./3.) )
01100     L = C0*DBLE(IQ)
01110     IF( L .LE. 2.D0*RCOFF ) THEN
01120       WRITE(NP,*) '****** N IS TOO SMALL.....'
01130       STOP
01140     END IF
01150     RCOFF2 = RCOFF**2
01160     L2 = L/2.D0
01170 C                                       --- PARAMETER (7) ---
01180     IX = 0
01190     CALL RANCAL( NRANMX, IX, RAN )
01200     NRAN    = 1
01210     NRANCHK = NRANMX - 5*N
01220 C                                       --- PARAMETER (8) ---
01230     DR      = 1.D0/DBLE( 20 )
01240     RMX     = 5.D0
01250     CALL RADIALR( R0, R )
01260     R0CHKSQ = R0(NR0MX)**2
01270     CRAD    = ( L**3/DBLE(N**2) ) / ( 4.D0*PI*DR )
01280 C
01290 C    ----------------------------------------------------------------
01300 C    ----------------  INITIAL CONFIGURATION  ------------------
01310 C    ----------------------------------------------------------------
01320 C
01330 C                               --- SET INITIAL POSITIONS ---
01340 CCC   OPEN(19,FILE='AA091.DATA',STATUS='OLD',TYPE='TEXT')
01350 CCC    READ(19,462)  N , L
01360 CCC    READ(19,464) (RX(I),I=1,N) , (RY(I),I=1,N) , (RZ(I),I=1,N)
01370 CCC   CLOSE(19,STATUS='KEEP')
01380 CCC    GOTO 7
01390 C
01400     CALL INIPOSIT( N , NDENS , L )
01410 C                                 --- CALCULATE ENERGY ---
01420   7 CALL ENECAL( N , RCOFF2 )
01430 C                                     --- PRINT OUT ---
01440     WRITE(NP,12) N, NDENS, TEMP, L, RCOFF, DLTA
01450     WRITE(NP,14) MCSMPLMX, NGRAPH1
01460     WRITE(NP,16) DR , RMX , NRMX
01470 C
01480 C                                 --- INITIALIZATION ---
01490     ETMXMX  = 0
```

```
01500 C
01510      RADCOUNT= 0
01520      DO 50 I=1,NRMX
01530         SUMRAD(I) = 0.D0
01540   50 CONTINUE
01550 C
01560 C     ----------------------------------------------------------------
01570 C     --------------   START OF MONTE CARLO PROGRAM   -------------
01580 C     ----------------------------------------------------------------
01590 C
01600      MCSMPL1 =MCSMPLST
01610      MCSMPL2 =MCSMPLMX
01620 C
01630      DO 500 MCSMPL = MCSMPL1 , MCSMPL2
01640 C
01650        DO 400 I=1,N
01660 C                            ----------  (1) CANDIDATE  ---------
01670 C
01680 C                                          +++ POSITION +++
01690 C
01700          RXCAN = RX(I) + DLTA*( 1.D0 - 2.D0*DBLE(RAN(NRAN)) )
01710          NRAN  = NRAN + 1
01720          RYCAN = RY(I) + DLTA*( 1.D0 - 2.D0*DBLE(RAN(NRAN)) )
01730          NRAN  = NRAN + 1
01740          RZCAN = RZ(I) + DLTA*( 1.D0 - 2.D0*DBLE(RAN(NRAN)) )
01750          NRAN  = NRAN + 1
01760          IF( RXCAN .GE. L   ) RXCAN = RXCAN - L
01770          IF( RXCAN .LT. 0.D0 ) RXCAN = RXCAN + L
01780          IF( RYCAN .GE. L   ) RYCAN = RYCAN - L
01790          IF( RYCAN .LT. 0.D0 ) RYCAN = RYCAN + L
01800          IF( RZCAN .GE. L   ) RZCAN = RZCAN - L
01810          IF( RZCAN .LT. 0.D0 ) RZCAN = RZCAN + L
01820 C
01830 C                            -------  (2) CALCULATE ENERGY -------
01840          PMX   = 0
01850          ETOTCAN= 0.D0
01860 C                                          +++ ENERGY +++
01870          DO 200 J=1,N
01880 C
01890            IF( J .EQ. I )              GOTO 200
01900 C
01910            RXIJ = RX(J) - RXCAN
01920            IF( RXIJ .GE. L2 ) RXIJ = RXIJ - L
01930            IF( RXIJ .LT. -L2 ) RXIJ = RXIJ + L
01940            IF( DABS(RXIJ) .GE. RCOFF )  GOTO 200
01950            RYIJ = RY(J) - RYCAN
01960            IF( RYIJ .GE.  L2 ) RYIJ = RYIJ - L
01970            IF( RYIJ .LT. -L2 ) RYIJ = RYIJ + L
01980            IF( DABS(RYIJ) .GE. RCOFF )  GOTO 200
01990            RZIJ = RZ(J) - RZCAN
02000            IF( RZIJ .GE.  L2 ) RZIJ = RZIJ - L
02010            IF( RZIJ .LT. -L2 ) RZIJ = RZIJ + L
02020            IF( DABS(RZIJ) .GE. RCOFF )  GOTO 200
02030 C
02040            RIJSQ = RXIJ**2 + RYIJ**2 + RZIJ**2
02050            IF( RIJSQ .GE. RCOFF2 )    GOTO 200
02060 C                         ++++++++++++++++++++++++++++++++++
02070 C                         PNUM(*):SAVE INTERACTIVE PARTICLE
02080 C                                NAMES.
02090 C                         ECAN(*):SAVE INTERACTION ENERGIES
02100 C                            *   :1,2,3,....,PMX
02110 C                         ++++++++++++++++++++++++++++++++++
02120 C
02130            PMX = PMX + 1
02140 C
02150            SR2  = 1.D0/RIJSQ
02160            SR6  = SR2*SR2*SR2
02170            EIJ  = SR6*( SR6-1.D0 )
02180 C
02190            PNUM(PMX) = J
02200            ECAN(PMX) = EIJ*4.D0
02210            ETOTCAN   = ETOTCAN + EIJ
```

```
C
  200      CONTINUE
C
           ETOTCAN = ETOTCAN*4.D0
C
C     --------  (3) JUDGEMENT ACCORDING TO METROPOLIS METHOD --------
C
           C3 = ETOTCAN - ETOT(I)
           IF( C3 .GE. 0.D0 )THEN
             IF( DBLE(RAN(NRAN)) .GE. DEXP(-C3/TEMP) )THEN
               NRAN = NRAN + 1
               GOTO 400
             END IF
             NRAN = NRAN + 1
           END IF
C                                          ++++++++++++++++++++++
C                                          CANDIDATES ARE ACCEPTED
C                                          ++++++++++++++++++++++
           NACCEPT = NACCEPT + 1
C
C                                 -------  (4) RENEW DATA  -------
C
           RX(I)   = RXCAN
           RY(I)   = RYCAN
           RZ(I)   = RZCAN
           ETOT(I) = ETOTCAN
C                                          ----- ENERGY DATA -----
C
C                                        ++++++++++++++++++++++++
C                                         FOR PARTICLES WHICH
C                                         INTERACT WITH PARTICLE I
C                                        ++++++++++++++++++++++++
           IF ( PMX .EQ. 0 )      GOTO 365
C
           DO 360 JJ=1,PMX
             J = PNUM(JJ)
             IF( ETMX(J) .EQ. 0 )  GOTO 355
             DO 350 JJJ=1,ETMX(J)
               IF( ETABLE(JJJ,J) .EQ. I ) THEN
                 ETOT(J)  = ETOT(J) - E(JJJ,J) + ECAN(JJ)
                 E(JJJ,J) = ECAN(JJ)
                 GOTO 360
               END IF
  350        CONTINUE
  355        ETMX(J)  = ETMX(J) + 1
             ETABLE(ETMX(J),J) = I
             E(     ETMX(J),J) = ECAN(JJ)
             ETOT(J)  = ETOT(J) + ECAN(JJ)
  360      CONTINUE
C                                    +++++++++++++++++++++++++++++
C                                     FOR PARTICLES WHICH BECOME
C                                     OUT OF RELATION TO PARTICLE I
C                                    +++++++++++++++++++++++++++++
  365      IF( ETMX(I) .EQ. 0 )       GOTO 383
C
           DO 380 JJ=1,ETMX(I)
             J = ETABLE(JJ,I)
             IF( PMX .EQ. 0 )         GOTO 373
             DO 370 JJJ=1,PMX
               IF( PNUM(JJJ) .EQ. J )  GOTO 380
  370        CONTINUE
  373        IF( ETMX(J) .EQ. 0 )     GOTO 380
             IC1 = ETMX(J)
             DO 375 JJJ=1,IC1
               IF( ETABLE(JJJ,J) .EQ. I ) THEN
                 ETOT(J)  = ETOT(J) - E(JJJ,J)
                 IF( JJJ .EQ. IC1 ) THEN
                   ETMX(J) = ETMX(J) - 1
                 ELSE
                   ETABLE(JJJ,J) = ETABLE(IC1,J)
                   E(     JJJ,J) = E(     IC1,J)
                   ETMX(J) = ETMX(J) - 1
```

```
02940                 END IF
02950                   GOTO 380
02960                 END IF
02970    375     CONTINUE
02980    380     CONTINUE
02990 C                             ++++++++++++++++++++++++++++++++++
03000 C                             RENEW DATA CONCERNING PARTICLE I
03010 C                             ++++++++++++++++++++++++++++++++++
03020    383     ETMX(I) = PMX
03030            IF( PMX .EQ. 0 )      GOTO 400
03040            DO 385 JJ=1,PMX
03050              ETABLE(JJ,I) = PNUM(JJ)
03060              E(JJ,I) = ECAN(JJ)
03070    385     CONTINUE
03080 C
03090 C
03100    400     CONTINUE
03110 C
03120 C                                ----- ENERGY OF SYSTEM -----
03130            C1 = 0.D0
03140            DO 450 J=1,N
03150              C1 = C1 + ETOT(J)
03160    450     CONTINUE
03170            MEANENE(MCSMPL) = REAL(C1)/REAL(2*N)
03180 C
03190 C                                 ----- RADIAL DISTRIBUTION -----
03200            IF( MCSMPL .GT. NRADIAL ) THEN
03210              RADCOUNT = RADCOUNT + 1
03220 C                                     --- ALONG Z-AXIS ---
03230            CALL RADIALCA( N, L, RADIAL, R0CHKSQ )
03240            DO 455 J=1,NRMX
03250              SUMRAD(J) = SUMRAD(J) + RADIAL(J)
03260    455     CONTINUE
03270            END IF
03280 C                          --- DATA OUTPUT FOR GRAPHICS (1) ---
03290 C
03300            IF( MOD(MCSMPL,NGRAPH1) .EQ. 0 ) THEN
03310              NOPT = NOPT + 1
03320              WRITE(NOPT,462)  N , L
03330              WRITE(NOPT,464) (RX(I),I=1,N),(RY(I),I=1,N),(RZ(I),I=1,N)
03340            END IF
03350 C
03360 C
03370 C                          --- CHECK OF MAXIMUM OF ETMX ---
03380            DO 490 J=1,N
03390              IF( ETMX(J) .GT. ETMXMX ) ETMXMX = ETMX(J)
03400    490     CONTINUE
03410 C
03420 C                      --- CHECK OF THE SUM OF RANDOM NUMBERS ---
03430 C
03440            IF( NRAN .GE. NRANCHK )THEN
03450              CALL RANCAL( NRANMX, IX, RAN )
03460              NRAN = 1
03470            END IF
03480 C                       ----- ADJUST MAXIMUM DISPLACEMENT ----
03490 C
03500          IF( MOD(MCSMPL,NCHK) .EQ. 0 ) THEN
03510            RATIO = REAL(NACCEPT)/REAL(N*NCHK)
03520            IF( RATIO .GT. 0.5 ) THEN
03530              DLTA = DLTA*1.05D0
03540            ELSE
03550              DLTA = DLTA*0.95D0
03560            END IF
03570            NACCEPT = 0
03580          END IF
03590 C
03600 C
03610 C
03620   500 CONTINUE
03630 C
03640 C     ---------------------------------------------------------------
03650 C     ---------------- END OF MONTE CARLO PROGRAM ------------------
```

```
03660 C      -----------------------------------------------------------------
03670 C
03680 C                                     --- PRINT OUT RESULTS ---
03690       CALL PRNTDATA( MCSMPL1 , MCSMPL2 , NP )
03700       WRITE(NP,612) ETMXMX , MCSMPL1 , MCSMPL2
03710 C
03720 C                          --- DATA OUTPUT FOR GRAPHICS (2) ---
03730       IC = 0
03740       WRITE(10,1012) N, IC, NDENS, TEMP, RCOFF, L
03750       WRITE(10,1013) DLTA, MCSMPLMX, SMPLSTEP
03760       WRITE(10,1014) ( MEANENE(I),I=SMPLSTEP,MCSMPLMX,SMPLSTEP )
03770 C
03780 C                          --- DATA OUTPUT FOR GRAPHICS (3) ---
03790       DO 1100 I=1,NRMX
03800          RADIAL(I) = SUMRAD(I)/DBLE(RADCOUNT)
03810  1100 CONTINUE
03820       WRITE(11,1102) NRMX , RADCOUNT
03830       DO 1105 I=1,NRMX
03840          WRITE(11,1104) I, R(I), RADIAL(I)
03850  1105 CONTINUE
03860 C
03870 C                                     --- PRINT OUT R.D.F. ---
03880       WRITE(NP,1106) RADCOUNT
03890       DO 1110 I=1,NRMX
03900          WRITE(NP,1108) I , R(I) , RADIAL(I)
03910  1110 CONTINUE
03920                                      CLOSE(9, STATUS='KEEP')
03930                                      CLOSE(10,STATUS='KEEP')
03940                                      CLOSE(11,STATUS='KEEP')
03950                                      CLOSE(21,STATUS='KEEP')
03960                                      CLOSE(22,STATUS='KEEP')
03970 C      ----------------------- FORMAT -----------------------------
03980    12 FORMAT(/1H ,'-------------------------------------------------'
03990       &         /1H ,'-            MONTE CARLO METHOD            -'
04000       &         /1H ,'-------------------------------------------------'
04010       &        //1H ,'N=',I4, 3X ,'NDENS=',F5.2, 2X ,'TEMP=',F5.2, 2X ,
04020       &          'L=',F6.2
04030       &         /1H ,'RCOFF=',F6.2, 2X ,'DLTA=',F7.4)
04040    14 FORMAT(/1H ,'MCSMPLMX=',I8, 2X ,'NGRAPH1=',I8)
04050    16 FORMAT(/1H ,'DR=',F6.3, 3X ,'RMX=',F6.3, 3X ,'NRMX=',I4/)
04060   462 FORMAT( I4 , F9.4 )
04070   464 FORMAT( (8F10.5) )
04080   612 FORMAT(///1H ,18X, 'MAXIMUM OF ETMX=',I7
04090       &          /1H ,18X, 'START OF MC SAMPLING STEP=',I7
04100       &          /1H ,18X, 'END  OF MC SAMPLING STEP=',I7/)
04110  1012 FORMAT( I7 , I3 , 4F8.4 )
04120  1013 FORMAT( F8.4 , 2I8 )
04130  1014 FORMAT( (5E16.9) )
04140  1102 FORMAT( 2I8 )
04150  1104 FORMAT( I6 , F8.4 , F10.5 )
04160  1106 FORMAT(///1H ,'------ RADIAL DISTRIBUTION ------'
04170       &          /1H ,'     SAMPLING NUMBER=',I7/)
04180  1108 FORMAT(1H ,'NR=',I4, 3X ,'R=',F6.3, 3X , 'R.D.F.=',F9.4)
04190       STOP
04200       END
04210 C***********************************************************************
04220 C*********************** SUBROUTINE *****************************
04230 C***********************************************************************
04240 C
04250 C**** SUB PRNTDATA ****
04260       SUBROUTINE PRNTDATA( MCSST, MCSMX, NP )
04270 C
04280       COMMON /BLOCK10/ MEANENE
04290 C
04300       PARAMETER( NNS=100000 )
04310 C
04320       INTEGER MCSST    , MCSMX    , NP
04330       REAL    MEANENE(NNS)
04340 C
04350       REAL    AMEANENE(10) , C0
04360       INTEGER IC , IMC(0:10) , JS , JE
04370 C
```

```
04380 C                              --- PRINT OUT PROCESS OF ENERGY ---
04390       IC = ( MCSMX-MCSST+1 )/50
04400       DO 20 I= MCSST-1+IC , MCSMX , IC
04410          WRITE(NP,10) I , MEANENE(I)
04420    20 CONTINUE
04430 C                                  --- MONTE CARLO STEP AVERAGE ---
04440       IC = ( MCSMX-MCSST+1 )/10
04450       DO 30 I=0,10
04460       IMC(I) = MCSST - 1 + IC*I
04470       IF( I .EQ. 10 ) IMC(I) =MCSMX
04480    30 CONTINUE
04490 C
04500 C
04510       DO 35 I=1,10
04520         AMEANENE(I) = 0.
04530    35 CONTINUE
04540 C
04550       DO 50 I=1,10
04560         JS = IMC(I-1) + 1
04570         JE = IMC(I)
04580         DO 40 J=JS,JE
04590           AMEANENE(I) = AMEANENE(I) + MEANENE(J)
04600    40   CONTINUE
04610    50 CONTINUE
04620 C
04630       DO 70 I=1,10
04640         C0        = REAL( IMC(I)-IMC(I-1) )
04650         AMEANENE(I) = AMEANENE(I)/C0
04660    70 CONTINUE
04670 C                                    --- PRINT OUT MEAN ENERGY ---
04680       WRITE(NP,75)
04690       DO 90 I=1,10
04700        WRITE(NP,80)I,IMC(I-1)+1,IMC(I), AMEANENE(I)
04710    90 CONTINUE
04720 C     ----------------------------------------------------------------
04730    10 FORMAT(1H ,'MCSMPL=',I5, 3X ,'MEAN ENERGY=',E12.5)
04740    75 FORMAT(//1H ,'----------------------------------------------'
04750       &      /1H ,'           MONTE CARLO AVERAGE                '
04760       &      /)
04770    80 FORMAT(1H ,'I=',I2, 2X ,'SMPLMN=',I5, 2X ,'SMPLMX=',I5, 2X,
04780       &           'MEAN ENERGY=',E12.5/)
04790                                                     RETURN
04800                                                     END
04810 C**** RADIALR ****
04820       SUBROUTINE RADIALR( R0, R )
04830 C
04840       IMPLICIT REAL*8 (A-H,O-Z)
04850 C
04860       INTEGER  SS
04870       PARAMETER( SS=200 )
04880 C
04890       COMMON /BLOCK12/ DR , RMX , CRAD , NRMX ,NR0MX
04900 C
04910       INTEGER  NRMX   , NR0MX
04920       REAL*8   R0(SS) , R(SS) , C0 , C1
04930 C
04940       C0   = DR/2.D0
04950 C                                    --- CALCULATE R(*) ---
04960       R0(1) = 0.8D0
04970       R(1)  = R0(1) + C0
04980 C
04990       DO 50 I=2,210
05000         R0(I) = R0(1) + DBLE(I-1)*DR
05010         C1    = R0(I) + C0
05020         IF( C1.GT.RMX )THEN
05030           NR0MX = I
05040           NRMX  = I-1
05050           GOTO 60
05060         END IF
05070         R(I) = C1
05080    50 CONTINUE
05090    60                                            RETURN
```

```
05100                                               END
05110 C**** RADIALCA ****
05120       SUBROUTINE RADIALCA( N, L, RADIAL, R0CHKSQ )
05130 C
05140       IMPLICIT REAL*8 (A-H,O-Z)
05150 C
05160       COMMON /BLOCK1/  RX  , RY  , RZ
05170       COMMON /BLOCK12/ DR  , RMX , CRAD, NRMX , NR0MX
05180       COMMON /BLOCK13/ R   , R0
05190 C
05200       INTEGER  NN , TT , SS
05210       PARAMETER( NN=2050 , TT=600 , SS=200 )
05220 C
05230       INTEGER  N
05240       REAL*8   RX(NN) , RY(NN) , RZ(NN) , L
05250       REAL*8   RXIJ  , RYIJ  , RZIJ  , RIJ , RIJSQ
05260 C
05270       INTEGER  NR0  , NRMX  , NR0MX
05280       REAL*8   R(SS) , R0(SS) , RADIAL(SS) , L2
05290 C
05300       L2 = L/2.D0
05310       DO 10 I=1,NRMX
05320         RADIAL(I) = 0.D0
05330    10 CONTINUE
05340 C                           --- CALCULATE RADIAL DIST. FUNC. ---
05350       DO 200 I=1,N
05360 C
05370 C
05380           DO 140 J=1,N
05390            IF( J .EQ. I ) GOTO 140
05400 C
05410            RXIJ = RX(J) - RX(I)
05420            IF( RXIJ .GE. L2 ) RXIJ = RXIJ - L
05430            IF( RXIJ .LT. -L2 ) RXIJ = RXIJ + L
05440            RYIJ = RY(J) - RY(I)
05450            IF( RYIJ .GE. L2 ) RYIJ = RYIJ - L
05460            IF( RYIJ .LT. -L2 ) RYIJ = RYIJ + L
05470            RZIJ = RZ(J) - RZ(I)
05480            IF( RZIJ .GE. L2 ) RZIJ = RZIJ - L
05490            IF( RZIJ .LT. -L2 ) RZIJ = RZIJ + L
05500 C
05510            RIJSQ= RXIJ**2 + RYIJ**2 + RZIJ**2
05520 C                              +++ OVER CHECK AREA(RADIAL) +++
05530            IF( RIJSQ .GE. R0CHKSQ ) GOTO 140
05540            RIJ = DSQRT(RIJSQ)
05550 C
05560            DO 130 NR0=2,NR0MX
05570             IF( R0(NR0).GT.RIJ ) THEN
05580               RADIAL(NR0-1) = RADIAL(NR0-1) + 1.D0
05590               GOTO 140
05600             END IF
05610   130      CONTINUE
05620 C
05630   140     CONTINUE
05640 C
05650 C
05660   200 CONTINUE
05670 C                           --- DIVIDE THE DATA BY SMALL VOLUME ---
05680       DO 210 I=1,NRMX
05690         RADIAL(I) = CRAD * RADIAL(I)/R(I)**2
05700   210 CONTINUE
05710       RETURN
05720       END
```

A11.7 Calculation of Resistance Functions (RESIST.FORT, RESIST2.FORT)

We here show programs for computing the resistance functions in terms of the algebraic equations explained in Appendix A10.   By the full program RESIST.FORT, the resistance functions X_{11}^{A*}, X_{12}^{A*}, X_{11}^{G*}, and X_{12}^{G*} are computed from the algebraic equations with $l=1$, $m=0$, $A_{lm}=1$, and $B_{lm}=0$, and X_{11}^{C*} and X_{12}^{C*} are evaluated for $l=1$, $m=0$, $A_{lm}=0$, and $B_{lm}=1$.   Also, by RESIST2.FORT, Y_{11}^{A*}, Y_{12}^{A*}, Y_{11}^{B*}, Y_{12}^{B*}, Y_{11}^{G*}, and Y_{12}^{G*} are computed for $l=m=1$, $A_{lm}=1$, and $B_{lm}=0$, and Y_{11}^{C*}, Y_{12}^{C*}, Y_{11}^{H*}, and Y_{12}^{H*} are evaluated for $l=m=1$, $A_{lm}=0$, and $B_{lm}=1$.   The subroutines necessary in RESIST2.FORT are the same as in RESIST.FORT.

```
(RESIST.FORT)

00010 C************************************************************************
00020 C*                 RESIST.FORT
00030 C*
00040 C*
00050 C*        ---- CALCULATION OF RESISTANCE FUNC. WHICH IS -----
00060 C*        ---- GIVEN BY S.KIM AND R.MIFFLIN           -----
00070 C*        ---- (PHYS. FLUIDS, 28(1985), 2033).        -----
00080 C*
00010 C*          (1) ALM=1 , BLM=0 , M=0  (IPATH=1 ) X11A,X12A,X11G,X12G
00020 C*          (3) ALM=0 , BLM=1 , M=0  (IPATH=3 ) X11C,X12C
00030 C*
00040 C*                             VER.1 BY A. SATOH, '96 1/12
00050 C************************************************************************
00060 C
00070 C----------------------------------------------------------------------
00080       IMPLICIT REAL*8 (A-H,O-Z)
00090 C
00100       PARAMETER( NN=1000 , NNN=667 )
00110       PARAMETER( NRADIAL=1000 , PI=3.141592653589793D0 )
00120 C
00130       COMMON /BLOCK1/ P0  , P1  , P2
00140       COMMON /BLOCK2/ PP1 , PP2
00150       COMMON /BLOCK3/ P0DEF0 , P0DEF1 , P0DEF2 , P1DEF1 , P1DEF2
00160       COMMON /BLOCK4/ AA  , XX  , BB
00170       COMMON /BLOCK5/ M  , S  , ALM , BLM
00180       COMMON /BLOCK6/ DIST, AANS1, BANS1, CANS1, AANS2, BANS2, CANS2
00190       COMMON /BLOCK7/ X11A , X12A , X11G , X12G
00280       COMMON /BLOCK8/ X11C, X12C
00290 C
00300       REAL*8  P0(2,0:NN)  , P1(2,0:NN)  , P2(2,0:NN)
00310       REAL*8  PP1(2,0:NN) , PP2(2,0:NN)
00320       REAL*8  P0DEF0(2,0:NN) , P0DEF1(2,0:NN) , P0DEF2(2,0:NN)
00330       REAL*8  P1DEF1(2,0:NN) , P1DEF2(2,0:NN)
00340       REAL*8  AA(NN,NN)  , XX(NN)  , BB(NN)
00350       REAL*8  DIST(NRADIAL)
00360       REAL*8  AANS1(2,NRADIAL), BANS1(2,NRADIAL), CANS1(2,NRADIAL)
00370       REAL*8  AANS2(2,NRADIAL), BANS2(2,NRADIAL), CANS2(2,NRADIAL)
00380       REAL*8  X11A(NRADIAL), X12A(NRADIAL)
00390       REAL*8  X11G(NRADIAL), X12G(NRADIAL)
00400       REAL*8  X11C(NRADIAL), X12C(NRADIAL)
00410 C
00420       REAL*8  CP0(2) , CP1(2) , CP2(2) , CPP1(2) , CPP2(2)
00430       REAL*8  CP0DEF0(2) , CP0DEF1(2) , CP0DEF2(2)
00440       REAL*8  CP1DEF1(2) , CP1DEF2(2)
00450       REAL*8  A1, B1, C1, D1, A2, B2, C2, D2, A3, B3, C3, D3
00460       REAL*8  AAWK(NNN,NNN) , XXWK(NNN) , BBWK(NNN)
00470       REAL*8  THETA1 , COS1 , R1 , THETA2 , COS2 , R2 , RR
00480       INTEGER NMX, NMX2, NMX3, INUM , ITREE , IPATH , ISTREET
00490 C
00500                                                NP=10
00510       OPEN(10,FILE='@aaa.data',STATUS='UNKNOWN')
00520       OPEN(11,FILE='aaa.data',STATUS='UNKNOWN')
00530 C                                --- PARAMETER (1) ---
00540       M  = 0
00550       IPATH = 1
00560 CCC   IPATH = 3
```

```
C                                             --- PARAMETER (2) ---
      E    = 1.0E-15
C                                             --- PARAMETER (3) ---
      IF( IPATH .EQ. 1 ) THEN
        ALM = 1.D0
        BLM = 0.D0
      ELSE IF( IPATH .EQ. 3 ) THEN
        ALM = 0.D0
        BLM = 1.D0
      END IF
C
C     -------------------------------------------------------------
      DO 400 ISTREET=1,2
C
      IF( ISTREET .EQ. 1 )  S = -1.D0
      IF( ISTREET .EQ. 2 )  S =  1.D0
C
      WRITE(NP,5) M , S , ALM , BLM
C
C                                            --- INITIALIZATION ---
      INUM = 0
C
C     -------------------------------------------------------------
C     ------------------ START OF MAIN ROUTINE ----------------------
C     -------------------------------------------------------------
C
C                         --- CHANGE DISTANCE BETWEEN PARTICLES ---
C                         +++     NMX MUST BE EVEN NUMBER      +++
CCC   DO 300 II = 20, 200
      DO 300 II = 21, 21
C
        INUM = INUM + 1
        RR   = DBLE(II)/10.D0
        NMX  = 60
        NMX2 = 2*NMX
        NMX3 = 3*NMX
C       ------------------------------------ START OF MAIN PART ------
C                                              --- CHANGE THETA ---
        DO 200 III=0,NMX-1
C
          I      = 3*III + 1
          THETA1 = DBLE(III)/DBLE(NMX-1)*PI
          COS1   = DCOS(THETA1)
          C1     = RR + COS1
          C2     = DSIN(THETA1)
          R2     = DSQRT( C1**2 + C2**2 )
          COS2   = - C1/R2
          R1     = 1.D0
C                                      --- CAL. LEGENDRE FUNCS ---
C                                              +++ (1) P(0,N) +++
          ITREE = 1
          CALL LEGENDRE( P0 , M , NMX , COS1 , ITREE )
          ITREE = 2
          CALL LEGENDRE( P0 , M , NMX , COS2 , ITREE )
C                                              +++ (2) P(1,N) +++
CCC        M1 = M + 1
CCC        ITREE = 1
CCC        CALL LEGENDRE( P1 , M1, NMX , COS1 , ITREE )
CCC        ITREE = 2
CCC        CALL LEGENDRE( P1 , M1, NMX , COS2 , ITREE )
C                                             +++ (3) PP(1,N) +++
          M1 = M + 1
          ITREE = 1
          CALL LEGENDR2( PP1 , M1, NMX , COS1 , ITREE )
          ITREE = 2
          CALL LEGENDR2( PP1 , M1, NMX , COS2 , ITREE )
C                                    +++ (4) DEF0. OF P(0,N) +++
          ITREE = 1
          CALL LEGEDEF0( P0DEF0 , PP1 , NMX , COS1 , ITREE )
          ITREE = 2
          CALL LEGEDEF0( P0DEF0 , PP1 , NMX , COS2 , ITREE )
C                                    +++ (5) DEF1. OF P(0,N) +++
```

```
            ITREE = 1
            CALL LEGEDEF1( P0DEF1 , P0 , NMX , COS1 , ITREE )
            ITREE = 2
            CALL LEGEDEF1( P0DEF1 , P0 , NMX , COS2 , ITREE )
C                                 +++ (6) DEF2. OF P(0,N) +++
CCC          ITREE = 1
CCC          CALL LEGEDEF2( P0DEF2 , P0 , NMX , COS1 , ITREE )
CCC          ITREE = 2
CCC          CALL LEGEDEF2( P0DEF2 , P0 , NMX , COS2 , ITREE )
C
            IF( IPATH .EQ. 3 )  GOTO 118
C                             --- CAL. COEFFICIENTS OF MATRIX ---
C                             +++     AA(I,J)*XX(I)= BB(I)    +++
C                                        +++ (1) FOR VZ +++
            DO 100 N=1,NMX
C
              CP0(1)     = P0(1,N)
              CP0(2)     = P0(2,N)
              CP0DEF1(1) = P0DEF1(1,N)
              CP0DEF1(2) = P0DEF1(2,N)
              A1 = A1CAL( N, M, COS1, COS2, R1, R2, CP0, CP0DEF1, S )
              B1 = B1CAL( N, M, COS1, COS2, R1, R2, CP0, CP0DEF1, S )
CCC            C1 = C1CAL( N, M, COS1, COS2, R1, R2, CP0, S )
C
              J  = 3*(N-1) + 1
              AA(I,J)   = A1
              AA(I,J+1) = B1
CCC            AA(I,J+2) = 0.D0
C
  100       CONTINUE
            BB(I) = 1.D0
C                                         +++ (2) FOR VR-VQ +++
            DO 110 N=1,NMX
C
              CP0(1)     = P0(1,N)
              CP0(2)     = P0(2,N)
              CP1(1)     = PP1(1,N)
              CP1(2)     = PP1(2,N)
              A2 = A2CAL( N, M, COS1, COS2, R1, R2, CP0, CP1, S )
              B2 = B2CAL( N, M, COS1, COS2, R1, R2, CP0, CP1, S )
CCC            C2 = C2CAL( N, M, COS1, COS2, R1, R2, CP1, S )
C
              J  = 3*(N-1) + 1
              AA(I+1,J)   = A2
              AA(I+1,J+1) = B2
CCC            AA(I+1,J+2) = C2
C
  110       CONTINUE
            BB(I+1) = 0.D0
            GOTO 200
C                                          +++ (3) FOR VQ +++
  118       DO 120 N=1,NMX
C
              CPP1(1)    = PP1(1,N)
              CPP1(2)    = PP1(2,N)
              CP0DEF0(1) = P0DEF0(1,N)
              CP0DEF0(2) = P0DEF0(2,N)
CCC            A3 = A3CAL( N, M, COS1, COS2, R1, R2, CPP1, S )
CCC            B3 = B3CAL( N, M, COS1, COS2, R1, R2, CPP1, S )
              C3 = C3CAL( N, M, COS1, COS2, R1, R2, CP0DEF0, S )
C
              J  = 3*(N-1) + 1
CCC            AA(I+2,J)   = 0.D0
CCC            AA(I+2,J+1) = 0.D0
              AA(I+2,J+2) = C3
C
  120       CONTINUE
            BB(I+2) = 1.D0
C
  200     CONTINUE
C         ------------------------------------ END OF MAIN PART ------
C
```

```
CCC       do 990 i=1,nmx3
CCC         write(np,*) '-------------------- i=',i
CCC         do 990 j=1,nmx3,6
CCC           write(np,988)i, aa(i,j),aa(i,j+1),aa(i,j+2),aa(i,j+3),
CCC  &                            aa(i,j+4),aa(i,j+5)
CC988         format(1h ,i3,6e12.5 )
CC990     continue
CCC       write(np,*) '+++++ BB +++++'
CCC       do 992 j=1,nmx3,6
CCC         write(np,988)j, bb(j),bb(j+1),bb(j+2),bb(j+3),
CCC  &                              bb(j+4),bb(j+5)
CC992     continue
C                                      --- CAL. INVERSE OF MATRIX ---
CCC    CALL TLEDD(NMX3,AA,BB,XX,NN,E,INDER)
       IF( IPATH .EQ. 1 )  GOTO 214
C                                         --- CAL. C1,C2,C3,... ---
       DO 210 I=1,N
         III = 3*I
         DO 205 J=1,N
           JJJ = 3*J
           AAWK(I,J) = AA(III,JJJ)
  205    CONTINUE
         BBWK(I) = BB(III)
  210  CONTINUE
       CALL TLEDD(NMX,AAWK,BBWK,XXWK,NNN,E,INDER)
C
       DO 212 J=1,N
         JJJ = 3*J
         XX(JJJ) = XXWK(J)
  212  CONTINUE
       GOTO 235
C                                       --- CAL. A1,B1,A2,B2,... ---
  214  DO 220 I=1,N
         II1 = 2*(I-1) + 1
         II2 = 3*(I-1) + 1
         DO 215 J=1,N
           JJ1 = 2*(J-1) + 1
           JJ2 = 3*(J-1) + 1
           AAWK(II1,JJ1)     = AA(II2,JJ2)
           AAWK(II1,JJ1+1)   = AA(II2,JJ2+1)
           AAWK(II1+1,JJ1)   = AA(II2+1,JJ2)
           AAWK(II1+1,JJ1+1) = AA(II2+1,JJ2+1)
  215    CONTINUE
         BBWK(II1)   = BB(II2)
         BBWK(II1+1) = BB(II2+1)
  220  CONTINUE
       CALL TLEDD(NMX2,AAWK,BBWK,XXWK,NNN,E,INDER)
C
       DO 230 J=1,N
         JJ1 = 2*(J-1) + 1
         JJ2 = 3*(J-1) + 1
         XX(JJ2)   = XXWK(JJ1)
         XX(JJ2+1) = XXWK(JJ1+1)
  230  CONTINUE
C
  235  WRITE(NP,250) RR, NMX,XX(1),XX(2),XX(3),XX(4),XX(5),XX(6)
C
       DIST(INUM)  = RR
       AANS1(ISTREET,INUM) = XX(1)
       BANS1(ISTREET,INUM) = XX(2)
       CANS1(ISTREET,INUM) = XX(3)
       AANS2(ISTREET,INUM) = XX(4)
       BANS2(ISTREET,INUM) = XX(5)
       CANS2(ISTREET,INUM) = XX(6)
C
  300 CONTINUE
C
  400 CONTINUE
C
C     ------------------------------------------------------------
C     --------------------- END OF MAIN ROUTINE ------------------
C     ------------------------------------------------------------
```

```
02730 C
02740 C                                    --- CAL. X11A,X12A,X11G,X12G ---
02750      IF( IPATH .EQ. 1 ) THEN
02760         DO 420 I=1,INUM
02770            X11A(I) =  (1.D0/3.D0)*( AANS1(1,I) + AANS1(2,I) )
02780            X12A(I) =  (1.D0/3.D0)*( AANS1(1,I) - AANS1(2,I) )
02790            X11G(I) = -(1.D0/4.D0)*( AANS2(1,I) + AANS2(2,I) )
02800            X12G(I) = -(1.D0/4.D0)*( AANS2(1,I) - AANS2(2,I) )
02810   420   CONTINUE
02820      END IF
02830 C                                         --- CAL. X11C,X12C ---
02840      IF( IPATH .EQ. 3 ) THEN
02850         DO 430 I=1,INUM
02860            X11C(I) =  0.5D0*( CANS1(1,I) + CANS1(2,I) )
02870            X12C(I) = -0.5D0*( CANS1(1,I) - CANS1(2,I) )
02880   430   CONTINUE
02890      END IF
02900 C                                         --- OUTPUT OF DATA ---
02910      WRITE(11,450) INUM, M, ALM, BLM
02920      DO 460 I=1,INUM
02930         IF( IPATH.EQ.1 ) WRITE(11,452) I, DIST(I), X11A(I), X12A(I),
02940      &                                X11G(I), X12G(I)
02950         IF( IPATH.EQ.3 ) WRITE(11,454) I, DIST(I), X11C(I), X12C(I)
02960   460 CONTINUE
02970 C                                         --- PRINT OUT ---
02980      WRITE(NP,*)
02990      DO 470 I=1,INUM
03000         IF( IPATH.EQ.1 ) WRITE(NP,466) DIST(I), X11A(I), X12A(I),
03010      &                                X11G(I), X12G(I)
03020         IF( IPATH.EQ.3 ) WRITE(NP,468) DIST(I), X11C(I), X12C(I)
03030   470 CONTINUE
03040 C
03050 C
03060      CLOSE(10,STATUS='KEEP')
03070      CLOSE(11,STATUS='KEEP')
03080 C
03090 C    --------------------------- FORMAT ---------------------------
03100    5 FORMAT(1H ,'----------------------------------------------'
03110      &      /1H ,'M=',I2, 2X ,'S=',F4.1, 2X ,'ALM=',F4.1, 2X ,
03120      &                     'BLM=',F4.1
03130      &      /1H ,'----------------------------------------------'/)
03140  250 FORMAT(1H ,'RR=',F6.3,'  NMX=',I4,
03150      &           ' A1=',E11.4,' B1=',E11.4,' C1=',E11.4
03160      &      /1H , 19X,     ' A2=',E11.4,' B2=',E11.4,' C2=',E11.4 )
03170  450 FORMAT( 2I4, 2F7.3 )
03180  452 FORMAT( I4, F7.3, 4E17.10 )
03190  454 FORMAT( I4, F7.3, 2E17.10 )
03200  466 FORMAT(1H ,'RR=',F6.3,' X11A=',E11.4, ' X12A=',E11.4,
03210      &           ' X11G=',E11.4, ' X12G=',E11.4 )
03220  468 FORMAT(1H ,'RR=',F6.3,' X11C=',E11.4, ' X12C=',E11.4 )
03230                                                          STOP
03240                                                          END
03250 C***********************************************************************
03260 C********************** SUBROUTINE ************************************
03270 C***********************************************************************
03280 C
03290 C**** SUB LEGENDRE ****
03300      SUBROUTINE LEGENDRE( P , M , NMX , X , ITREE )
03310 C
03320      IMPLICIT REAL*8 (A-H,O-Z)
03330 C
03340      PARAMETER( NN=1000 )
03350 C
03360      REAL*8  P(2,0:NN)
03370 C                           +++ M MUST BE SMALLER THAN 4 +++
03380      IF( M .EQ. 0 ) THEN
03390         P(ITREE,0) = 1.D0
03400         P(ITREE,1) = X
03410         NST = 2
03420      ELSE IF( M .EQ. 1 ) THEN
03430         C0   = DSQRT( 1.D0 - X**2 )
03440         P(ITREE,1) = C0
```

```
        P(ITREE,2) = 3.D0*X*C0
        NST  = 3
      ELSE
        C0   = ( 1.D0 - X**2 )
        P(ITREE,2) = 3.D0*C0
        P(ITREE,3) = 15.D0*X*C0
        NST  = 4
      END IF
C
      DO 100 N=NST,NMX
        C0    = DBLE( N - M )
        C1    = X*DBLE( 2*N - 1 )
        C2    = DBLE( N + M - 1 )
        P(ITREE,N)  = (C1/C0)*P(ITREE,N-1) - (C2/C0)*P(ITREE,N-2)
  100 CONTINUE
                                                       RETURN
                                                       END
C**** SUB LEGENDR2 ****
      SUBROUTINE LEGENDR2( PP , M , NMX , X , ITREE )
C
      IMPLICIT REAL*8 (A-H,O-Z)
C
      PARAMETER( NN=1000 )
C
      REAL*8  PP(2,0:NN)
C                                           +++ M MUST BE 1 OR 2 +++
      IF( M .EQ. 1 ) THEN
        PP(ITREE,1) = 1.D0
        PP(ITREE,2) = 3.D0*X
        NST  = 3
      ELSE
        PP(ITREE,2) = 3.D0
        PP(ITREE,3) = 15.D0*X
        NST  = 4
      END IF
C
      DO 100 N=NST,NMX
        C0    = DBLE( N - M )
        C1    = X*DBLE( 2*N - 1 )
        C2    = DBLE( N + M - 1 )
        PP(ITREE,N)  = (C1/C0)*PP(ITREE,N-1) - (C2/C0)*PP(ITREE,N-2)
  100 CONTINUE
                                                       RETURN
                                                       END
C**** SUB LEGEDEF0 ****
      SUBROUTINE LEGEDEF0( P0DEF0 , PP1 , NMX , X , ITREE )
C
      IMPLICIT REAL*8 (A-H,O-Z)
C
      PARAMETER( NN=1000 )
C
      REAL*8  P0DEF0(2,0:NN) , PP1(2,0:NN)
C
      P0DEF0(ITREE,0) = 0.D0
      DO 100 N=1,NMX
        P0DEF0(ITREE,N) = PP1(ITREE,N)
  100 CONTINUE
                                                       RETURN
                                                       END
C**** SUB LEGEDEF1 ****
      SUBROUTINE LEGEDEF1( P0DEF1 , P0 , NMX , X , ITREE )
C
      IMPLICIT REAL*8 (A-H,O-Z)
C
      PARAMETER( NN=1000 )
C
      REAL*8  P0DEF1(2,0:NN) , P0(2,0:NN)
C
      P0DEF1(ITREE,0) = 0.D0
      DO 100 N=1,NMX
        C0        = DBLE(N) * X
        C1        = DBLE(N)
```

```
04170          P0DEF1(ITREE,N) = - C0*P0(ITREE,N) + C1*P0(ITREE,N-1)
04180    100 CONTINUE
04190                                                          RETURN
04200                                                          END
04210 C**** SUB LEGEDEF2 ****
04220       SUBROUTINE LEGEDEF2( P0DEF2 , P0 , NMX , X , ITREE )
04230 C
04240       IMPLICIT REAL*8 (A-H,O-Z)
04250 C
04260       PARAMETER( NN=1000 )
04270 C
04280       REAL*8  P0DEF2(2,0:NN) , P0(2,0:NN)
04290 C
04300       P0DEF2(ITREE,0) = 0.D0
04310       DO 100 N=1,NMX
04320         C00      = 1.D0/ DSQRT( 1.D0 - X**2 )
04330         C0       = DBLE(N) * C00 * X
04340         C1       = DBLE(N) * C00
04350         P0DEF2(ITREE,N) = - C0*P0(ITREE,N) + C1*P0(ITREE,N-1)
04360    100 CONTINUE
04370                                                          RETURN
04380                                                          END
04390 C**** SUB LEGEDEF3 ****
04400       SUBROUTINE LEGEDEF3( P1DEF1 , P1 , P0 , NMX , X ,ITREE )
04410 C
04420       IMPLICIT REAL*8 (A-H,O-Z)
04430 C
04440       PARAMETER( NN=1000 )
04450 C
04460       REAL*8  P1DEF1(2,0:NN) , P1(2,0:NN) , P0(2,0:NN)
04470 C
04480       DO 100 N=1,NMX
04490         C0       = DBLE( N*(N+1) ) * DSQRT( 1.D0-X**2 )
04500         C1       = X
04510         P1DEF1(ITREE,N) = - C0*P0(ITREE,N) + C1*P1(ITREE,N)
04520    100 CONTINUE
04530                                                          RETURN
04540                                                          END
04550 C**** SUB LEGEDEF4 ****
04560       SUBROUTINE LEGEDEF4( P1DEF2 , PP1 , P0 , NMX , X , ITREE )
04570 C
04580 C
04590       IMPLICIT REAL*8 (A-H,O-Z)
04600 C
04610       PARAMETER( NN=1000 )
04620 C
04630       REAL*8  P1DEF2(2,0:NN) , PP1(2,0:NN) , P0(2,0:NN)
04640 C
04650       DO 100 N=1,NMX
04660         C0       = DBLE( N*(N+1) )
04670         C1       = X
04680         P1DEF2(ITREE,N) = - C0*P0(ITREE,N) + C1*PP1(ITREE,N)
04690    100 CONTINUE
04700                                                          RETURN
04710                                                          END
04720 C#### FUN A1CAL ####
04730       DOUBLE PRECISION FUNCTION A1CAL( N,M,X1,X2,R1,R2,CP,CPDEF1,S)
04740 C
04750       IMPLICIT REAL*8 (A-H,O-Z)
04760 C
04770       REAL*8  CP(2) , CPDEF1(2)
04780 C
04790       C1   = DBLE(N+1)/DBLE(4*N-2)
04800       C2   = DBLE(N-2)/DBLE( N*(4*N-2) )
04810       CC1  = 1.D0/R1**N
04820       CC2  = 1.D0/R2**N
04830       ANS1 = C1*CC1*X1*CP(1) - C2*CC1*CPDEF1(1)
04840       ANS2 = C1*CC2*X2*CP(2) - C2*CC2*CPDEF1(2)
04850       A1CAL = ANS1 - S*ANS2
04860                                                          RETURN
04870                                                          END
04880 C#### FUN B1CAL ####
```

```
      DOUBLE PRECISION FUNCTION B1CAL( N,M,X1,X2,R1,R2,CP,CPDEF1,S)
C
      IMPLICIT REAL*8 (A-H,O-Z)
C
      REAL*8  CP(2) , CPDEF1(2)
C
      C1     = DBLE(N+1)
      CC1    = 1.D0/R1**(N+2)
      CC2    = 1.D0/R2**(N+2)
      ANS1   = -C1*CC1*X1*CP(1) + CC1*CPDEF1(1)
      ANS2   = -C1*CC2*X2*CP(2) + CC2*CPDEF1(2)
      B1CAL  = ANS1 - S*ANS2
                                                       RETURN
                                                       END
C#### FUN C1CAL ####
      DOUBLE PRECISION FUNCTION C1CAL( N,M,X1,X2,R1,R2,CP1,S)
C
      IMPLICIT REAL*8 (A-H,O-Z)
C
      REAL*8  CP1(2)
C
      CC1    = 1.D0/R1**(N+1)
      CC2    = 1.D0/R2**(N+1)
      ANS1   = CC1*CP1(1)
      ANS2   = CC2*CP1(2)
      C1CAL  = DBLE(M)*( ANS1 - S*ANS2 )
                                                       RETURN
                                                       END
C#### FUN A2CAL ####
      DOUBLE PRECISION FUNCTION A2CAL( N,M,X1,X2,R1,R2,CP0,CP1,S)
C
      IMPLICIT REAL*8 (A-H,O-Z)
C
      REAL*8  CP0(2) , CP1(2)
C
      C1    = DBLE(N+1)/DBLE(4*N-2)
      C2    = DBLE(N-2)/DBLE( N*(4*N-2) )
      SN1   = 1.D0
      SN2   = R1/R2
      C11   = 1.D0/R1**N
      C22   = 1.D0/R2**N
      ANS1  = C1*C11*SN1*CP0(1)
     &             + C2*C11*SN1*( X1*CP1(1) + DBLE(M)*CP0(1) )
      ANS2  = C1*C22*SN2*CP0(2)
     &             + C2*C22*SN2*( X2*CP1(2) + DBLE(M)*CP0(2) )
      A2CAL = ANS1 + S*ANS2
                                                       RETURN
                                                       END
C#### FUN B2CAL ####
      DOUBLE PRECISION FUNCTION B2CAL( N,M,X1,X2,R1,R2,CP0,CP1,S)
C
      IMPLICIT REAL*8 (A-H,O-Z)
C
      REAL*8  CP0(2) , CP1(2)
C
      C1    = DBLE(N+1)
      SN1   = 1.D0
      SN2   = R1/R2
      C11   = 1.D0/R1**(N+2)
      C22   = 1.D0/R2**(N+2)
      ANS1  = -C1*SN1*CP0(1) - SN1*( X1*CP1(1) + DBLE(M)*CP0(1) )
      ANS2  = -C1*SN2*CP0(2) - SN2*( X2*CP1(2) + DBLE(M)*CP0(2) )
      B2CAL = C11*ANS1 + C22*S*ANS2
                                                       RETURN
                                                       END
C#### FUN C2CAL ####
      DOUBLE PRECISION FUNCTION C2CAL( N,M,X1,X2,R1,R2,CP,S)
C
      IMPLICIT REAL*8 (A-H,O-Z)
C
      REAL*8  CP(2)
C
```

```
05610       C1    = 1.D0/R1**(N+1)
05620       C2    = 1.D0/R2**(N+1)
05630       ANS1  = C1*CP(1)
05640       ANS2  = C2*CP(2)
05650       C2CAL = - ( ANS1 + S*ANS2 )
05660                                              RETURN
05670                                              END
05680 C#### FUN A3CAL ####
05690       DOUBLE PRECISION FUNCTION A3CAL( N,M,X1,X2,R1,R2,CPP1,S)
05700 C
05710       IMPLICIT REAL*8 (A-H,O-Z)
05720 C
05730       REAL*8  CPP1(2)
05740 C
05750       C1    = DBLE(N-2)/DBLE( N*(4*N-2) )
05760       C11   = 1.D0/R1**N
05770       C22   = 1.D0/R2**N
05780       ANS1  = C1*C11*CPP1(1)
05790       ANS2  = C1*C22*CPP1(2)
05800       A3CAL = - DBLE(M)*( ANS1 + S*ANS2 )
05810                                              RETURN
05820                                              END
05830 C#### FUN B3CAL ####
05840       DOUBLE PRECISION FUNCTION B3CAL( N,M,X1,X2,R1,R2,CPP1,S)
05850 C
05860       IMPLICIT REAL*8 (A-H,O-Z)
05870 C
05880       REAL*8  CPP1(2)
05890 C
05900       C11   = 1.D0/R1**(N+2)
05910       C22   = 1.D0/R2**(N+2)
05920       ANS1  = C11*CPP1(1)
05930       ANS2  = C22*CPP1(2)
05940       B3CAL = DBLE(M)*( ANS1 + S*ANS2 )
05950                                              RETURN
05960                                              END
05970 C#### FUN C3CAL ####
05980       DOUBLE PRECISION FUNCTION C3CAL( N,M,X1,X2,R1,R2,CPDEF2,S)
05990 C
06000       IMPLICIT REAL*8 (A-H,O-Z)
06010 C
06020       REAL*8  CPDEF2(2)
06030 C
06040       CC1   = 1.D0/R1**(N+1)
06050       CC2   = 1.D0/R2**(N+1)
06060       SN1   = 1.D0
06070       SN2   = R1/R2
06080       ANS1  = CC1*SN1*CPDEF2(1)
06090       ANS2  = CC2*SN2*CPDEF2(2)
06100       C3CAL = ANS1 + S*ANS2
06110                                              RETURN
06120                                              END
06130 C-------------------------------------------------------------------
06140 C-                      TLEDD
06150 C-
06160 C-          THE SOLUTION OF LINEAR EQUATION OF TYPE
06170 C-              AX=B      A- N*N  MATRIX
06180 C-          BY  GAUSSIAN ELIMINATION
06190 C-------------------------------------------------------------------
06200 C
06210 C**** SUB TLEDD ***
06220       SUBROUTINE   TLEDD(N,A,B,X,NN,E,INDER)
06230 C
06240       REAL*8    A,B,X,E,PIV,PMAX,EPS,F,AA,WORK
06250       DIMENSION  A(NN,N) , B(N) , X(N)
06260       INDER=0
06270       NM1   = N-1
06280       IF(NM1)                40,41,42
06290    40 INDER = -1000
06300       WRITE(6,640)   N
06310   640 FORMAT(1H0,8H *** N =,I5,5X,' N SHOULD BE POSITIVE IN TLEDD')
06320       RETURN
```

```
 41 X(1)  =  B(1)/A(1,1)
    IF(A(1,1).EQ.0.0D0)  GO TO   21
    RETURN
 42 IF(N.LE.NN)       GO TO    43
    INDER =  -1001
    WRITE(6,642)   N,NN
642 FORMAT(1H0,8H *** N =,I5,5X,8H *** NN=,I5,
   &/1H ,'     NN SHOULD BE GREATER THAN OR EQUAL TO N IN TLEDD')
    RETURN
 43 PIV  =  0.0D0
    DO    1     I=1,N
    DO    1     J=1,N
    IF(DABS(A(I,J)).LE.PIV)    GO TO  1
    PIV   =  DABS(A(I,J))
    L     =  I
    M     =  J
  1 CONTINUE
    IF(PIV.EQ.0.0D0)         GO TO  21
    EPS  =  E*PIV
    DO   2   K=1,NM1
    KK   = K+1
    X(K) = M
    IF(K.EQ.L)   GO TO  4
    DO   3   J=K,N
    WORK  =  A(K,J)
    A(K,J)=  A(L,J)
  3 A(L,J)=  WORK
    WORK  =  B(K)
    B(K)  =  B(L)
    B(L)  =  WORK
  4 IF(K.EQ.M)    GO TO  6
    DO   5   I=1,N
    WORK  =  A(I,K)
    A(I,K)=  A(I,M)
  5 A(I,M)=  WORK
  6 PMAX  =  1.0D0/A(K,K)
    PIV   =  0.0D0
    DO   7   I=KK,N
    AA    =  A(I,K)*PMAX
    DO   8   J=KK,N
    A(I,J)=  A(I,J)-A(K,J)*AA
    WORK  =  DABS(A(I,J))
    IF(WORK.LE.PIV)   GO TO  8
    PIV   =  WORK
    M     =  J
    L     =  I
  8 CONTINUE
  7 B(I)  =  B(I)-B(K)*AA
    IF(PIV.LE.EPS)          GO TO  22
  2 CONTINUE
    GO TO    10
 22 DO  20    I=KK,N
    IF(DABS(B(I)).GT.EPS)   GO TO  25
    B(I)   =  0.0D0
 20 X(I)  =  I
    INDER =  N-KK+1
    WRITE(6,610)    K
    L     =  N-K+2
    B(K)  =  B(K)/A(K,K)
    IF(K.EQ.1)   GO TO  15
    GO TO   16
 21 K     =  0
    KK    =  1
 25 INDER =  -(N-KK+1)
    WRITE(6,611)    K
    RETURN
610 FORMAT(1H0,'***LINEAR EQUATION AX=B TO TLEDD IS ILL CONDITI'
   &,'ON OR INDETERMINATE WITH RANK=',I4
   &/1H ,4X,'SOLUTION CAN BE EXPRESSED BY B+A*Y WHERE'
   &, ' Y IS AN ARBITRARY VECTOR'                        )
611 FORMAT(1H0,'***LINEAR EQUATION AX=B TO TLEDD IS ILL CONDITI'
   &,'ON OR INCONSISTENT  WITH RANK=',I4,
```

```
07050     & 38H.   RETURN WITH NO FURTHER CALCULATION   )
07060 C
07070    10 B(N)  =  B(N)/A(N,N)
07080       L    = 2
07090    16 DO  11    II=L,N
07100       I    = N+1-II
07110       F    = B(I)
07120       IM1  = I+1
07130       DO  12   J=IM1,N
07140    12 F    = F-A(I,J)*B(J)
07150    11 B(I) = F/A(I,I)
07160    15 DO  13   II=2,N
07170       I    = N+1-II
07180       M    = X(I)
07190       WORK = B(I)
07200       B(I) = B(M)
07210    13 B(M) = WORK
07220       IF(INDER.NE.0)        GO TO  30
07230    17 DO  14   I=1,N
07240    14 X(I)  = B(I)
07250       RETURN
07260    30 DO  50   J=KK,N
07270       DO  52  II=1,K
07280       I    = K-II+1
07290       F    = A(I,J)
07300       IM1  = I+1
07310       IF(II.EQ.1)    GO TO  52
07320       DO  51  M=IM1,K
07330    51 F    = F+A(I,M)*A(M,J)
07340    52 A(I,J)=  -F/A(I,I)
07350    50 CONTINUE
07360       DO  31   I=1,N
07370       DO  32   J=1,N
07380    32 IF(I.GE.KK.OR.J.LT.KK)   A(I,J)=0.0D0
07390    31 IF(I.GE.KK)     A(I,I)=1.0D0
07400    33 M    = X(K)
07410       IF(M.EQ.K)   GO TO  35
07420       DO  34   I=KK,N
07430       WORK = A(K,I)
07440       A(K,I)= A(M,I)
07450    34 A(M,I)= WORK
07460    35 K    = K-1
07470       IF(K.NE.0)   GO TO  33
07480       GO TO  17
07490       END
```

(RESIST2.FORT)

```
00010 C********************************************************************
00020 C*             RESIST2.FORT
00030 C*
00040 C*
00050 C*        ---- CALCULATION OF RESISTANCE FUNC. WHICH IS -----
00060 C*        ---- GIVEN BY S.KIM AND R.MIFFLIN            -----
00070 C*        ---- (PHYS. FLUIDS, 28(1985), 2033).         -----
00080 C*
00010 C*           (2) ALM=1 , BLM=0 , M=1  (IPATH=2 ) Y11A,Y12A,Y11B,Y12B
00020 C*                                     Y11G,Y12G
00030 C*           (4) ALM=0 , BLM=1 , M=1  (IPATH=4 ) Y11C,Y12C,Y11H,Y12H
00040 C*
00050 C*                         VER.1 BY A. SATOH, '96 1/12
00060 C********************************************************************
00070 C
00080 C--------------------------------------------------------------------
00090       IMPLICIT REAL*8 (A-H,O-Z)
00100 C
00110       PARAMETER( NN=1000 , NNN=667 )
00120       PARAMETER( NRADIAL=1000 , PI=3.141592653589793D0 )
00130 C
00140       COMMON /BLOCK1/ P0 , P1 , P2
```

```
00150      COMMON /BLOCK2/ PP1 , PP2
00160      COMMON /BLOCK3/ P0DEF0 , P0DEF1 , P0DEF2 , P1DEF1 , P1DEF2
00170      COMMON /BLOCK4/ AA  , XX  , BB
00180      COMMON /BLOCK5/ M  , S , ALM , BLM
00190      COMMON /BLOCK6/ DIST, AANS1, BANS1, CANS1, AANS2, BANS2, CANS2
00280      COMMON /BLOCK7/ Y11A, Y12A , Y11B , Y12B , Y11G, Y12G
00290      COMMON /BLOCK8/ Y11C, Y12C , Y11H , Y12H
00300 C
00310      REAL*8  P0(2,0:NN)  , P1(2,0:NN)  , P2(2,0:NN)
00320      REAL*8  PP1(2,0:NN) , PP2(2,0:NN)
00330      REAL*8  P0DEF0(2,0:NN) , P0DEF1(2,0:NN) , P0DEF2(2,0:NN)
00340      REAL*8  P1DEF1(2,0:NN) , P1DEF2(2,0:NN)
00350      REAL*8  AA(NN,NN) , XX(NN) , BB(NN)
00360      REAL*8  DIST(NRADIAL)
00370      REAL*8  AANS1(2,NRADIAL), BANS1(2,NRADIAL), CANS1(2,NRADIAL)
00380      REAL*8  AANS2(2,NRADIAL), BANS2(2,NRADIAL), CANS2(2,NRADIAL)
00390      REAL*8  Y11A(NRADIAL), Y12A(NRADIAL)
00400      REAL*8  Y11B(NRADIAL), Y12B(NRADIAL)
00410      REAL*8  Y11G(NRADIAL), Y12G(NRADIAL)
00420      REAL*8  Y11C(NRADIAL), Y12C(NRADIAL)
00430      REAL*8  Y11H(NRADIAL), Y12H(NRADIAL)
00440 C
00450      REAL*8  CP0(2) , CP1(2) , CP2(2) , CPP1(2) , CPP2(2)
00460      REAL*8  CP0DEF0(2) , CP0DEF1(2) , CP0DEF2(2)
00470      REAL*8  CP1DEF1(2) , CP1DEF2(2)
00480      REAL*8  A1, B1, C1, D1, A2, B2, C2, D2, A3, B3, C3, D3
00490      REAL*8  AAWK(NNN,NNN) , XXWK(NNN) , BBWK(NNN)
00500      REAL*8  THETA1 , COS1 , R1 , THETA2 , COS2 , R2 , RR
00510      INTEGER NMX, NMX2, NMX3, INUM , ITREE , IPATH , ISTREET
00520 C
00530                                                          NP=10
00540      OPEN(10,FILE='@bbb.data',STATUS='UNKNOWN')
00550      OPEN(11,FILE='bbb.data',STATUS='UNKNOWN')
00560 C                                        --- PARAMETER (1) ---
00570      M  = 1
00580      IPATH = 2
00590 CCC   IPATH = 4
00600 C                                        --- PARAMETER (2) ---
00610      E  = 1.0E-15
00620 C                                        --- PARAMETER (3) ---
00630      IF( IPATH .EQ. 2 ) THEN
00640        ALM = 1.D0
00650        BLM = 0.D0
00660      ELSE IF( IPATH .EQ. 4 ) THEN
00670        ALM = 0.D0
00680        BLM = 1.D0
00690      END IF
00700 C
00710 C      -------------------------------------------------------------
00720      DO 400 ISTREET=1,2
00730 C
00740      IF( ISTREET .EQ. 1 ) S = -1.D0
00750      IF( ISTREET .EQ. 2 ) S =  1.D0
00760 C
00770      WRITE(NP,5) M , S , ALM , BLM
00780 C
00790 C                                       --- INITIALIZATION ---
00800      INUM = 0
00810 C
00820 C      -------------------------------------------------------------
00830 C      ------------------ START OF MAIN ROUTINE ---------------------
00840 C      -------------------------------------------------------------
00850 C
00860 C                      --- CHANGE DISTANCE BETWEEN PARTICLES ---
00870 C                      +++     NMX MUST BE EVEN NUMBER      +++
00880 CCC   DO 300 II = 20, 200
00890      DO 300 II = 21, 21
00900 C
00910       INUM = INUM + 1
00920       RR  = DBLE(II)/10.D0
00930      NMX  = 60
00940      NMX2 = 2*NMX
```

```
      NMX3 = 3*NMX
C     ------------------------------------ START OF MAIN PART ------
C                                         --- CHANGE THETA ---
      DO 200 III=0,NMX-1
C
        I      = 3*III + 1
        THETA1 = DBLE(III)/DBLE(NMX-1)*PI
        COS1   = DCOS(THETA1)
        C1     = RR + COS1
        C2     = DSIN(THETA1)
        R2     = DSQRT( C1**2 + C2**2 )
        COS2   = - C1/R2
        R1     = 1.D0
C                                   --- CAL. LEGENDRE FUNCS ---
C                                          +++ (1) P(0,N) +++
        M1 = M - 1
        ITREE = 1
        CALL LEGENDRE( P0 , M1 , NMX , COS1 , ITREE )
        ITREE = 2
        CALL LEGENDRE( P0 , M1 , NMX , COS2 , ITREE )
C                                          +++ (2) P(1,N) +++
        ITREE = 1
        CALL LEGENDRE( P1 , M, NMX , COS1 , ITREE )
        ITREE = 2
        CALL LEGENDRE( P1 , M, NMX , COS2 , ITREE )
C                                         +++ (3) PP(1,N) +++
        ITREE = 1
        CALL LEGENDR2( PP1 , M, NMX , COS1 , ITREE )
        ITREE = 2
        CALL LEGENDR2( PP1 , M, NMX , COS2 , ITREE )
C                                         +++ (4) PP(2,N) +++
        M1 = M + 1
        ITREE = 1
        CALL LEGENDR2( PP2 , M1, NMX , COS1 , ITREE )
        ITREE = 2
        CALL LEGENDR2( PP2 , M1, NMX , COS2 , ITREE )
C                                  +++ (5) DEF2. OF P(1,N) +++
        ITREE = 1
        CALL LEGEDEF4( P1DEF2 , PP1 , P0 , NMX , COS1 , ITREE )
C
        ITREE = 2
        CALL LEGEDEF4( P1DEF2 , PP1 , P0 , NMX , COS2 , ITREE )
C
C                           --- CAL. COEFFICIENTS OF MATRIX ---
C                           +++    AA(I,J)*XX(I)= BB(I)    +++
C                                          +++ (1) FOR VZ +++

        DO 100 N=1,NMX
C
          CPP1(1)    = PP1(1,N)
          CPP1(2)    = PP1(2,N)
          CP1DEF2(1) = P1DEF2(1,N)
          CP1DEF2(2) = P1DEF2(2,N)
          A1 = A1CAL( N, M, COS1, COS2, R1, R2, CPP1, CP1DEF2, S )
          B1 = B1CAL( N, M, COS1, COS2, R1, R2, CPP1, CP1DEF2, S )
          C1 = C1CAL( N, M, COS1, COS2, R1, R2, CPP1, S )
C
          J  = 3*(N-1) + 1
          AA(I,J)   = A1
          AA(I,J+1) = B1
          AA(I,J+2) = C1
C
  100   CONTINUE
        IF( IPATH .EQ. 2 )  BB(I) = 0.D0
        IF( IPATH .EQ. 4 )  BB(I) = 1.D0
C                                       +++ (2) FOR VR-VQ +++
        DO 110 N=1,NMX
C
          CPP1(1)     = PP1(1,N)
          CPP1(2)     = PP1(2,N)
          CPP2(1)     = PP2(1,N)
          CPP2(2)     = PP2(2,N)
```

```
01670              A2 = A2CAL( N, M, COS1, COS2, R1, R2, CPP1, CPP2, S )
01680              B2 = B2CAL( N, M, COS1, COS2, R1, R2, CPP1, CPP2, S )
01690              C2 = C2CAL( N, M, COS1, COS2, R1, R2, CPP2, S )
01700 C
01710              J   = 3*(N-1) + 1
01720              AA(I+1,J)   = A2
01730              AA(I+1,J+1) = B2
01740              AA(I+1,J+2) = C2
01750 C
01760   110      CONTINUE
01770              IF( IPATH .EQ. 2 ) BB(I+1) = 0.D0
01780              IF( IPATH .EQ. 4 ) BB(I+1) = 0.D0
01790 C                                          +++ (3) FOR VQ +++
01800            DO 120 N=1,NMX
01810 C
01820              CPP1(1)    = PP1(1,N)
01830              CPP1(2)    = PP1(2,N)
01840              CP1DEF2(1) = P1DEF2(1,N)
01850              CP1DEF2(2) = P1DEF2(2,N)
01860              A3 = A3CAL( N, M, COS1, COS2, R1, R2, CPP1, S )
01870              B3 = B3CAL( N, M, COS1, COS2, R1, R2, CPP1, S )
01880              C3 = C3CAL( N, M, COS1, COS2, R1, R2, CP1DEF2, S )
01890 C
01900              J   = 3*(N-1) + 1
01910              AA(I+2,J)   = A3
01920              AA(I+2,J+1) = B3
01930              AA(I+2,J+2) = C3
01940 C
01950   120      CONTINUE
01960              IF( IPATH .EQ. 2 ) BB(I+2) = 1.D0
01970              IF( IPATH .EQ. 4 ) BB(I+2) = -COS1
01980 C
01990   200  CONTINUE
02000 C     ---------------------------------- END OF MAIN PART ------
02010 C
02020 CCC      do 990 i=1,nmx3
02030 CCC       write(np,*) '-------------------- i=',i
02040 CCC       do 990 j=1,nmx3,6
02050 CCC         write(np,988)i, aa(i,j),aa(i,j+1),aa(i,j+2),aa(i,j+3),
02060 CCC  &                        aa(i,j+4),aa(i,j+5)
02070 CC988        format(1h ,i3,6e12.5 )
02080 CC990     continue
02090 CCC      write(np,*) '+++++ BB +++++'
02100 CCC      do 992 j=1,nmx3,6
02110 CCC       write(np,988)j, bb(j),bb(j+1),bb(j+2),bb(j+3),
02120 CCC  &                       bb(j+4),bb(j+5)
02130 CC992     continue
02140 C                              --- CAL. INVERSE OF MATRIX ---
02150          CALL TLEDD(NMX3,AA,BB,XX,NN,E,INDER)
02160 C
02170   235  WRITE(NP,250) RR, NMX,XX(1),XX(2),XX(3),XX(4),XX(5),XX(6)
02180 C
02190          DIST(INUM)  = RR
02200          AANS1(ISTREET,INUM) = XX(1)
02210          BANS1(ISTREET,INUM) = XX(2)
02220          CANS1(ISTREET,INUM) = XX(3)
02230          AANS2(ISTREET,INUM) = XX(4)
02240          BANS2(ISTREET,INUM) = XX(5)
02250          CANS2(ISTREET,INUM) = XX(6)
02260 C
02270   300 CONTINUE
02280 C
02290   400 CONTINUE
02300 C
02310 C     ----------------------------------------------------------------
02320 C     ------------------- END OF MAIN ROUTINE ----------------------
02330 C     ----------------------------------------------------------------
02340 C
02350 C                      --- CAL. Y11A,Y12A,Y11B,Y12B,Y11G,Y12G ---
02360       IF( IPATH .EQ. 2 ) THEN
02370         DO 420 I=1,INUM
02380         Y11A(I) = (1.D0/3.D0)*( AANS1(1,I) + AANS1(2,I) )
```

```
02390          Y12A(I) = -(1.D0/3.D0)*( AANS1(1,I) - AANS1(2,I) )
02400          Y11B(I) = -( CANS1(1,I) + CANS1(2,I) )
02410          Y12B(I) =  ( CANS1(1,I) - CANS1(2,I) )
02420          Y11G(I) = -(1.D0/4.D0)*( AANS2(1,I) + AANS2(2,I) )
02430          Y12G(I) =  (1.D0/4.D0)*( AANS2(1,I) - AANS2(2,I) )
02440   420  CONTINUE
02450       END IF
02460 C                                  --- CAL. Y11C,Y12C,Y11H,Y12H ---
02470       IF( IPATH .EQ. 4 ) THEN
02480        DO 430 I=1,INUM
02490          Y11C(I) =  0.5D0*( CANS1(1,I) + CANS1(2,I) )
02500          Y12C(I) =  0.5D0*( CANS1(1,I) - CANS1(2,I) )
02510          Y11H(I) = -(1.D0/8.D0)*( AANS2(1,I) + AANS2(2,I) )
02520          Y12H(I) = -(1.D0/8.D0)*( AANS2(1,I) - AANS2(2,I) )
02530   430  CONTINUE
02540       END IF
02550 C                                       --- OUTPUT OF DATA ---
02560       WRITE(11,450) INUM, M, ALM, BLM
02570       DO 460 I=1,INUM
02580        IF( IPATH.EQ.2 ) WRITE(11,452) I, DIST(I), Y11A(I), Y12A(I),
02590      &                            Y11B(I), Y12B(I)
02600        IF( IPATH.EQ.2 ) WRITE(11,453)  Y11G(I), Y12G(I)
02610        IF( IPATH.EQ.4 ) WRITE(11,454)  I, DIST(I), Y11C(I), Y12C(I),
02620      &                            Y11H(I), Y12H(I)
02630   460 CONTINUE
02640 C                                       --- PRINT OUT ---
02650       WRITE(NP,*)
02660       DO 470 I=1,INUM
02670        IF( IPATH.EQ.2 ) WRITE(NP,466) DIST(I), Y11A(I), Y12A(I),
02680      &                     Y11B(I), Y12B(I), Y11G(I), Y12G(I)
02690        IF( IPATH.EQ.4 ) WRITE(NP,468) DIST(I), Y11C(I), Y12C(I),
02700      &                            Y11H(I), Y12H(I)
02710   470 CONTINUE
02720 C
02730 C
02740       CLOSE(10,STATUS='KEEP')
02750       CLOSE(11,STATUS='KEEP')
02760 C
02770 C    --------------------------- FORMAT ---------------------------
02780     5 FORMAT(1H ,'------------------------------------------------'
02790      &      /1H ,'M=',I2, 2X ,'S=',F4.1, 2X ,'ALM=',F4.1, 2X ,
02800      &                      'BLM=',F4.1
02810      &      /1H ,'------------------------------------------------'/)
02820   250 FORMAT(1H ,'RR=',F6.3,' NMX=',I4,
02830      &           ' A1=',E11.4,' B1=',E11.4,' C1=',E11.4
02840      &      /1H , 19X,      ' A2=',E11.4,' B2=',E11.4,' C2=',E11.4 )
02850   450 FORMAT( 2I4, 2F7.3 )
02860   452 FORMAT( I4, F7.3 , 4E17.10 )
02870   453 FORMAT( 2E17.10 )
02880   454 FORMAT( I4, F7.3 , 4E17.10 )
02890   466 FORMAT(1H ,'RR=',F6.3,' Y11A=',E11.4, ' Y12A=',E11.4,
02900      &                     ' Y11B=',E11.4, ' Y12B=',E11.4
02910      &      /1H , 9X ,     ' Y11G=',E11.4, ' Y12G=',E11.4 )
02920   468 FORMAT(1H ,'RR=',F6.3,' Y11C=',E11.4, ' Y12C=',E11.4,
02930      &                     ' Y11H=',E11.4, ' Y12H=',E11.4 )
02940                                                         STOP
02950                                                         END
02960 C****************************************************************
02970 C********************** SUBROUTINE ******************************
02980 C****************************************************************
02990 C
03000 C#### FUN A1CAL ####
03010       DOUBLE PRECISION FUNCTION A1CAL( N,M,X1,X2,R1,R2,CP,CPDEF1,S)
03020 C
03030       IMPLICIT REAL*8 (A-H,O-Z)
03040 C
03050       REAL*8  CP(2) , CPDEF1(2)
03060 C
03070       C1   = DBLE(N+1)/DBLE(4*N-2)
03080       C2   = DBLE(N-2)/DBLE( N*(4*N-2) )
03090       SN1  = 1.D0
03100       SN2  = R1/R2
```

```
      CC1   = 1.D0/R1**N
      CC2   = 1.D0/R2**N
      ANS1  = SN1*( C1*CC1*X1*CP(1) - C2*CC1*CPDEF1(1) )
      ANS2  = SN2*( C1*CC2*X2*CP(2) - C2*CC2*CPDEF1(2) )
      A1CAL = ANS1 - S*ANS2
                                                            RETURN
                                                            END
C#### FUN B1CAL ####
      DOUBLE PRECISION FUNCTION B1CAL( N,M,X1,X2,R1,R2,CP,CPDEF2,S)
C
      IMPLICIT REAL*8 (A-H,O-Z)
C
      REAL*8  CP(2) , CPDEF2(2)
C
      C1     = DBLE(N+1)
      SN1   = 1.D0
      SN2   = R1/R2
      CC1    = 1.D0/R1**(N+2)
      CC2    = 1.D0/R2**(N+2)
      ANS1   = SN1*( -C1*CC1*X1*CP(1) + CC1*CPDEF2(1) )
      ANS2   = SN2*( -C1*CC2*X2*CP(2) + CC2*CPDEF2(2) )
      B1CAL  = ANS1 - S*ANS2
                                                            RETURN
                                                            END
C#### FUN C1CAL ####
      DOUBLE PRECISION FUNCTION C1CAL( N,M,X1,X2,R1,R2,CP1,S)
C
      IMPLICIT REAL*8 (A-H,O-Z)
C
      REAL*8  CP1(2)
C
      SN1   = 1.D0
      SN2   = R1/R2
      CC1    = 1.D0/R1**(N+1)
      CC2    = 1.D0/R2**(N+1)
      ANS1   = SN1*CC1*CP1(1)
      ANS2   = SN2*CC2*CP1(2)
      C1CAL  = DBLE(M)*( ANS1 - S*ANS2 )
                                                            RETURN
                                                            END
C#### FUN A2CAL ####
      DOUBLE PRECISION FUNCTION A2CAL( N,M,X1,X2,R1,R2,CP0,CP1,S)
C
      IMPLICIT REAL*8 (A-H,O-Z)
C
      REAL*8  CP0(2) , CP1(2)
C
      C1    = DBLE(N+1)/DBLE(4*N-2)
      C2    = DBLE(N-2)/DBLE( N*(4*N-2) )
      SN1   = 1.D0
      SN2   = R1/R2
      SN1SQ = SN1**2
      SN2SQ = SN2**2
      C11   = 1.D0/R1**N
      C22   = 1.D0/R2**N
      ANS1  = SN1SQ*( C1*C11*CP0(1)
     &                + C2*C11*( X1*CP1(1) + DBLE(M)*CP0(1) ) )
      ANS2  = SN2SQ*( C1*C22*CP0(2)
     &                + C2*C22*( X2*CP1(2) + DBLE(M)*CP0(2) ) )
      A2CAL = ANS1 + S*ANS2
                                                            RETURN
                                                            END
C#### FUN B2CAL ####
      DOUBLE PRECISION FUNCTION B2CAL( N,M,X1,X2,R1,R2,CP0,CP1,S)
C
      IMPLICIT REAL*8 (A-H,O-Z)
C
      REAL*8  CP0(2) , CP1(2)
C
      C1    = DBLE(N+1)
      SN1   = 1.D0
      SN2   = R1/R2
```

```
      SN1SQ = SN1**2
      SN2SQ = SN2**2
      C11   = 1.D0/R1**(N+2)
      C22   = 1.D0/R2**(N+2)
      ANS1  = SN1SQ*( -C1*CP0(1) - ( X1*CP1(1) + DBLE(M)*CP0(1) ) )
      ANS2  = SN2SQ*( -C1*CP0(2) - ( X2*CP1(2) + DBLE(M)*CP0(2) ) )
      B2CAL = C11*ANS1 + C22*S*ANS2
                                                       RETURN
                                                       END
C#### FUN C2CAL ####
      DOUBLE PRECISION FUNCTION C2CAL( N,M,X1,X2,R1,R2,CP,S)
C
      IMPLICIT REAL*8 (A-H,O-Z)
C
      REAL*8  CP(2)
C
      C1     = 1.D0/R1**(N+1)
      C2     = 1.D0/R2**(N+1)
      SN1    = 1.D0
      SN2    = R1/R2
      SN1SQ  = SN1**2
      SN2SQ  = SN2**2
      ANS1   = SN1SQ*C1*CP(1)
      ANS2   = SN2SQ*C2*CP(2)
      C2CAL  = - ( ANS1 + S*ANS2 )
                                                       RETURN
                                                       END
C#### FUN A3CAL ####
      DOUBLE PRECISION FUNCTION A3CAL( N,M,X1,X2,R1,R2,CPP1,S)
C
      IMPLICIT REAL*8 (A-H,O-Z)
C
      REAL*8  CPP1(2)
C
      C1    = DBLE(N-2)/DBLE( N*(4*N-2) )
      C11   = 1.D0/R1**N
      C22   = 1.D0/R2**N
      ANS1  = C1*C11*CPP1(1)
      ANS2  = C1*C22*CPP1(2)
      A3CAL = - DBLE(M)*( ANS1 + S*ANS2 )
                                                       RETURN
                                                       END
C#### FUN B3CAL ####
      DOUBLE PRECISION FUNCTION B3CAL( N,M,X1,X2,R1,R2,CPP1,S)
C
      IMPLICIT REAL*8 (A-H,O-Z)
C
      REAL*8  CPP1(2)
C
      C11   = 1.D0/R1**(N+2)
      C22   = 1.D0/R2**(N+2)
      ANS1  = C11*CPP1(1)
      ANS2  = C22*CPP1(2)
      B3CAL = DBLE(M)*( ANS1 + S*ANS2 )
                                                       RETURN
                                                       END
C#### FUN C3CAL ####
      DOUBLE PRECISION FUNCTION C3CAL( N,M,X1,X2,R1,R2,CPDEF2,S)
C
      IMPLICIT REAL*8 (A-H,O-Z)
C
      REAL*8  CPDEF2(2)
C
      CC1    = 1.D0/R1**(N+1)
      CC2    = 1.D0/R2**(N+1)
      ANS1   = CC1*CPDEF2(1)
      ANS2   = CC2*CPDEF2(2)
      C3CAL  = ANS1 + S*ANS2
                                                       RETURN
                                                       END
```

LIST OF SYMBOLS

Latin Alphabet

a	particle radius
c	lattice speed (14.3)
d	particle diameter
D	diffusion coefficient
$\mathbf{D}$	diffusion matrix
$\mathbf{e}$	unit vector denoting particle direction
E	energy (thermodynamic)
$\mathbf{E}$	rate-of-strain tensor
f	Maxwellian distribution (Chap.2)
$f_\alpha, f_\alpha^{(eq)}$	single-particle distributions (14.3)
$\mathbf{f}_i$	force acting on particle i
$\mathbf{f}_{ij}$	force acting on particle i by particle j
F	Helmholtz free energy
$F_x, F_y, F_z, \mathbf{F}$	forces
$\mathbf{F}_{ij}^C$	conservative force (14.2)
$\mathbf{F}_{ij}^D$	dissipative force (14.2)
$\mathbf{F}_{ij}^R$	random force (14.2)
g	radial distribution
$g^{(2)}$	pair correlation function
G	Gibbs free energy
$\mathbf{h}$	unit vector denoting magnetic field direction
H	Hamiltonian
$\mathbf{H}$	magnetic field (H=|$\mathbf{H}$|)
$\mathbf{I}$	unit matrix or unit tensor
$\mathbf{J}$	Oseen tensor
k	Boltzmann's constant
$\mathbf{k}$	reciprocal vector
K	kinetic energy
m	mass
	magnitude of magnetic moment
$\mathbf{m}, \mathbf{m}_i$	magnetic moments (m=|$\mathbf{m}$| or |$\mathbf{m}_i$|)

M	mass
$\mathbf{M}$	magnetization (A7)
	mobility matrix
n	number density of particles
$\mathbf{n}$	unit vector normal to surface
	unit vector denoting particle direction
$\mathbf{n}_i$	unit vector denoting particle direction
N	number of particles
p, P	pressure
$\mathbf{p}, \mathbf{p}_1, \mathbf{p}_2, \ldots$	momentum vectors
p_{ij}	transition probability
Pe	Péclet number
P_n^m	associated Legendre function
$\mathbf{q}, \mathbf{q}_1, \mathbf{q}_2, \ldots$	position vectors
q_i	electric charge of particle i (11.7)
r	particle radius
	magnitude of vector $\mathbf{r}$
r_{coff}	cutoff radius for forces or interaction energies between particles
$\mathbf{r}_i$	position of paretic i
$\mathbf{r}_{ij}$	$=\mathbf{r}_i - \mathbf{r}_j$ ($r_{ij}=\|\mathbf{r}_{ij}\|$)
$R, R_1, R_2, \ldots$	uniform random numbers
$\mathbf{R}$	resistance matrix
Re	Reynolds number
R_H	nondimensional number denoting strength of magnetic filed relative to viscous shear force
R_m	nondimensional number denoting strength of magnetic particle-particle interaction relative to viscous shear force
R_V	nondimensional number denoting strength of particle-particle interaction due to steric layers relative to viscous shear force
s	eccentricity of ellipse (Chap. 5)
S	entropy
$\mathbf{S}, \mathbf{S}_1, \mathbf{S}_2$	stresslets
t	time
$\mathbf{t}_{ij}$	unit vector denoting direction of line connecting particles i and j
T	temperature
$\mathbf{T}, \mathbf{T}_1, \mathbf{T}_2$	torques

T_x, T_y, T_z	components of torque
$\mathbf{u}$	flow velocity
u_i	interaction energy
u_{ij}	interaction energy between particles *i* and *j*
U, U_i	potential energies
$\mathbf{U}$	flow velocity
$\mathbf{v}$, $\mathbf{v}_i$	velocity vectors of particle
V	volume
w_D, w_R	weighting functions (14.2)
$w_\alpha^{(8)}$, $w_\alpha^{(18)}$	weighting factors (14.3)
W	probability density function
	internal virial (A8)
x^a, x^c, x^g, x^k	mobility functions
X^A, X^C, X^G, X^K	resistance functions
y^a, y^b, ..., y^k	mobility functions
Y^A, Y^B, ..., Y^K	resistance functions
z^k, Z^K	mobility and resistance functions, respectively
Z	partition function (Chap.2)
Z_K	momentum integral (Chaps.2 and 3)
Z_U	configuration integral (Chaps.2 and 3)

Greek Alphabet

$\dot{\gamma}$	shear rate
$\mathbf{\Gamma}$	velocity gradient tensor
δ	thickness of steric layer of colloidal particle
$\delta(*)$	Dirac delta function
δ_{ij}	Kronecker's delta
$\boldsymbol{\delta}_x$, $\boldsymbol{\delta}_y$, $\boldsymbol{\delta}_z$	fundamental vectors
ε	permittivity (3.7)
ε, σ	constants of Lennard-Jones potential (A3)
$\boldsymbol{\varepsilon}$	alternate tensor or Eddington's epsilon
$\dot{\varepsilon}$	strain rate (4.3)
η, η^{eff}, η_s, η_{yx}, η^D	shear viscosity
θ, φ	zenithal and azimuthal angles of polar coordinate system
θ_{ij}	stochastic variable

λ	nondimensional parameter representing strength of magnetic particle-particle interactions
λ_V	nondimensional parameter representing strength of repulsive interactions due to overlap of steric layers
Λ	thermal de Broglie wavelength (Chap.2 and 3.4)
μ	chemical potential (Chap.2 and 3.4)
μ_0	permeability of free space
ν	kinematic viscosity (14.3)
ξ	nondimensional parameter representing strength of magnetic particle-field interactions
	$=(1-s^2)^{1/2}$ (Chap.5)
ξ_i	friction coefficient
Ξ	grand partition function (Chap.2 and 3.4)
ρ	mass density
	charge density (A6)
	probability density function
	multivariate normal distribution
ρ_i	probability density function
σ	constant of Lennard-Jones potential (A3)
τ	relaxation time (14.3)
$\boldsymbol{\tau}$, $\boldsymbol{\tau}^{eff}$	stress tensors
$\boldsymbol{\tau}^K$, $\boldsymbol{\tau}^U$	kinetic and potential parts of stress tensor (14.2)
φ	azimuthal angle of polar coordinate system
φ_V	volumetric fraction
χ	probability density function (Chap.2)
ψ	transition probability
Ψ	orientational distribution (Chap.15)
$\boldsymbol{\omega}$	angular velocity of particle
Ω	grand potential (Chap.2)
$\boldsymbol{\Omega}$	angular velocity of flow field

Superscripts

*	denotes nondimensional quantities
t	denotes matrix or tensor transpose
(eq)	denotes equilibrium quantities

Subscripts

1, 2, α, β	denote quantities related to particles
α	denotes quantities related to site velocity vector (14.3)

Mathematical Operators

∇	nabla
$\partial/\partial\mathbf{r}_i$	gradient with respect to particle positions
$\partial/\partial\mathbf{v}_i$	gradient with respect to particle velocities
$\partial/\partial t$	partial derivative with respect to time
$\langle * \rangle$	average or ensemble average

Abbreviations

AF	additivity of forces (12.1)
AIHI	approximation of ignoring hydrodynamic interactions between particles(12.1)
AV	additivity of velocities (12.1)
BD	Brownian dynamics
DPD	dissipative paretic dynamics (14.2)
LBM	lattice Boltzmann method (14.3)
MC	Monte Carlo
MC step	Monte Carlo step
MD	molecular dynamics
SD	Stokesian dynamics

INDEX